Praise for *A SAFE HAVEN*

"*A Safe Haven* is an outstanding achievement. The Radoshes succeed in debunking the many myths surrounding President Truman's policies toward Palestine and Zionism, and answer lingering questions concerning his decision-making on the crucial issue of Jewish statehood. This is certain to become an essential work for students, journalists, and statesmen–indeed, anyone interested in understanding Israel's origins."

–Michael B. Oren, ambassador of Israel to the United States, and author of *Power, Faith, and Fantasy: America in the Middle East, 1776 to the Present* and *Six Days of War: June 1967 and the Making of the Modern Middle East*

"[A] revelatory account of Truman's vital contributions to Israel's founding. . . . The story of how President Truman resisted pressures both in and out of government to ignore or undermine the aborning state of Israel is told by the Radoshes with an elegance informed by thorough research." –Martin Peretz, *Wall Street Journal*

"The story of Truman's integral role in the birth of Israel . . . is recounted by the Radoshes . . . with good anecdotal color . . . and impressive detail. . . . The creation of Israel remains a remarkable, odds-defying story that bears retelling from different angles. And at a time when Washington and Jerusalem find themselves at odds and some are questioning the future of the alliance, it's worth recalling how the relationship began, and the role of the straight-talking haberdasher from Missouri who worked so hard to make it happen."

–Jonathan D. Tepperman, *New York Times Book Review*

"The state of Israel might well not have ever existed without the support of the United States government; and that support depended largely, and at moments entirely, on the thoughts and efforts of President Harry S. Truman. Allis and Ronald Radosh have written a thorough, powerful, and often surprising account of a fascinating political history, covering everything from diplomacy at the highest levels to the backroom machinations of left-wing Manhattan. It is one of the great stories in modern history, with a seemingly unlikely but steadfast hero in Truman–a book which will absorb anyone who cares about how the world we know came to be."

–Sean Wilentz, Sidney and Ruth Lapidus Professor of History at Princeton University, and author of *The Rise of American Democracy: Jefferson to Lincoln* and *The Age of Reagan: A History*

"[This history] of Harry S. Truman's role in the partition of Palestine and the establishment of the state of Israel offers a useful and clear account of a complex chain of events. . . . [Truman] remained enormously proud of his support for Israel for the rest of his life; *A Safe Haven* helps readers understand why."

–Walter Russell Mead, *Foreign Affairs*

"Allis and Ronald Radosh have given us an invaluable and compelling account of President Harry S. Truman's fight against enormous and unscrupulous opposition from within his own administration, in order to make sure America played a public role in the creation of Israel." –Arthur Herman, *National Review*

"This is an excellent examination of a presidential decision that has had immense historical consequences." –Jay Freeman, *Booklist*

"In their deeply engaging study of Truman and the foundation of Israel, the Radoshes capture the dramatic intersection of momentous millennial aspirations and the thrilling intricacies of political intrigue with remarkable narrative skill."

–Ron Rosenbaum, author of *Explaining Hitler* and *The Shakespeare Wars*

"An extremely well-researched literary effort . . . which gives the reader a full picture, not only of the events being recorded but also of the motivations that drove actions and events. . . . Its historical relevance is most haunting." –Alan Jay Gerber, *Jewish Star*

"The authors, using archival documents that have never been used in a book . . . examine all points of view in an outstandingly researched and very readable work of history. . . . Allis and Ronald Radosh in *A Safe Haven* have written an outstanding work of history that resonates to this day, sixty-one years after the creation of Israel."

–David M. Kinchen, *Huntington News*

"[An] illuminating new history. . . . *A Safe Haven* offers an expert case study of just how complicated . . . policy-making in a democracy can be." –Adam Kirsch, *Tablet Magazine* and *The New Republic*

"In this very timely, very well-researched work, Allis and Ronald Radosh cover the beginnings of what has become an ever-enduring challenge of American foreign policy–the creation of Israel. . . . In the end, [they] credit President Harry S. Truman with a sense of purpose and determination not readily evident to his contemporaries." –Alonzo L. Hamby, *The Weekly Standard*

"What the Radoshes book shows is how a skillful politician was able to overcome entrenched bureaucratic and other vested interests to use his rightful and legitimate presidential power to help bring about a Jewish state in Palestine. . . . [An] enthralling book."

–Martin Rubin, *Washington Times*

About the Author

ALLIS RADOSH has taught at Sarah Lawrence College and the City University of New York, and served as a program officer at the National Endowment for the Humanities. RONALD RADOSH, professor emeritus of history at the City University of New York and adjunct senior fellow at the Hudson Institute, is the author or coauthor of fourteen books, including *The Rosenberg File.* He has written for *The New Republic, National Review,* the *New York Times,* and the *Los Angeles Times,* among many other publications. This is the second book they have written together. They live in Martinsburg, West Virginia.

A SAFE HAVEN

A SAFE HAVEN

Harry S. Truman
and
the Founding of Israel

Allis Radosh
AND Ronald Radosh

NEW YORK • LONDON • TORONTO • SYDNEY • NEW DELHI • AUCKLAND

HARPER PERENNIAL

A hardcover edition of this book was published in 2009 by HarperCollins Publishers.

FIRST HARPER PERENNIAL EDITION PUBLISHED 2010.

Designed by Jaime Putorti

Library of Congress Cataloging-in-Publication Data is available upon request.

ISBN 978-0-06-059464-0

10 11 12 13 14 DIX/RRD 10 9 8 7 6 5 4 3 2 1

For Greta and Harry Freundlich
and Henry Gruenberg

CONTENTS

INTRODUCTION

Harry S. Truman never expected or wished to be president of the United States. A senator from Missouri, Truman was picked at the Democratic Party's 1944 convention to run as Roosevelt's vice president. He was a compromise choice. The labor movement and the left wing of the party favored the candidacy of Henry A. Wallace, Roosevelt's "progressive" vice president. Wallace had written a New Deal bestseller, *The Century of the Common Man*, which became a textbook for the hopes and dreams of the strong left-wing component of the New Deal. Wallace, however, had some major handicaps for a party seeking a candidate who might be elevated to the presidency. He was a devotee of a mystical guru, and letters he had written to him threatened to be made public by the Republican opposition. Even more important, Wallace had what historian Alonzo Hamby accurately describes as "ideological rigidity," which later on would reveal itself as an ever increasing tendency to look too kindly on the Soviet Union, and to even embrace American Communists as allies of liberal Democrats. In 1944, more than a few loyal Democratic supporters of FDR began to view Wallace as an unguided missile, too close to becoming president.[1]

Party leaders shifted to the little-known senator from Missouri, much to the consternation of his wife, Bess, who wished to avoid the limelight and was more than content with her husband's current position. Roosevelt kept Truman at arm's length as vice president, as he ran the country with the counsel and support of his trusted longtime advisers. When Roosevelt died of a stroke at his vacation home in Warm Springs, Georgia, Truman found himself immediately confronted with some of the most pressing problems facing the nation. Eventually, the public as well as professional historians would rate him as one of the great American presidents. Few would have predicted that when Truman entered the office, or in fact, when he left in political disgrace with the lowest popularity polls of any American President to date.

History, as we know, has been kinder to him than his own country's contemporaries had been. Forgotten by most were his inept domestic decisions, such as his misguided call to draft striking railroad workers into the Army to force them to keep working. Even more unpopular was his announcement that to prevent a steel strike in 1952, he was using executive authority to seize the factories and make them government property. The courts ruled that unconstitutional, but the damage had been done. With the plants back in private hands, the country went through a fifty-three-day strike. When the political pollsters asked the public to rate the president, only 32 percent gave him a good rating.

Things looked no better when one turned to events abroad. China had gone Communist during Truman's watch, as Mao Tse-tung defeated the Kuomintang Nationalists and took power on the Chinese mainland. Mao's victory led Republicans to put the blame on the Truman administration. The supporters of Senator Joseph McCarthy of Wisconsin and Pat McCarran of Nevada began to gain strength by raising the question: "Who lost China?" In Korea, Truman brought the United States into battle after North Korea crossed that 38th parallel, which divided South from North Korea. American troops quickly pushed back the invading North Koreans, but events turned for the worse when Chinese "volunteers" entered the fight on the side of the North. The Korean War dragged on, as the public grew weary and the troops filled with disdain at fighting a seemingly endless series of battles to take small hills. Truman had fired the popular General Douglas MacArthur for threatening civilian leadership of the military. Mac-

Arthur made known his plan to extend the war across the Yalu River into China, thereby threatening Soviet intervention on China's side and a potential new world war. The war became a stalemate, and lingered on until the new Eisenhower administration negotiated a truce that holds to our own time.

Dwight D. Eisenhower, whose leadership during World War II propelled him to major attention as a potential candidate, gained the Republican nomination and easily defeated the Democratic candidate, Adlai Stevenson in 1952. The election was widely seen as a personal repudiation of Truman. A saying of the day made the rounds, "To Err is Truman." The nation seemed to forget Truman's many foreign policy triumphs–the Marshall Plan for reconstruction of Western Europe; the creation of NATO to protect the West against potential Soviet aggression; and the decision to drop atomic bombs on Hiroshima and Nagasaki, thereby shortening the war and saving thousands of American lives. Above all, Truman began the slow reversal of a soft policy towards Stalin and the Soviet Union. The Soviets were moving towards the creation of a new Cold War, and at the Potsdam Conference and after, Truman began a major reorientation of America's position in the world. No one has explained it better than the historian Wilson D. Miscamble. "Under Truman's leadership," he writes, "the United States moved to a level of world engagement and assumed international commitments far beyond anything that Roosevelt had conceived."[2]

All these actions took guts and leadership. Yet Truman would leave office unpopular and repudiated. In 1962, the Harvard historian Arthur Schlesinger, Sr., asked seventy-five people to rank the American presidents. How satisfied Truman must have felt to learn that historians, ten years after he had left the White House, declared him a "near great" president. Truman was put into the same category as his own personal hero, President Andrew Jackson. The reason he received a high rating was because of his series of triumphs, from the Truman Doctrine to protect Greece and Turkey from Communist aggression, to the Berlin airlift, the Point Four program for economic aid to Third World nations, all policies that Schlesinger called "landmarks in an assumption of global responsibilities."[3] Subsequent polls of historians, as well as the public, have consistently shown that Truman's standing remains high.

While American policy toward Palestine, including Truman's decision to recognize Israel, did not turn up in any of the polls as one of the accomplishments that led him to be highly regarded, it ranked among the hardest he had to deal with. On most other critical issues, and in the days of the early Cold War there were many, Truman relegated day-to-day policy to his secretaries of state, first Edward Stettinius, then James F. Byrnes, and finally General George C. Marshall. But on the issue of the future of Palestine, Truman was often at odds with his own Defense and State departments. The State Department's Near East Division was staffed by a group of men whose experience and outlook led them to emulate British policy and tilt toward the Arabs. They were vehemently opposed to the creation of a Jewish state in Palestine, which they thought would be a disaster. Much of Truman's handling of policy questions relating to Palestine and the Middle East involved removing this group's decision-making abilities on the issue, and placing them under his control in the White House.

Palestine was, his daughter Margaret pointed out, "the most difficult dilemma of his entire administration." In the midst of an emerging Cold War with the Soviets, and scores of international crises, from the sealing off of Berlin by the Russians to the Czech coup that put the Communists into power, there were various events that demanded the president's immediate attention. Among these, Palestine was the one constant crisis. He had done his best to deal with it, Margaret Truman wrote, but "his best was probably not good enough." As the days and months went by, the depths of the problem seemed to defy a solution. On one day in July 1946, Truman told his wife, Bess, that he had "the most awful day I've ever had." He held a Cabinet luncheon "and spent two solid hours discussing Palestine and got nowhere."[4] In fact, during the next two years Truman would have many such days. A plain-talking and straightforward man who valued loyalty and friendship, Truman would find his values tested and his decisions second-guessed.

Truman became president as the concentration camps were being liberated, exposing the full horror of the Nazis' crimes against the

Jews. The Jewish survivors, in pitiable condition, were among the refugees housed in European Displaced Persons camps. Those in Germany and Austria were the responsibility of the United States. This situation could not go on indefinitely; they would have to be settled somewhere. But where? When asked where they wanted to go, most answered "We want to go to Palestine." The British, the mandatory power in Palestine, was refusing their entry, limiting Jewish immigration to a small quota. Although Truman found Britain a major ally when it came to dealing with the Soviets and fighting the Cold War, he could not say the same regarding Britain's relations with the United States when it came to Palestine. And in Palestine, there were constant skirmishes and fighting between the Yishuv (the Jewish community), the Arabs of Palestine who vehemently opposed what they considered to be Jewish interlopers, and the British who held the Mandate.

At home, Truman received conflicting advice from his State Department and the Jewish community. The Jews had prospered in pluralistic and democratic America. For many it was their Zion. But after the Holocaust, the assimilationists became Zionists. The inability of the Western democracies to save their brethren from the gas chambers and their refusal, even now, to admit the survivors, strengthened the Zionist argument that the Jews, persecuted throughout history in the Diaspora, would only be safe once they had their own country. American Jews did not want this other country for themselves, but rather for the suffering Jews across the seas. A large segment of American public opinion came to support them.

Harry Truman was insecure about many things when he became president, but he was confident he could handle the issue of Palestine in a just way. He did not anticipate the maelstrom he was about to enter. In his memoir, Truman wrote that his challenge was to create neither an Arab nor a Jewish policy, but "an American policy [that] . . . aimed at the peaceful solution of a world trouble spot . . . based on the desire to see promises kept and human misery relieved."[5] That task would consume him from the day he became president to the day he recognized Israel on May 14, 1948, and beyond. The story of why he made the decisions and took the actions he did is the subject of our book.

A SAFE HAVEN

ONE

FDR'S LEGACY: PALESTINE, THE JEWS, AND THE ARABS

Franklin D. Roosevelt and his new vice president, former Senator Harry S. Truman of Missouri, were sworn in as the nation's top executive officers on January 20, 1945. FDR made history that day, becoming the first man to be inaugurated for a fourth term as president and the first president-elect to be inaugurated at the South Portico of the White House. Unlike most inaugurations, the event was toned down, coming as it did in the midst of a brutal world war.

Because the inaugural took place in the middle of the war, as *The New York Times* reported, there were no "long, colorful parades, silk hats, resplendent uniforms, fireworks and receptions." The nation did not witness an inaugural parade at all, as the president knew that such a display would be considered an insult to the half-frozen soldiers in tanks and foxholes on Europe's battlefields. Instead, fewer than five thousand invited guests–far less than the eight thousand originally expected–sat on the snow-covered White House lawn, gritting their teeth to shield themselves from the bitter Washington cold.[1]

The president, weak from a tough twelve years of guiding the nation through depression and war, and physically in pain from his polio, nevertheless strode to his balcony podium, helped by his son, Colonel James Roosevelt. James offered his father a cape to shield him, but, turning it down, the president spoke bareheaded, wearing a dark business suit. The other guests, including Harry Hopkins, Bernard Baruch, governors, and cabinet officers, all wore heavy overcoats. The audience was respectfully somber, pausing only to applaud when the new vice president, Harry S. Truman, took the oath of office administered by the outgoing holder of that office, Henry A. Wallace.[2]

Anxious to get started, the new vice president slipped away from the one event scheduled for some of the distinguished guests, a post-inaugural luncheon at the White House. Hitching a ride to Capitol Hill from the sergeant-at-arms of the Senate, Truman sat at his desk, reading through his voluminous mail and thinking about the issues he would be dealing with in his new position.[3]

Among the immediate decisions to be made by the administration was planning for the Allied coalition's final blow that would lead to the collapse of the Axis powers. Once that was out of the way, top priority would be given to the creation of a new world organization, which the president in particular hoped would become a mechanism to guarantee a peaceful postwar world. Beneath the surface, wartime tensions with the United States' Soviet ally were destined to come to a head, creating a dangerous situation that might sully the dream of postwar unity. At home FDR faced having to shift the nation's economy from wartime production to peacetime stability and continued full employment, thereby fulfilling his promise to realize his famous "Four Freedoms," including freedom from want and poverty.

As Roosevelt and Truman took their oaths, the full scope of the war's horrific death toll and looming refugee crisis were yet to be fully revealed. Historians would eventually estimate that some 36.5 million Europeans had died as a result of the war between 1939 and 1945. Over half those, 19 million, were civilian casualties.[4] Of these, no one group suffered proportionately more than the Jews. What made the war against the Jews unique, wrote the Zionist leader Meyer Weisgal (who became the Zionist leader Chaim Weizmann's personal representative in the United States), was that this genocide was not for a "ruthless seizure of territory," it was "for the sake of extermination as a principle."[5]

By 1942, Adolf Hitler's campaign of legal restrictions, expulsions, and pogroms against the Jews had escalated to the Jews' physical destruction. During the war, however, many still hoped that stories about Hitler's Final Solution were grossly exaggerated. Those who sought the truth, if they looked, were able to learn about the terrible reality. Stories buried inside *The New York Times* were beginning to tell a very grim tale. Dr. Joseph Schwartz, the European director of the American Jewish Joint Distribution Committee, wrote that perhaps only 500,000 out of Europe's 6 million Jews had escaped destruction by emigration and that "only one to one and a half million of Europe's 6 million Jews were now left on the Continent.[6] "At the worst of times," wrote Weisgal, "we had never believed that the diabolic mind of Hitler would translate into those facts. . . . The worst of our forebodings paled before even the partial revelation."[7] By the final tally, one third of the entire Jewish people and almost two thirds of European Jewry had been destroyed.

The situation of the Jewish masses, though never very good, had looked brighter during the waning days of World War I. Britain, seeking to gain the Jews' support for the Allied cause, held out the promise to them of Palestine, the homeland Diaspora Jews had longed for since they had been defeated by the Romans in 70 C.E., almost 2,000 years before. More immediately, the British war cabinet hoped the incentive would cause Russian Jews to encourage their country to stay in the war and for American Jews to work for rapid entry of U.S. troops into the battle.[8]

The idea of a British protectorate over a Jewish homeland in Palestine was not only attractive to the Jews but would give Britain a presence in the Middle East once Turkey was defeated and its Ottoman Empire dismantled. Not least, the concept of the Jews' return to their ancient homeland as prophesied in the Bible appealed to the faith of many of Britain's restorationist statesmen. Among them was Lord Arthur James Balfour, Britain's secretary of state for foreign affairs. In 1917, Balfour had negotiated an agreement with Dr. Chaim Weizmann, lending British support to the establishment of a "National Home for the Jewish people" in Palestine.[9]

Before proceeding, the British wanted to secure American support for the plan. They were successful when President Woodrow Wilson, encouraged by Supreme Court Justice Louis D. Brandeis–the leader of American Zionism at the time–approved a draft of the declaration. Wilson claimed he "had been influenced by a desire to give the Jews their rightful place in the world; a great nation without a home is not right" and then added, "My personal hope is that all the Jews will make good and eventually found a Jewish State."[10] Despite Wilson's remark, exactly what a "National Home" meant would become a subject of major debate.

This unheard-of compact between a sovereign state and a stateless people, which met great resistance from many of Britain's wealthier assimilated Jews, who did not want to be perceived as having dual loyalty, probably would not have been achieved without the efforts of Dr. Chaim Weizmann. Born in 1874 in the little town of Motol in czarist Russia's Pale of Settlement, Weizmann, the son of a timber merchant, experienced the oppression of Eastern European Jewry firsthand. At eighteen, like thousands of Russian Jewish students in what he likened to "a sort of educational stampede," he left for the West.[11] After earning his doctorate in Germany in chemical engineering, he settled in Manchester, England, where he was able to pursue his two lifelong interests, science and Zionism. The former led him to develop a process for manufacturing synthetic rubber (acetone) that was needed to make changes in the production of naval guns, greatly benefiting Great Britain during World War I and earning the gratitude of the British establishment.[12] At home in the culture of Eastern European Jewry as well as that of the West, Weizmann spoke fluent English, German, Yiddish, French, Hebrew, Russian, and Italian.[13]

Weizmann's rise as a Zionist leader was aided by his passionate, almost messianic belief in his cause, his eloquence, and his great charm. "Weizmann was an overwhelming figure," wrote the Zionist leader Nahum Goldmann. "Anyone who encountered him fell under his spell."[14] He was, Justice Felix Frankfurter later wrote, "among the great men of our time." When he spoke, he seemed taller than he already was, giving the impression of a man who "had that something that makes a difference, that makes a great man," who had "something electric about him."[15] Balfour would not be the last statesman to be influenced by Weizmann and his Zionist vision.

After the war, the League of Nations awarded Britain a mandate over Palestine. Using the spirit of the Balfour Declaration as its cornerstone, the Mandate charged Britain with facilitating the creation of a Jewish national home. It established a Jewish Agency to administer Jewish Palestine and supervise its relations with both Britain and world Jewry. Other countries soon gave their approval. By 1922, fifty-two governments had endorsed the major goals of world Zionism.[16] However, Arab protests led the British to issue a White Paper reassuring them that although the Jews were to be in Palestine as a matter of right, not of sufferance, the Jewish national home did not necessitate a Jewish majority or the establishment of a Jewish state in Palestine.

Weizmann recognized that the state he so desired could not be created by decree. "Even if all the governments of the world gave us a country," he wrote, "it would be a gift of words, but if the Jewish people will go and build Palestine, the Jewish state will become a reality and a fact."[17] One project especially dear to Weizmann was the creation of the Hebrew University in Jerusalem, which he hoped would be a "partial solution" for Russian Jewish youth, who were excluded from Russian schools."[18] He found success when the university was inaugurated on Mount Scopus in Jerusalem in 1925. Balfour himself attended the opening ceremony and had tears in his eyes when he saw the progress that the Jews of Palestine had made.[19]

There had always remained a small Jewish presence in Palestine, but now the Yishuv, as the Jewish community in Palestine was called, grew in size. At the end of the First World War, there were 50,000 Jews in Palestine. But under the terms of the British Mandate, along with British protection, immigration to Palestine began to increase. By 1928, the Jewish population numbered 160,000 and made up 20 percent of the total population, with 70 percent Muslim and 10 percent Christian. By the time the Second World War came to an end, the Jewish population of Palestine had exploded to 600,000, while the Arab population increased to twice that number.[20]

By 1945, the Yishuv had its own schools, public services, trade unions, and army. It was, for all purposes, a "state within a state."[21] Yet, due to Arab resistance, including deadly riots, British pledges to aid in the development of the Jewish national home had become a burden. In 1939, Britain adopted a White Paper that limited Jewish immigration to 75,000 over a five-year period, to be continued only with the consent of the Arabs, who were unlikely to grant it. Britain also sought to halt the growth of the Yishuv by limiting land sales to Jews. This meant that at a time when many Jews were trying to escape from Hitler's clutches, the door to Palestine was closed to them, condemning multitudes to death. While the Arabs saw the White Paper as a promise to stop the Zionists, whom they viewed as intruders taking over their land, the Jews saw it as a betrayal of the Balfour Declaration and the Mandate established by the League of Nations. The League's Permanent Mandates Commission refused to sanction it, saying that Britain's new policy was not in accordance with their interpretation of the Palestine Mandate.[22]

While Hitler proceeded to annihilate Europe's Jews with diabolical efficiency, the American Jewish community became by default the largest and wealthiest in the world. Even the most assimilated Jewish American, who might barely identify as a Jew, realized that it could very well be him- or herself in Hitler's gas chambers. And many of their relatives indeed were. The guilt was enormous, especially torturous for those who had friends or family frantically writing for help to get out of Europe. America's Jews, never very united, cast about for a solution.

One response was to place their faith in Franklin D. Roosevelt. The overwhelming majority of American Jews voted for and loved Franklin D. Roosevelt–they were, after all, part of the labor/liberal coalition that gave the president continued electoral victories in the North and the East. FDR had personally given assurances to Jewish leaders. He had told Brooklyn congressman Emanuel Celler that after the war ended, British prime minister Winston Churchill was going to abrogate the British White Paper on Palestine, and allow Jews to go to Palestine without restriction. Celler concluded that Roosevelt "would supply the impetus that would open the gates of Palestine to the Jews fleeing murder."[23] Roosevelt, moreover, ran on the Democratic Party platform that promised the Jews the establishment of a Jewish commonwealth in Palestine. In a letter to Senator Robert F. Wagner of New York, FDR said that he was convinced that the American people gave their support to the establishment of a Jewish "commonwealth" in Palestine and pledged that, if reelected, "I shall help to bring about its realization."[24]

Blame for not rescuing more Jews from Hitler's clutches has been laid at FDR's feet. Today, many historians and observers have concluded that FDR could have done more to save Europe's Jews and that his policies have been found wanting. Scholars have criticized the president's handling of his refugee policy in waiting until 1944 to establish the War Refugee Board, deciding not to bomb the railroad tracks leading to Auschwitz, and failing to aid the Jewish passengers fleeing Hitler on board the ocean liner *St. Louis* when they were denied entry into Cuba and forced to return to Europe in 1939. In short, it has been argued Roosevelt had done too little too late.[25] At the time, however, most Jews believed they had a friend in FDR. They had the utmost confidence in their commander in chief. He regularly advised them to be patient, to put their demands on hold, and that the best way to help Europe's Jews was to win the war as quickly as possible.

A fervent believer in FDR's promises was Weizmann's contemporary Rabbi Stephen Wise, whose reputation rested on his willingness to champion the cause of Zionism when few young Reform rabbis in America would risk their future careers on a cause so unpopular with the Reform Movement.[26] Wise was chief rabbi of a new Reform congregation, the Free Synagogue, which he founded in New York City in 1907. Among the many leadership roles in the Zionist movement that he took on were the presidencies of both the World and American Jewish Congresses.

Eliahu Epstein (Elath), a representative of the Jewish Agency in Washington and later Israel's first ambassador to the United States, regarded Wise as the dominant figure among America's Jewish leadership when he met him in 1945. A man with "personal charm, brilliance in conversation, and great talent for public relations," Wise was one of the only Zionist leaders with direct access to the president. However, his very closeness to Roosevelt, Epstein observed, "often prevented his being firm enough in his demands."[27] Wise completely believed in FDR's promises; the Jews had a "great, good friend" in Washington who was thinking about them and planning for their future.[28]

Hitler's escalating war against Europe's Jewish population, however, led many American Jews to take a more militant stance. They were frustrated with the Jewish establishment's inability to make FDR intervene. Instead of relying on the personal diplomacy of men such as Weizmann and Wise, many Zionists were attracted to leaders with a more activist program, such as that advocated by Rabbi Abba Hillel Silver. Born in Lithuania in 1893, Silver came from a long line of rabbis. After gaining his own rabbinical degree in 1915, he joined the Temple Tifereth-Israel in Cleveland, Ohio, the largest Reform congregation in the United States. An active and committed Zionist, Silver loathed the quiet diplomacy of his counterpart Stephen Wise, with whom he vied for control of the Zionist movement.

To his regret, in 1942 Weizmann had encouraged Silver to take a leading role in the Zionist movement. Now he was changing the rules on Zionism's elder statesmen. Silver was not only confrontational and unyielding, but, unlike Weizmann and Wise, he was not charming. Many people who had contact with him found him abrasive and did not particularly like him, a group that unfortunately included both FDR and Truman. But Silver was a very effective political strategist and organizer. He was an inspired speaker whose "baritone-voiced oratory," a *New York Times* reporter wrote, "would do credit to a Shakespearean actor."[29]

In 1942, the year Hitler's Final Solution was made known to the world, six hundred American Zionists met at New York City's Biltmore Hotel. The presence of Chaim Weizmann and the Palestinian leader David Ben-Gurion added to the importance of the event and signified that the United States had replaced Great Britain as the Zionist movement's center of gravity. At the conference Silver launched an effective attack on the opponents of political Zionism. Silver argued that the only solution for the Jews was to establish their own state and that this should be their top priority. By the end, the attendees unanimously adopted what came to be known as the Biltmore Declaration, which stated for the first time (at least officially among American Zionists) that "Palestine be established as a Jewish Commonwealth."

Silver, accused by many of his opponents of being a Republican, actually denied belonging to either party. He believed that both major political parties should be made to vie for Jewish support.[30] It was a mistake, he thought, to be in the pocket of the Democratic Party and be taken for granted. To the chagrin of the Democrats, Silver proceeded to form a close alliance with Republican Senator Robert A. Taft of Ohio, a strong supporter of Zionist goals.

Silver also argued that the Zionists' focus should be on creating a climate of public opinion in the United States that would move non-Zionist Jews to support the goal of establishing an independent Jewish state in Palestine. They should reach out and educate the non-Jewish majority, convince journalists, and put pressure on Congress and through it the administration. In a democracy, the people, Silver and his followers hoped, would have to be listened to.

This ambitious effort was coordinated by an umbrella group of American Zionist organizations, the American Zionist Emergency Council (AZEC), which had been formed in 1939 with Wise as its leader. In 1943, with Wise's encouragement, Silver and Wise became cochairmen in the hope that the organization would become more effective. AZEC's two main goals in 1943 and 1944 became pressuring government and media figures to help rescue the remnant of Europe's Jewry and convincing them at the same time that Palestine had to be opened to receive them. Their hope was that such settlement would eventually build support for creating a Jewish state.[31]

AZEC then set about organizing a conference consisting of democratically elected delegates from organized American Jewry. Not all were Zionists. On August 29, 1943, the American Jewish Conference opened at the Waldorf-Astoria hotel in New York City. More than five hundred delegates were present, representing sixty-five different national organizations. In one of his most electrifying speeches, Silver convinced his audience to support the Biltmore Declaration. His success now proved that the majority of organized American Jewry was behind the effort to bring about a Jewish state in Palestine.[32]

The lone dissenter at the conference was the American Jewish Committee (AJC), a small organization founded in 1906 that fought anti-Semitism in America and worked to secure Jewish rights abroad. Made up mostly of affluent East Coast Jews of German descent, the AJC had a much greater influence than its numbers suggested, largely because of the economic and social position of its membership. Represented by Judge Joseph M. Proskauer, the head of its anti-Zionist wing, the AJC seceded from the American Jewish Conference after its confirmation of Biltmore, claiming that no one organization had the right to speak for all Jews on the important topic of a Jewish homeland. Although Proskauer would later change his mind, during the war he sought to convince FDR that "the idea of a state based on religion was anachronistic and against the spirit of the age." It would produce a theocracy, contrary to American principles. Worse, he thought the demand for a Jewish state was "an admission . . . of despair." Jews should fight for the right to live anywhere and go back to their homes after the war. The State Department, which also opposed the Zionist solution, found the AJC a useful counterweight to the growing popularity of the Zionist movement.[33]

In January 1944, it looked as though the Zionists' campaign was getting results. Zionist groups worked to have resolutions introduced in the Senate and House of Representatives supporting the establishment of a Jewish state in Palestine and the abolishment of the British White Paper. But their hopes were dashed when the resolutions, which had substantial support in both houses of Congress, were opposed by Secretary of State Cordell Hull, Secretary of War Henry L. Stimson, and FDR, who argued that their passage would hurt the war effort.[34]

The war had not as yet been won, they said, and passage of the resolution by Congress might throw a monkey wrench into operations in the Middle East, which was a major supply route for the European theater. Both the War Department and the State Department were pressing the case that Arab nations would be alienated by such a resolution, and their support was needed for victory in the war against the Germans. The Middle East Supply Center was located in Cairo, as was the Persian Gulf Command, through which Lend-Lease to the Soviet Union passed. And should the resolution lead to conflict between Arabs and Jews, troops might have to be diverted from the European theater to control the disturbances.[35]

However, in June and July, Silver's strategy of making the parties compete for Jewish support achieved success. Both the Republican and Democratic Parties' conventions included pro-Zionist planks, which were then endorsed by the two parties' presidential candidates, Thomas E. Dewey and Franklin D. Roosevelt.

When Stimson finally withdrew his objections to the shelved congressional resolutions in October, they were resurrected and introduced in the Senate again by a bipartisan resolution bearing the names of Senators Robert F. Wagner of New York and Robert Taft of Ohio, who was nicknamed by his conservative supporters "Mr. Republican." Once again, problems arose when the resolution was brought before the Senate Foreign Relations Committee. There were objections to the phrase "Jewish Commonwealth." Roosevelt, who wanted room to maneuver with the Arabs after the war, agreed with Secretary of State Edward R. Stettinius, Jr., who had told him that "putting the resolution through now would just stir things up."[36]

Indeed, the administration was angry that the Zionist leaders had pushed for the reintroduction of the resolution without consulting the White House. Judge Samuel Rosenman, FDR's special counsel and unofficial adviser on Jewish affairs, tried to intercede. Rosenman believed in going through official channels and did not care for the tactics of "extreme Zionists" such as Silver.[37] Meeting with the Jewish Agency's representative, Nahum Goldmann, Rosenman complained that no one had even told him or Supreme Court Justice Felix Frankfurter, a major figure working for the Zionist cause behind the scenes, about the resolution. How could he help the Zionists or give them advice, he asked, "when they are neither informed of, nor consulted about . . . the resolution?" Had they done so, Rosenman continued, they would have seen "that such a step was impossible at the moment, and such a resolution could not have been passed."[38]

Wise had been informed of the president's request that the pro-Zionist congressional resolution be put off for the time being and he had tried to forestall it. Roosevelt had said that if Congress could just give him more time, when he was at Yalta, he would be able to "unravel this whole situation on the ground."[39]

Silver, however, was determined to press ahead on the resolution. He was furious at Wise and those in AZEC who supported him, who had gone to Washington, willingly collaborating with the State Department to keep Congress from "expressing itself in favor of the Jewish Commonwealth." It was not the "charm, winsomeness or eloquence" of individual leaders that would help them, Silver said, nor the promise of support for a candidate in an election. Such intercession was an anachronism. Putting it as sharply as possible and alluding to Wise, Silver bellowed in a speech, "It is too late for Court Jews."[40]

Roosevelt, as one might imagine, resented Silver. "The President," Rosenman informed Jewish Agency Representative Nahum Goldmann, "is becoming antagonized by Dr. Silver's takeover of the Zionist political leadership, instead of Dr. Wise." It was crazy to shift leadership to "a man like Dr. Silver whom the President dislikes." Moreover, Silver's tactics were getting to FDR, who was "antagonized by the fact that all the hostile [Republican] elements in Congress are poised to make Zionist speeches." The result was ominous for the Zionists, Rosenman said. FDR, "who always thought of Palestine as a noble and idealistic venture, is beginning to think of it as a nuisance." Rosenman's advice was that "Dr. Wise should appear more on the Washington scene and assume active political leadership."[41]

Turf battles soon arose between AZEC and the Jewish Agency. When Goldmann had the temerity to see Stettinius without clearing it with Silver, Silver was furious. Calling Goldmann, who functioned as sort of an international ambassador of the future Jewish state representing both world and American Jewry, "an international gigolo" and a "damn nuisance," Silver threatened to resign.[42]

Silver's childhood friend and associate Emanuel Neumann appealed to Weizmann for help. It was true, Neumann wrote, that Silver "lacks the flexibility required in dealing with internal situations. But his firmness and boldness, and if you like, his inflexibility are also the qualities which made him such an undaunted champion of our cause in the world at large."[43] Weizmann should not abandon the difficult Silver, Neumann told Weizmann, because whatever his faults, he had "revolutionized the American Zionist movement from a weak conglomerate, unaware of its own potential and somewhat uncertain of the direction it should take, into a political force to be reckoned with."[44]

FDR's wish regarding Silver and Wise appeared to be coming true. A split in the Zionist ranks was inevitable. Silver resigned from AZEC in December 1944, forfeiting the leadership to Wise and his allies, only to be brought back at the demand of the members, who appreciated the fiery Silver.

One group that was not going to wait for Roosevelt to save the Jews of Europe was the "revisionist" Vladimir Jabotinsky and his followers, who arrived in America from Palestine in 1940 and began a dramatic crusade. Jabotinsky died soon after, and his mantle was taken up by Hillel Kook. The nephew of Palestine's chief rabbi, Kook was born in 1915 in Lithuania, but his family soon moved to Palestine. By the time he was a young man, Kook had become a member of the Irgun, the underground militia in Palestine founded by Jabotinsky that was a split off from the Haganah. At the Irgun leader's suggestion, Kook moved to the United States, where he took the name Peter Bergson, because, he said, he did not want to cause his family any embarrassment.[45]

Bergson quickly learned that in America, the way to gain attention was to use the techniques so well developed by Hollywood and Broadway, as well as the American advertising industry. He became a master of fund-raising, created new committees that included Jews and non-Jews, and took out full-page ads in the nation's leading newspapers. Gaining the support of the playwright Ben Hecht, Bergson got Billy Rose to produce an extravaganza in 1943, *We Will Never Die,* which packed Madison Square Garden and toured the nation. One of the actors appearing in it was a young novice named Marlon Brando. More established performers included Edward G. Robinson and Paul Muni. An estimated 100,000 people saw the company in its tour of major cities.[46]

Bergson got the attention of the White House, Congress, and American Jews when he agitated for a brash new proposal, the creation of a Jewish army, which would allow Jews in Palestine and Europe to contribute to the Allied war effort. His technique for gaining support was described well by Jabotinsky's son Ari: "We bought a page in *The New York Times* and advertised the Committee for a Jewish Army just as you would advertise Chevrolet Motors or Players cigarettes." Fund-raising was so successful, young Jabotinsky continued, that "we became the best known Jewish organization among the Gentiles."[47]

Bergson created yet another new group in 1943, which he called the Emergency Committee to Save the Jewish People of Europe. His genius was to focus the group's attention on one goal, in this case actions to save Jews from Hitler. Consciously downplaying the issue of Palestine and its independence, Bergson got major American figures from all points of view to sign on, including the conservative radio personality Lowell Thomas, the left-wing producer Erwin Piscator, and the pro-Communist intellectual Mary Van Kleeck.[48]

FDR took notice, and let it be known to the official Zionist leadership that he was disturbed by Bergson's dramatic advertisements in *The Washington Post* for his Emergency Committee to save the Jewish People of Europe. They were "doing a great deal of harm," the president thought, and the "time had come for the responsible Jewish groups to come out with a statement . . . making it clear that they have no connection with the Jewish Army Committee group or its statements." He wanted either the Zionists or the American Jewish Committee to take action and to inform all members of Congress. FDR even met with *Washington Post* publisher Phil Graham and asked him to stop running Bergson's ads.

Wise agreed, and when Harold Ickes, the secretary of the interior, accepted the position of honorary chairman of the Emergency Committee's Washington Division, Wise wrote to him, "I wish I could have seen you before you gave your consent. I know that your aim is to save Jews, but why tie up with an organization which talks about saving Jews, gets a great deal of money for saving them, but in my judgment, has not done a thing which may result in the saving of a single Jew."[49]

When Bergson organized a march on Washington of four hundred rabbis on October 6, 1943, to demand immediate action by the government to rescue European Jews and to open Palestine to them, the president, on the advice of Stephen Wise and Samuel Rosenman, refused to meet with them. Rosenman told FDR that "the group behind this petition [was] not representative of the most thoughtful elements in Jewry" and that he had tried without success "to keep the horde from storming Washington." Moreover, he told the chief, "The leading Jews of his acquaintance opposed this march on the Capitol."[50]

After the rabbis' march on the capital, when Rosenman met with Silver and complained about Bergson, the rabbi told him that "our failure to get anywhere with the present administration . . . was responsible for the success of the strategy and tactics of the Bergson group." An hour later, Silver met with the inventive Bergson, who told him about yet another new project: the American League for a Free Palestine. The underlying idea, said Bergson, was that Palestine was already a Jewish state that needed to be freed from enemy occupation, like Holland or France. Silver brought up the possibility of the Zionists and the Bergson group cooperating, but Bergson "had no reaction." Silver ended the meeting by telling the young man that since his group had failed at both his campaign for the Jewish army and the rescuing of European Jewry, "he should reconsider the validity of his technique on the American scene."[51]

Bergson further alienated America's Jewish leadership when he bought a former embassy building in Washington and made it the headquarters of his new group, the Hebrew Committee of National Liberation. Declaring it "the government-in-exile" for the yet-to-be-created Jewish state, Bergson claimed his new committee was the "authentic representative of Palestine Jewry and those stateless Jews hoping to immigrate to the Holy Land." This was a direct challenge to the World Zionist Organization and the Jewish Agency, "which considered themselves the sole legitimate representatives of the Palestine Jewish community." Bergson also tried to allay the fears of many assimilated Jews, who worried that they would be accused of dual loyalty if a Jewish state should arise. He called for a distinction to be made between "Hebrews," which he said referred to Jewish residents of Palestine and European Jewish refugees, and "Jews," which referred to Diaspora Jews, indicating only their religion. But this distinction further angered Zionist leaders "since it undermined the basic Zionist tenet of international Jewish solidarity."[52]

Bergson, nevertheless, gained a large group of congressional supporters who met with him and endorsed his organizations, despite the State Department's opposition and that of mainstream Zionists, who considered him dangerous and sought to marginalize him. Yet Bergson managed to touch a raw nerve on the part of American Jews, because he alone appeared to be doing something to stop the slaughter of Europe's Jews. Bergson ironically forced the established Zionist groups to meet him halfway, by changing their tactics and assuming a more militant position.[53]

Prior to 1943, the United States had no clearly defined policy toward Palestine and generally regarded it as a British responsibility. FDR had been postponing decisions on the issues of the Jewish refugees and U.S. support for the creation of a Jewish commonwealth in Palestine. Despite his promises to Jewish leaders and unbeknown to them, the president had pursued a policy that tried to please the Arabs, the Department of State, and the Jews simultaneously. His favored technique was to assure representatives of the various groups that he stood with them, giving all a false hope that the president shared their agenda.[54]

Up until the end of the war, the State Department had functioned like a nineteenth-century exclusive club. This sedate, unhurried workplace was housed in an "ornate building, dating from the 1870s, with high ceilings, white-painted swinging doors, and long corridors paved with black and white marble." There, Palestine came under the jurisdiction of the Office of Near Eastern and African Affairs, which was headed by Wallace S. Murray, who regarded the Zionists as intruders in an otherwise calm Near East and was alarmed at their increasing success in swaying the American public to support the creation of a Jewish state. Arab leaders too were anxiously watching the situation and started to question the U.S. government's position.[55]

But this state of affairs could not last forever. Acting on FDR's order in 1943, Murray sent Colonel Harold B. Hoskins to the region. Hopkins spoke fluent Arabic and was told by State to consult various Arab leaders, including King Saud of Saudi Arabia. What the Arabs feared above all, Hoskins reported, was that a Jewish Palestine would be forced on them by the Great Powers. Such a result, Hoskins argued, could not be attained "without outside assistance from British or British and American military forces." Because the Arabs so strongly opposed a Jewish state, he advised Roosevelt to make it "very clear to the American people . . . that only by military force can a Zionist State in Palestine be imposed upon the Arabs." It was therefore important that support for aid to the persecuted European Jews not be tied up with the Zionist effort to create a Jewish state in Palestine. One thing had to be done: a simple statement should be issued that "any post-war decisions will be taken only after full consultation with both Arabs and Jews."[56]

A similar report came from former Secretary of War Patrick Hurley, whom FDR also sent to the Near East on a fact-finding mission. Hurley's acerbic report amounted to a frontal assault on the Zionist arguments. Jewish leaders outside Palestine, Hurley argued, "view the Zionist program with a degree of distrust and alarm." Moreover, Arabs were not anti-Semitic, showing hostility only "toward the Jewish claim that they are the 'chosen people' and hence entitled . . . to special privileges." The Arabs saw the establishment of a Jewish state as a conduit for imperialism in the area. Hurley suggested that Jews should accept the compromise offered by the prime minister of Iraq, who argued that they should accept an Arab Federation including Palestine, Transjordan, Lebanon, Syria, and Iraq, with an Arab majority and a Jewish minority that had "autonomous rights" in districts where Jews were a majority.

Hurley was most disturbed that David Ben-Gurion, the Palestinian Zionist leader, was going everywhere saying that the United States supported a Jewish state in Palestine. Hurley asked that the president clarify the issue and point out that the United States consented only to the British Mandate for a Jewish national home that would not trespass on the rights of non-Jews in Palestine. Moreover, Hurley claimed, the Jews were seeking a state that could be imposed only by removing Arabs already there by force, even if the Jews had once had a historical claim to Palestine themselves. If the United States supported such a claim, Hurley argued, Mexico could then argue that it was entitled to restitution for the United States' having taken much of what was once its land and was now in the American West and Southwest.[57]

Such reports led Roosevelt to tell his friend Senator Robert F. Wagner, Jr., New York's leading Democrat, about his fears. There were about half a million Jews in Palestine, he wrote the senator, and on the other side were "seventy million Mohammedans who want to cut their throats the day they land." His main goal was thus to avoid a massacre, and he hoped the situation could be resolved through negotiations. Anything done such as action toward a Jewish state, FDR implied, would only "add fuel to the flames." American hopes were one thing. "If we talk about them too much," he cautioned Wagner, "we will hurt fulfillment." Better to keep discussions on the matter under the radar.[58]

The result of both missions was to be found in the president's reply to Ibn Saud on May 26, 1943, a note that echoed Hurley's conclusions and even used Hoskins's exact words. No decision regarding Palestine would be taken, Roosevelt assured Ibn Saud, "without full consultation with both Arabs and Jews." That statement quickly became the cornerstone and mantra of U.S. policy toward Palestine for the next few years.[59]

The loosely worded pledge to King Saud, however, became a virtual political football. When American Zionists elicited a positive response to their demands from the White House, as they invariably would, the Arab states and leaders immediately protested. The Division of Near Eastern Affairs at State would then reassure the Arabs that policy had not changed; nothing would be done without consulting them and looking out for their interests. The evidence indicates that FDR approved this strategy and that reassuring messages to the Arabs had not been sent out by State against his wishes.[60]

It was clear that the president was highly ambivalent, giving assurances to both the Arabs and the Jews about the possibilities for the future. Foreign policy was one thing; pacifying a key component of the coalition that had given him electoral victory was another. In March 1944, FDR had reiterated his good intentions to Wise and Silver. Emerging from a White House meeting, the two rabbis told the press that FDR had told them the U.S. government had not approved the British White Paper of 1939 limiting Jewish immigration to Palestine. The president was happy, they reported, "that the doors of Palestine are today open to Jewish refugees, and that when future decisions are reached, full justice will be given to those who seek a Jewish national home."[61] The Zionist leaders presented this as a major victory, receiving cheers from the audience when they spoke at a Zionist dinner in New York that evening.

They were unaware, of course, that almost immediately, Secretary of State Cordell Hull had sent out a reply to the Arab states. In FDR's comments to the rabbis, Hull stressed, "the president pointed out that the statement mentioned only a Jewish National Home, not a Jewish Commonwealth, and added that although the United States had never approved of the White Paper, it had never disapproved of it either."[62] As usual, the president carefully pursued his policy of obfuscation and again was successful at keeping all the adversaries at bay.

FDR, believing in his personal charm and powers of persuasion, was convinced that if he could just sit down with King Saud of Saudi Arabia, whom he considered to be the leader of the Arabs, he would be able to iron out the Palestine situation.[63] He had told his cabinet that after the wartime meeting of the United States, Britain, and the Soviet Union at Yalta, he would meet with Ibn Saud and "try to settle the Palestine situation." Back in Washington, the economist and White

House consultant Herbert Feis was amazed that Roosevelt "cherished the illusion that presumably he, and he alone, as head of the United States, could bring about a settlement–if not a reconciliation–between Arabs and Jews." Feis remembered thinking as he left the White House after hearing the president talk about Middle Eastern affairs, "I've read of men who thought they might be King of the Jews and other men who thought they might be King of the Arabs, but this is the first time I've listened to a man who dreamt of being King of both the Jews and the Arabs."[64]

It is difficult to know what FDR truly believed. The president told Rosenman that the issue of Palestine could be settled by "letting the Jews in to the limit that the country will support them–with a barbed wire fence around the Holy Land." Rosenman mused back that it might work: "if the fence was a two-way one to keep the Jews in and the Arabs out."[65]

In his diary on November 10, 1944, the handsome, gray-haired, ex–steel executive Edward R. Stettinius, Jr., now acting secretary of state, wrote that in a private meeting with FDR, the president had told him he was "confident" that he "will be able to iron out the whole Arab-Jewish issue on the ground where he can have a talk." The president's own position, he told the Stettinius, was that "Palestine should be for the Jews and no Arabs should be in it, and he has definite ideas on the subject. It should be exclusive Jewish territory."[66] Not even the most militant Zionists, it must be said, had ever called for such a drastic solution to the Palestine issue.

Two months later the president told Stettinius, who had replaced the ailing Cordell Hull as secretary of state, that he planned to meet with Ibn Saud during his trip to Yalta. Stettinius told him that he thought this would be an excellent idea "because sooner or later we would have to take a definite position in regard to the Arabian-Jewish difficulties and to the Jewish national home in Palestine." FDR then told him his plan: he was going to take a map with him showing the "Near Eastern area as a whole and the relationship of Palestine to the area and on that basis to point out to Ibn Saud what an infinitesimal part of the whole area was occupied by Palestine and that he could not see why a portion of Palestine could not be given to the Jews without harming in any way the interests of the Arabs with the understanding, of course, that the Jews would not move into adjacent parts of the Near East from Palestine."[67]

Roosevelt hoped that a Jewish commonwealth in Palestine would emerge that would aid the economic development of Arab lands. Undersecretary of State Sumner Welles, for one, claimed in his memoir that Roosevelt thought a Jewish state could be a model of social policy and would help raise living standards throughout the Middle East. He had anticipated, Welles wrote, that once a Palestinian Jewish commonwealth was created, "far-reaching projects for irrigation, power development and the construction of communications" could be carried out. Perhaps the Arab countries would find this an inducement and their people and leaders would "overcome racial antagonism."[68]

This vision came in part from Walter Clay Lowdermilk, who was assistant chief of the Soil Conservation Service of the United States. A former Rhodes Scholar, Lowdermilk was a religious Methodist who had first visited Palestine in 1938, when he had sought to gather information that might help him understand what factors had led to the devastating dust bowl that had hit the American heartland in the 1930s.

Lowdermilk's report, later turned into the best-selling book *Palestine: Land of Promise,* argued that engineering miracles such as the TVA had proved that "modern engineers can harness wild waters to produce cheap power for industry and scientific agriculture to make over waste lands into fields, orchards and gardens, to support populous and thriving communities." Surveying the Jordan Valley in the Near East, Lowdermilk argued that the "combination of natural factors and a concentration of resources" there made it viable for a new, far-reaching reclamation project that would surpass what the TVA had done in the Tennessee Valley. As Lowdermilk put it, "The Holy Land can be reclaimed from the desolation of long neglect and wastage and provide farms, industry and security for possibly five million Jewish refugees from the persecutions and hatreds of Europe in addition to the 1,800,000 Arabs and Jews already in Palestine and Trans-Jordan."[69]

Lowdermilk's conclusions delighted the Zionists. Here was a scientific study that offered evidence that Palestine could support a much larger Jewish population. Moreover, Lowdermilk was obviously sympathetic to the Zionist goals. "Some place must be found," he emphasized, to "reinstate the Jews long without a country among the peoples of the earth." And that place could only be Palestine. "They have nowhere else to go."[70]

Lowdermilk was impressed by what the Jews had already accomplished. They have "demonstrated the finest reclamation of old lands that I have seen in three continents," he wrote. And "they have done this by the application of science, industry and devotion to the problems of reclaiming lands, draining swamps, improving agriculture and livestock and creating new industries," all done against "great odds and with sacrificial devotion to the ideal of redeeming the Promised Land."[71]

On the other hand, when he took an 18,000-mile car trip through Arab lands he found neglected and desolate wasteland. "The Arabs," Lowdermilk concluded, "have not the genius or ability to restore the Holy Land to its possibilities." He did not call for a formal Jewish homeland, but his implication was clear. Settling millions of Jewish refugees in Palestine, he concluded, "would erect an eternal memorial to our victory in this world struggle for democracy and world freedom." The Jews had to become "custodians of a new Palestine."[72]

Despite what the president had told some of his confidants, he had his doubts about the viability of Lowdermilk's analysis. His hesitation was strengthened by the State Department. Writing to the president in January, Stettinius warned him that other credible experts showed "that from a purely technical standpoint there are serious obstacles to" Lowdermilk's plan.[73] Lowdermilk's chief opponent was Dr. Isaiah Bowman, the president of Johns Hopkins University, who headed a State Department advisory group. Not only was Lowdermilk incorrect, Bowman argued, but Palestine could not support even its current population satisfactorily. Bowman's arguments influenced even Eleanor Roosevelt, who was otherwise more than sympathetic to the plight of Europe's Jews.[74] If such was the case, it was obviously out of the question for Zionists to advocate a major increase of Jewish immigrants, since it would be virtually impossible for them to sustain themselves.

Before leaving for Yalta, Roosevelt met with Wise and asked him for his opinion of Lowdermilk's optimistic projections for the development of the Jordan Valley. Wise admitted that some had called it impractical but that TVA chief David Lilenthal saw it as "extremely practical and desirable." The president was also concerned about Arab fears that the Jews would "seek to infiltrate the surrounding Arab countries." In truth, Wise responded, the reverse was taking place: the Arabs were migrating to Palestine in order to take advantage of its rapid development. Most important, Wise assured FDR, "the Jews have not the slightest desire or intention to colonize the Arab lands outside Palestine." Jews living in Arab lands would undoubtedly themselves leave for Palestine, since Jews in Iraq, Syria, and Yemen were ready to go at a moment's notice. As for the president's fear that the Soviets would oppose a Jewish state, Wise informed Roosevelt that President Edvard Beneš of Czechoslovakia, whom he called a "Zionist of long standing," had personally been told by Joseph Stalin that if Britain and the United States had no objections to the creation of a Jewish commonwealth in Palestine, neither would Russia.[75]

In preparing for Yalta, the president also asked James Landis, his director for Middle East operations, what he might say to King Saud regarding "a *rapprochement* to the Palestine problem." Ibn Saud felt strongly about the topic, Landis told FDR, and was adamant that he would not accept any middle ground. Unless the president was ready to offer new and far-reaching proposals, he suggested that FDR steer clear of the issue altogether. Indeed, Landis admitted that for the past twenty years, the State Department in effect had had "no policy" on the Palestine issue at all, considering it a British responsibility. To develop a new one, it had to start from the assumption that the goal of a Jewish commonwealth in Palestine had to be entirely given up, since "the political objective implicit in the Jewish state idea will never be accepted by the Arab nations." Such a stance might indeed be inconsistent with both Balfour and the Atlantic Charter, Landis advised. Nevertheless, one had to accept the clear "political limitations" facing the United States. His hope was that a Jewish home, rather than a state or commonwealth, could be found acceptable to both Arabs and Jews.[76]

Reinforcing Landis's opinion of King Saud's attitude toward the idea of a Jewish state, the future U.S. ambassador to Saudi Arabia, William E. Eddy, forwarded another missive. He reported that King Saud had told him that "if America should choose in favor of the Jews, who are accursed in the Koran as enemies of the Muslims until the end of the world, it will indicate to us that America has repudiated her friendship with us."[77] Stettinius reiterated to the president that Ibn Saud "regards himself as a champion of the Arabs of Palestine and would himself feel it an honor to die in battle in their cause."[78]

Now the State Department offered its own proposal: Palestine should become an international territory under trusteeship of the British, with a charter granted it by the new United Nations. This arrangement would supersede all previous agreements on Palestine, including those promised by the British in the Balfour Declaration.[79] Trusteeship would remain the State Department's favored policy for Palestine.

Two days after the inauguration, the president left on the long trip to the Yalta Conference at a Black Sea resort, a journey that would only drain the energy of the already ill leader. The main task addressed at the conference centered on the dismemberment, disarming, and demilitarization of defeated Germany. The Big Three–the United States, Britain, and the Soviet Union–agreed in principle to dismantle Germany, to reestablish the so-called Curzon Line as the border between Poland and the Soviet Union, to recognize Josip Broz Tito's authority in Yugoslavia, and to give the Soviets access to the Baltic in the former German port city of Königsberg. Most important, Stalin and Roosevelt agreed in principle to the holding of free elections in Poland and Eastern Europe, in the lands that had been liberated from Nazism. The situation of the Jews, it appeared, had been pushed aside.

At Yalta, concern for the Jewish refugees and their future in Palestine was not taken up by the Big Powers' delegates in their official deliberations. However, Roosevelt tried to find out where Stalin stood on Zionism. After dinner on February 10, Roosevelt asked Stalin if he supported the Zionist program. Stalin answered warily: yes, in principle, but he recognized the difficulty of solving the Jewish problem. "The Soviet attempt to establish a Jewish home at Birobidzhan," the Soviet dictator explained, "had failed because the Jews scattered to other cities. Only some small groups had been successful at farming." FDR then told him he was going to see Ibn Saud after the Yalta Conference. What, Stalin asked, was he going to give Ibn Saud when he saw him? He had thought of only one concession, the president quipped to Stalin; "to give Ibn Saud the six million Jews in the United States." FDR was being facetious, but Stalin evidently did not realize it. That would be difficult, Stalin replied seriously, since Jews were "middlemen, profiteers and parasites."[80] Roosevelt, however, came away from the conversation pleased that Stalin did not seem opposed to Zionist aims in Palestine.

The president had been duly warned by the State Department and his advisers about the implacability of the Arab states to Zionist demands, as well as the hostile position taken by King Saud. Ever the optimist, FDR still believed that his charm and commitment to negotiation could work and that he would be able to make a breakthrough that would be acceptable to all sides.

It was in this context that the president decided, at the end of the Yalta Conference, to meet with Emperor Haile Selassie of Ethiopia, King Farouk of Egypt, and, most important, King Saud of Saudi Arabia. Palestine was not the only subject on the agenda. Roosevelt was fully informed about the importance of the Saudi oil reserves. These were considered essential to American security, given the awareness that both increased demand for oil in the West, along with the dwindling of domestic reserves, made the securing of Middle Eastern oil a key necessity of an American foreign policy at the war's end. He knew that oil industry executives in particular were waiting for the war's conclusion, at which time they looked forward to exploiting the vast oil reserves found in Arabia. While the war was still on, FDR had to be careful not to antagonize the Arab leaders, since a major supply line to the Soviet ally ran through Arab territory on the Persian Gulf.

Already American interests in Saudi oil were being developed. The

Arabian American Oil Company (ARAMCO), *The New York Times* reported, which had invested $100 million in Saudi Arabia, was expected to expand the amount tenfold in the next decade. American wartime airfields were being surveyed for use in the future by commercial aircraft, and American banking firms were studying the potential for opening bank branches.[81]

One can only be impressed by the insistence of the physically weak president to carry out this mission. After the strenuous eight-day meeting with Stalin and Churchill at Yalta, FDR took a five-and-a-half-hour flight to Egypt, where he boarded a Navy cruiser, *The Quincy,* at Great Bitter Lake on the Suez Canal. That body of water, the historian Frank Freidel pointed out, was "called Mara in biblical times, and was an ironic location for Roosevelt's debate with Ibn Saud over the fate of Jewish refugees, for (as he knew) Moses had stopped there with the Israelites on the flight from Egypt toward the promised land."[82] The next day, he entertained both King Farouk and Emperor Haile Selassie of Ethiopia, waiting anxiously for the really important meeting–the one scheduled the following day: February 14, 1945.

The president went all out to impress Ibn Saud. *The New York Times* reported in a banner headline that the warship had been made into an "Arab Court in Miniature." The display was "without precedent in American naval annals." The ship's crew had spread dozens of thick Oriental carpets on the deck and set up a royal tent in front of the forward gun turret.[83] Ibn Saud "lived in it as he were making a pilgrimage somewhere in the vast desert regions of Arabia."[84] Aware of the eating habits of the Muslim monarch, they built a sheep pen at the fantail, large enough to feed the entire royal party.

The meeting was difficult for both leaders. The ailing king could not walk up the gangplank and was hoisted onto the ship in a lifeboat. Roosevelt's polio, bad health, and the stress of the Yalta Conference had left him weak and tired. He waited to receive the king sitting on an armchair put up under the *Quincy*'s forward guns. The monarch's chair, unlike that of the president, was a gilt armchair, on which he sat for most of his meetings, "guarded by barefoot Nubian soldiers." Ibn

Saud's party also included, the diplomat Charles E. "Chip" Bohlen observed, a royal astrologer, a coffee server, and "nine miscellaneous slaves, cooks, porters and scullions."[85]

The president, realizing the high stakes, tried to charm the Saudi monarch. He was "in top form," his translator William A. Eddy wrote, "as a charming host, witty conversationalist, with the spark and light in his eyes and that gracious smile which always won people over whenever he talked with them as a friend."[86] As a guest of the president, Ibn Saud waited for his host to bring up topics for discussion. Roosevelt immediately introduced the problem of the Jewish refugees' plight, for which he had obvious sympathy. FDR expressed his admiration for their role in developing the land in Palestine. But he prefaced his comments by stating that he had "a serious problem in which he desired the King's advice and help," namely, the rescue and settlement of the remnant of European Jewry. "What would the King suggest?" FDR asked.[87]

Simply introducing the topic of the Jews met with a negative reaction from Ibn Saud, who told the president that "in his opinion the Jews should return to live in the lands from which they were driven." If that was impossible because their homes were destroyed, the answer was to give them "living space in the Axis countries which oppressed them." Moreover, he was not impressed with FDR's claim that Jewish farmers had developed the land in Palestine. That had been accomplished with American and British capital, he claimed, and only the Jews would gain from the prosperity. The truth was, Ibn Saud continued, that Arabs and Jews "could never cooperate, neither in Palestine, nor in any other country." Arabs would rather die, he told FDR, "than yield their land to the Jews."[88] As far as he was concerned, the Arabs had done no wrong to Europe's Jews. The cost of helping them had to be carried out exclusively by the defeated Germans. Said the monarch, "Make the enemy and the oppressor pay; that is how we Arabs wage war."[89] The king's expectation was that the United States would give him full support.

FDR, who had come to the meeting hoping to convince the king to accept some kind of compromise, decided immediately that it was best to assure Ibn Saud that he would get precisely the support he expected from the United States. He would never, he assured the concerned monarch, help the Jews at the expense of the Arabs. King Saud immediately seemed relieved.

The president then explained to Ibn Saud that he was talking as the chief executive but it was impossible for him to prevent speeches made on behalf of the Zionist project or to stop resolutions in favor of a Jewish homeland passed by Congress. He should not confuse these outbursts, FDR emphasized, with U.S. policy. His own future policy, he assured King Saud, would be one friendly to the interests of the Saudi monarchy.[90] His chief adviser, Harry Hopkins, was not happy. On the *Quincy,* Hopkins had remained in his cabin, too sick to attend the meeting itself. When he heard the outcome, Hopkins thought that the president's ill health had led him to be "overly impressed" with Ibn Saud and had caused him to abandon his earlier pro-Zionist position far too easily.[91]

British Prime Minister Winston Churchill was riled at the news of FDR's meeting with Ibn Saud. An American-Saudi alliance, he feared, might be a potential threat to British dominance in the Middle East. Hence he decided to meet the three leaders that the president had just seen. Churchill had made his own promises to the Zionists, and he considered himself to be one, if somewhat inconsistently. During a visit to Palestine in 1921, as colonial secretary, Churchill had been impressed by the achievements of the Jewish pioneers in Palestine. After visiting Rishon Le-Zion, one of the oldest Jewish agricultural villages, founded in 1882, he enthusiastically told the House of Commons, "Anyone who has seen the work of the Jewish colonies . . . will be struck by the enormous productive results which they have achieved." "From the most inhospitable soil, surrounded on every side by barrenness and the most miserable form of cultivation," he was driven during his visit "into a fertile and thriving country estate, where the scanty soil gave place to good crops and good cultivation, and then to vineyards and finally to the most beautiful, luxurious orange groves, all created in 20 or 30 years by the exertions of the Jewish community who live there."[92]

Churchill had his own plans for Palestine after the war. In 1943, he told Chaim Weizmann that he wanted Britain to offer Ibn Saud the leadership of a Middle Eastern Arab Federation, for which he would be paid £20 million a year. In exchange, the king would have to give his support for a Jewish state in Palestine.[93] Later Churchill proclaimed the 1939 White Paper a "gross breach of faith" and told Chaim Weizmann that after victory he would see to it that Jewish refugees would be able to go to Palestine. The promise of a Jewish homeland, Churchill said, was "an inheritance left to him by Lord Balfour and he was not going to change his attitude." After Hitler's defeat, he added, "the allies will have to establish the Jews in the position where they belong." The Jews and the Zionists should not worry, the prime minister assured Weizmann, because they had a "wonderful case." He hadn't changed his views, and he promised that "he would bite deep into the problem and it is going to be the biggest plum of the war."[94]

But Churchill fared no better with Ibn Saud than Roosevelt had. He began the meeting by reminding the Saudi monarch that Great Britain had "supported and subsidized" him for twenty years and had made his reign possible "by fending off potential enemies." Therefore Britain, Churchill argued, was entitled to ask for Ibn Saud's help "in the problem of Palestine, where a strong Arab leader can restrain fanatical Arab elements, insist on moderation in Arab councils, and effect a realistic compromise with Zionism." Churchill expected both sides to make concessions, and he expected the Saudi monarch to do his part.

What Churchill suggested, Ibn Saud answered, would not help the Allies or Britain, "but [was] an act of treachery to the Prophet and all believing Muslims which would wipe out my honor and destroy my soul. I could not acquiesce in a compromise with Zionism much less take any initiative." Furthermore, "the British and their Allies would be making their own choice between 1. a friendly and peaceful Arab world, and 2. a struggle to the death between Arab and Jew if unreasonable immigration of Jews to Palestine is renewed. In any case, the formula must be one arrived at by and with Arab consent."[95] Churchill, Rabbi Stephen Wise commented, had the same problem that FDR faced: he was unable to make an impression on King Saud. "It may have been his own fault," Wise later wrote Weizmann, "because he set out to browbeat the old man for two hours, after which he changed his tack and was sweet as honey with him, but nothing availed."[96]

On March 1, FDR's acquiescence to Ibn Saud was made clear when he reported on Yalta in a speech to Congress. The country could see the toll his trip had taken. The president, despite his polio, had always given his speeches standing up. He delivered this one sitting down, apologizing for his "unusual posture" but telling his nationwide audience and the members of Congress that "I know you will realize it makes it a lot easier for me in not having to carry about ten pounds of steel around the bottom of my legs, and also because I have just completed a 14,000 mile trip."[97]

Truman commented years later that all could see that "the famous Roosevelt manner and delivery were not there. And he knew it." Yet both houses of Congress were "awed by his dramatic display of sheer will power and courage."[98] In the speech, FDR made the comment that would shock the Zionist activists, as well as the entire nation, whose citizens overwhelmingly favored a Jewish homeland in Palestine. At the meeting with Ibn Saud, he noted, "I learned more about the whole problem of Arabia–the Moslems–the Jewish problem–by talking to Ibn Saud for five minutes than I could have learned in the exchange of two or three dozen letters."[99] In saying this, he was undoubtedly referring to the extremely negative–some would say fanatical–response of the Wahhabi king on the subject of a Jewish homeland in Palestine.

The American Jewish community and all of its representative groups were horrified. FDR had thrown in the comment toward the end of the speech, almost an aside to a presentation of what had been attained at Yalta. The pro-Zionist American Jewish Congress accused the president of reneging on his preelection endorsement of the Democratic Party's platform. Congressman Emanuel Celler bitterly complained that of all the great issues discussed by the Big Three at Yalta, not one of the world's leaders had seen fit even to mention the catastrophe overtaking Europe's Jewry. Celler was especially irked that the president had gone out of his way to meet with Ibn Saud but had had no time for the Jewish leaders in Palestine. As Celler wrote in his autobiography, "I listened with dismay and disbelief to his implied repudiation of the Jewish claim to Palestine."[100] Senator Edwin C. Johnson (D.–Colorado), a member of the pro-Zionist American Christian Palestine Committee, commented that "the choice of a desert king as an expert on the Jewish question is nothing short of amazing." Johnson added that he thought even FDR's dog, his beloved Fala, "would be more of an expert."[101] Only Sumner Welles, the president's assistant secretary of state and his good friend, stood by his boss. Calling the charge that FDR had backed down on his support of a Jewish Palestine "malicious," Welles tried what today would be called spin: FDR, he wrote, "did not modify in one iota the basic principles that he had constantly supported . . . the kind of settlement . . . that would provide the Jews with their promised national homeland."[102]

Realizing the harm his comments had caused in the Jewish community, FDR tried to minimize the damage. Once again, he turned to the Zionist leader he most trusted, Rabbi Stephen Wise. The rabbi in turn provided a detailed account of their meeting to Chaim Weizmann. Wise began his meeting with the president by "congratulating him upon a gloriously successful mission," referring to the Yalta Agreement. Knowing full well that Wise's central concern was the Zionist project, Roosevelt mournfully and candidly acknowledged, "I have had a failure. The one failure of my mission was with Ibn Saud." The president felt especially bad about this, he told Wise, because he had "arranged the whole meeting with him for the sake of your cause." FDR continued, "I tried to approach the Jewish question a number of times. Every time I mentioned the Jews he would shrink and give me some such answer as this–'I am too old to understand new ideas!' When, for example, the President began to tell him about what we have done for Palestine through irrigation and the planting of trees, Ibn Saud's answer was, 'My people don't like trees; they are desert dwellers. And we have water enough without irrigation.' The President added, 'I have never so completely failed to make an impact upon a man's mind as in his case.' "

The rabbi accepted FDR's explanation, telling Weizmann that Roosevelt had seen Ibn Saud "for our sake" but had come out of the meeting fearful that if the king united the Arab armies, he could easily defeat all the Jews of Palestine. The president did give Wise some good news: "Stalin is all right and he is with us," and as far as Churchill went, there had been no change "with regard to Zionist plans." Moreover, he assured Wise, Churchill was equally concerned with the plight of European Jewish refugees who sought entry to Palestine but were being sent elsewhere against their will. The prime minister had said he would simply avoid talking about the harshness of the White Paper and "let the Jews come in."

Finally, the president told Wise that he was seeking another way to deal with the Saudi ruler. "He is seventy-five years old," the president commented, "and has swollen ankles, so that perhaps we had better wait until he goes!" Wise "instantly demurred" and, given the pressing situation of the Jews, told the president that he had to act quickly and in a firm manner. Then FDR thought that perhaps there was "another way of dealing with the problem. Since we cannot move I.S., and since the other Arab chiefs will go along with him, I have been thinking about the plan of our putting the case up to the first meeting of the Council of the United Nations, whenever it meets." Ever loyal to FDR and his intentions, Wise concluded that "the President is dead in earnest about this, and that if we were to agree with him, he would feel that he and Churchill must put it up to their associates and urge the imposition of a Jewish State on I.S. and his associates by the Council of the United Nations." "The President," Wise confidently told Weizmann, "remains our friend as much as ever."[103]

On the issue of Saudi Arabia and the Middle East, the State Department was firm in its belief that Arab goodwill would lead to increased American influence, proven especially by King Saud's preference for American investment in Arabia.[104] Near Eastern Office Chief Wallace Murray's feathers were therefore ruffled when he heard that the president had told Rabbi Wise that he still favored a Jewish state in Palestine. Warning the president that such a step would have "serious repercussions in the Middle East," Murray noted that it could even lead to bloodshed as well as "endanger the security of our immensely valuable oil concession in Saudi Arabia." Finally, he argued, Stalin was actually opposed to a Jewish state, and continued American endorsement of the Zionist goal by the president might result in throwing the entire Arab world into the arms of the Russians.[105]

The president, bombarded by Wise and the Zionists on the one hand and the opponents of Zionism in the State Department and Arab leaders on the other, felt increasingly gloomy about the prospects for any kind of Middle East settlement. At a luncheon he held with the first lady and Colonel Harold Hoskins, he reiterated sadly how opposed King Saud was to Jewish concerns, even with a proposal floated by the president to settle Jews in Libya. Eleanor was as usual more sympathetic to the plight of the dispossessed Jews and tried to lift her husband's gloom by noting the "wonderful work that had been done by the Zionists in certain parts of Palestine." Her comments did not impress the president. Except along the coastal plain, he replied, "Palestine looked extremely rocky and barren to him as he flew over it." But the Zionists were stronger, the first lady responded, and "were perhaps willing to risk a fight with the Arabs at Palestine."

That, precisely, is what upset FDR. "There were many more Arabs than Jews in and around Palestine," the president noted, "and in the long run, he thought these numbers would win out." Hoskins recalled that he had been attacked by American Zionists when he had said in 1943 that the only way to establish a Jewish state in Palestine would be by force. Now Hoskins asked the president if he agreed with this conclusion, and he said he did. The answer, Hoskins still maintained, lay in State's plan for a trusteeship and the establishment of Palestine as an international territory sacred to Muslims, Christians, and Jews. FDR concurred and "said he thought such a plan might well be given to the" newly created United Nations.[106]

A favorable conclusion to the Palestine issue seemed to lie far in the future. In the meantime, Roosevelt sought both to contain the damage from his spontaneous comment about Ibn Saud and to assure

American Zionists that his true sympathies lay with their cause. In his heart, Roosevelt may well have thought he had done what he could on their behalf.

All of FDR's contradictory positions and vacillations created havoc at the State Department. State's deputy director of Near Eastern and African affairs, Paul H. Alling (who was actually the nominal chief of the Division of Near Eastern Affairs at State),[107] thought that FDR's indecision was producing political instability in the Near East and having a negative effect on America's standing in the entire region. Even worse was the president's meeting with Wise and Silver, which on March 9 had allowed the two to claim at a press conference that FDR's promises to them "appeared to affirm the President's support of the Zionist position." And when FDR wrote Senator Robert F. Wagner of New York that he gave his support to the Democratic Party's Palestine plank–one sympathetic to Zionist aims–it further raised doubts for the Arabs about the veracity of Roosevelt's pledges to them.

The very last meeting the president held before departing for Warm Springs, Georgia–where he was to get some much-needed rest–was with American Jewish Committee leaders Jacob Blaustein and Joseph Proskauer. The president told them that he had previously warned Rabbi Wise "that he gravely feared a continuation of the agitation for a Jewish state," because it might cause a world war as well as well as strife in Palestine. What he did not tell Wise, and what he did tell Proskauer and Blaustein, was that after talking with Ibn Saud, he had been "frightened," and he wanted them "to talk to the Zionist leaders about it."[108] Most important, FDR told Blaustein and Proskauer of his "belief that the project of a Jewish state in Palestine was, under present conditions, impossible of accomplishment" and that the Jews' objective should be to secure liberal immigration into Palestine and work for the rights of Jews in all nations through the auspices of the new United Nations. Admitting that his new view was "somewhat at variance with his public utterances," Roosevelt explained that he had "learned a great deal at Yalta" and hence supported the American Jewish Committee's attempts to "moderate the sharpness of the propaganda of the extreme Zionists."[109]

FDR's view of the Palestine situation was ambiguous. At times he seemed sympathetic to the creation of a Jewish state; at others, he seemed to suggest that it could not be built without using military force, a commitment he did not want to make. When Harry S. Truman assumed the presidency, his only knowledge of what FDR thought would come from his public statements on Palestine. As on other issues, Truman had been left out of the loop. Now he would find that Palestine would become one of his most difficult problems.

TWO

TRUMAN INHERITS A PROBLEM

Yalta and the meeting with Ibn Saud exhausted the ailing Roosevelt. As usual, he sought solace and rest at what everyone called the Southern White House, his retreat and health spa at Warm Springs, Georgia. William D. Hassett, FDR's correspondence secretary, accompanied the president. The sixty-one-year-old staff member and former journalist kept a daily diary of his White House years.[1]

The president had scheduled many events for the future–a state dinner for the regent of Iraq, a trip to San Diego to visit his grandchildren, and a short stop at Hyde Park before returning to the capital. None of these was to be. "In the quiet beauty of the Georgia spring," Bill Hassett wrote in his diary, "like a thief in the night, came the day of the Lord." On April 12, 1945, the president suffered a massive cerebral hemorrhage and died.[2]

The nation, as well as many of the president's own staff, was shocked. Today we have full knowledge of the extent of FDR's ill health. At the time, his true condition was carefully sanitized by his own medical staff. Rear Admiral Ross McIntyre, the president's personal physician–who knew the truth about FDR's condition–told the press that his death had come "out of a clear sky."[3] Even Judge Samuel Rosenman, FDR's special counsel, was somewhat surprised. Rosenman had met with Roosevelt each night when he was at the White House. When FDR returned from Yalta, Rosenman thought the president looked sick. Yet when he got the news about Roosevelt's hemorrhage and death he was unprepared, because FDR had always bounced back when he went to Warm Springs.[4]

But others had feared the worst. Vice President Truman's old army buddy (and chief administrative assistant to Truman when he became president), Ed McKim, had been shocked by FDR's appearance when Truman had taken him to a White House reception for the Hollywood film *Wilson*. Wondering if the president had long to live, McKim told Truman to take a long look at the White House, since he would soon be living there. "I'm afraid you're right, Eddie," Truman responded, "and it scares the hell out of me."[5] Truman was actually alarmed at FDR's condition. When he met with Roosevelt, a very rare occurrence, the president's "hands shook so badly . . . that he could not get the cream from the pitcher into his coffee," and even speech was difficult. "I'm very much concerned about him," Truman confided to his daughter, Margaret.[6] And when Truman met the president a week after he returned from Yalta, he was stunned. "His eyes were sunken," Truman recalled. "His magnificent smile was missing from his careworn face. He seemed a spent man."[7]

It was a rainy day back in Washington that sad day in April. Truman had spent it listening to a rather dreary Senate debate about a Mexican-American water treaty. When the Senate adjourned, Truman walked over to the office of the speaker of the House, Sam Rayburn. Waiting for him there were Rayburn and the House parliamentarian, Lewis Deschler. Sitting down for an end-of-the-day drink, they were joined by congressional liaison James M. Barnes. The vice president was looking forward to his favorite form of relaxation, a good game of poker at the Statler Hotel, where his friend Ed McKim was staying. Before Truman arrived, the speaker took a phone call from Steve Early, FDR's assistant. He had asked Rayburn to tell Truman to phone him immediately at the White House when he arrived.

By the time Truman walked in, the men had forgotten about Early's call. The vice president poured a bourbon and shone in the presence of the seasoned politicos. Deschler then remembered that Early had phoned. Wasn't Truman supposed to return the call right away?

Truman dialed the White House and asked to speak to Early. When he hung up, he looked pale and exclaimed, "Jesus Christ and General Jackson." Early had told him to come to the White House immediately and enter through the main Pennsylvania Avenue entrance. Grabbing his coat on the way out, he said, "Boys, this is in the room, something must have happened."[8]

Eluding his Secret Service retinue, Truman ran through the Capitol basement building to where his car and driver were waiting, stopping in his office to grab his hat. The vice president had his driver go as fast as possible in the evening rush-hour traffic. Truman speculated about why he had suddenly been summoned. He didn't want to think the worst. Perhaps he had to meet FDR, who had secretly returned from Warm Springs to visit the grave of one of his good friends, the Episcopal bishop of Arizona, who had been buried in Washington that day. Perhaps, he thought, the president wanted him to undertake some special mission to Congress. He entered the White House at approximately 5:25 P.M. Taken to the second floor, he saw the first lady, Eleanor Roosevelt, sitting with Steve Early. "Harry," Mrs. Roosevelt calmly told him, "the President is dead." Truman found himself unable to speak. "Is there anything I can do for you?" he finally asked. She immediately replied, "Is there anything *we* can do for you? For you are the one in trouble now."

Before long, Secretary of State Stettinius entered the room. Truman asked that the cabinet be assembled as quickly as possible. He was soon joined by Speaker Rayburn, the House majority leader, John W. McCormack, the minority leader, Joseph W. Martin, and Truman's wife, Bess, and daughter, Margaret. Rayburn searched for a Bible on which Truman could take the oath of office. All that could be found was a weathered, simple one that sat in the desk drawer of the Senate usher, Howell Crim. At 7:09 P.M.–Truman committed the exact moment to memory, looking at the clock beneath Woodrow Wilson's portrait in the cabinet room–Chief Justice Harlan Stone administered the oath. "Standing erect," one reporter wrote, "with his sharp features taut and looking straight ahead through his large round glasses," Harry Truman "became the thirty-second President of the United States."

The ceremony took one minute.[9] "His face was grave," the *Times* columnist and Washington insider Arthur Krock reported, "but his lips were firm and his voice was strong."[10]

Truman made his first presidential decision when he met with his cabinet a short time later. The press, and the nation, wanted to know if he would go through with plans for the San Francisco Conference to plan the new United Nations that was scheduled for April 25. Tom Connally, the chairman of the Senate Foreign Relations Committee, was certain it would be postponed. Truman, by his own candid account, was rather uninformed about the issues pertaining to American foreign affairs, but he understood the need to let it proceed. The United Nations was crucial to the goal of winning the peace, and the United States' European allies needed to be assured that the country was in good hands and that FDR's policies would continue as planned.

Having settled that issue, Truman turned to his cabinet. He would be "President in his own right," he informed them. He wanted their thoughts, and they should feel free to differ with him, but he alone would make the final decisions.[11] Jonathan Daniels, the press secretary, would shortly announce Truman's first goal. He promises, Daniels informed the world, "to prosecute the war on both fronts, East and West, with all the vigor we possess to a successful conclusion." The president would be at his desk by nine the next morning.[12] The many tough issues facing the country would now be his problems.

For a nation still in shock, the very idea of Harry S. Truman taking over as chief executive and commander in chief was simply unacceptable. Many saw him as a product of the Tom Pendergast political machine in Truman's native Missouri, as a politician who only by accident had become the leader of the most powerful nation on Earth. Truman himself knew he was not prepared. The few times he had met with Roosevelt, he had never been made privy to the nation's top secrets or the considerations FDR had faced when making foreign policy decisions. Thus FDR's secretary of war, Henry L. Stimson, stayed behind as the cabinet adjourned. He had something important to let the new president know. There was a new explosive of immense power being developed, he told Truman. Ironically, the Soviet dictator, Joseph Stalin, who had learned long ago about the A-bomb's development from Soviet agents in the Manhattan Project, had known about it long before Harry S. Truman.

Many had doubts about Truman's capacity to do the job. Yet once they saw him in action, they had to acknowledge the great contrast between Truman and FDR. At age sixty, just two years younger than FDR, Truman exuded vitality. He stood five feet, ten inches and weighed 167 pounds, with a compact frame. George Elsey, who worked in the White House Map Room–the heart of the government's top secret intelligence and communications center–recalled that "it was hard to believe he was only two years younger than the gaunt man in the wheelchair whose hands trembled."[13]

The new president not only knew that he had a great deal of catching up to do, he read regularly in the press how millions of Americans regarded Roosevelt as indispensable. Immediately, Truman sought to reassure Americans of both his good intentions and his capabilities. For that purpose, he turned to a journalist who was familiar with his roots, the editor of *The Kansas City Star,* Roy Roberts. He and another Missouri journalist, Duke Shoop, were the only members of the press who were given access to the president his first day in office.

Truman always felt the most comfortable with men from his home state. Americans were already commenting that Truman looked just like any average man on the street. That, Roberts said, was precisely his advantage. Unlike FDR, who was an upper-class patrician, Truman knew the meaning of hard work, debt, and struggle and the responsibilities of leadership during wartime. "He has plowed corn . . . he sold haberdashery, and failed at it. He worked in a mailing room. He fought bravely in his country's war. Then he started climbing the ladder in politics, with a political machine, his sponsor, as the worst handicap to overcome in any possible climb to the Presidency." As he was a true man of the people, Roberts implied, the shift of power would move from the capital to the Missouri River. The viewpoint in the White House would no longer be that of the nation's East Coast elites. In the New Deal days, the gospel at the Roosevelt White House had been the New York City left-wing paper *P.M.*, and the liberal newsweeklies *The Nation* and *The New Republic*. Most of the country had never heard of them, Roberts wrote, yet "they were bibles in Washington, D.C." The journalist predicted that Truman would willingly shift power from the executive branch to the people's representatives in Congress. He would soon develop his own Truman administration and Truman program; the postwar climate the new president envisioned would move the country slightly to the right. Truman's innate Missouri conservatism would temper FDR's desire to please the left wing of his party.[14]

Harry S. Truman woke up at 6:30 A.M. on his first morning as president. After his usual morning walk–his "constitutional," as he would always refer to it–and breakfast, he left for the White House in order to arrive at 9 A.M., as he had promised. His new office, so recently occupied by FDR, was filled with the late president's mementos and paraphernalia. There were Roosevelt's ship models and ship prints, items foreign to Truman, whose own past was that of Missouri farm fields. And Truman as yet had no staff of his own. His only personal aide was his old vice presidential secretary, Matthew J. Connelly, whom he would soon make his new appointments secretary.[15]

In the afternoon he received a detailed memo from Secretary of State Stettinius, presenting a summary of the major world issues he would have to deal with. The long list included relations with Britain, the problems emerging between the United States and the Soviet Union, the question of how to deal with Poland and the Balkan areas, the issue of how to deal with Germany, Austria, and Italy at the war's end, and other less pressing matters.

Later, he drove to the Capitol to meet with leaders of both houses of Congress, seeking their cooperation and bipartisan support. As he left, turning to both reporters and Senate pages who had lined up to applaud him, Truman made what would become one of his best-known statements: "Boys," he said, "if you ever pray, pray for me now. . . . When they told me yesterday [about FDR's passing] I felt like the moon, the stars and all the planets had fallen on me."[16]

Truman, as vice president and a man of the Senate, had a good grasp of domestic policy issues. But on the all-important and increasingly sticky areas surrounding foreign affairs, he was a greenhorn. "Everyone," Admiral William D. Leahy said, "including Truman himself, knew that in the field of international relations he had much to learn."[17] Above all, the developing conflicts with Joseph Stalin and the Soviet Union stood as the most important. He would quickly hear, from men like Secretary of State Edward Stettinius, Admiral Leahy, Ambassador to Russia W. Averell Harriman, Russian expert Charles E. ("Chip") Bohlen, and Secretary of War Henry Stimson, that tensions between the two wartime allies were beginning to heat up and the always tenuous alliance was beginning to slowly unravel.

What came to be called the Cold War would hover over the nation for half a century. In its earliest days, Truman would have to learn how to balance FDR's hopes for a cooperative postwar relationship with the Soviets with the growing realization that Stalin's expansionist goals would make such cooperation almost impossible to achieve. FDR had thought that his personal diplomacy would maintain a postwar alliance composed of the United States, Britain, the Soviet Union, and China. Truman would be forced to adopt a different policy, based on a show of firmness and a refusal to make unwarranted concessions to the Russians. Communism and democracy, he understood, were bound to be in fundamental opposition and conflict; no charm or personal negotiating could succeed in taming Soviet ambitions.[18]

Truman was a quick learner. Admiral Leahy had quickly gotten to work gathering papers that the president would have to read, in order to make good policy judgments. These included summaries prepared by the Joint Chiefs of Staff, as well as other policy papers from various government departments. He was pleasantly surprised to see that within a few days, Truman had thoroughly digested them and was catching up on what had to be done to win the war. But it was a daunting task. When the Democratic financier and wealthy businessman Abe Feinberg came to see Truman at the White House, he found him in what he called a "semi-dazed" state. Explaining why he looked that way, Truman told Feinberg, "When I came into this office I asked for all the documents that I should know about." Feinberg found them piled all around the president's desk, standing higher than the desk from the floor.[19]

Aside from the critical foreign policy issues, Truman was busy deciding whom to appoint to the important presidential staff positions. Among those he retained from FDR's staff were two Jewish men who would become instrumental in dealing with the Palestine issue: Judge Samuel I. Rosenman and David K. Niles.

Rosenman was born in San Antonio, Texas, in 1896 to Russian Jewish immigrants. His father was a clothing manufacturer who moved his family to New York City when his son was eight years old. After attending Columbia College and Law School, Rosenman was elected to the New York State Assembly. He began his career with FDR during the 1928 campaign for governor of New York. When Roosevelt won, Rosenman stayed on as a speechwriter and counsel. Elected in 1932 to the New York State Supreme Court, Rosenman in his spare time served Roosevelt in different capacities when he became president, including that of speechwriter. Retiring from the bench in 1943, he moved directly to the White House staff, as special counsel to the president.[20] FDR appreciated Rosenman, whom he nicknamed "Sammy the Rose" but called Sam, for his intellect, judicious judgment, and writing abilities. He was a confidant of the president and worked in every Roosevelt political campaign, drafted nearly every speech, and coined the term "New Deal."[21]

Truman realized that Rosenman's two-decades-long relationship with Franklin Roosevelt would serve him well. His very presence became a reassuring symbol for old New Dealers who worried that Truman would not carry on the policies that had made FDR beloved to the American public. They worried that, as some people expected, Truman would be far more conservative than Roosevelt. If some had those fears, they were ill advised. Since 1944, Truman had become a major figure in the liberal wing of the Democratic Party. He had always had good relations with both the labor movement and blacks, the two key constituencies that had stood with FDR in the northern centers.[22] Rosenman also became an important, though unofficial, adviser to Truman on the Jewish situation.

David K. Niles had served FDR as an adviser on patronage and minority-group issues and New York City politics. He also served as a liaison with the Jews, who were pressing FDR on the issues of rescue, refugees, and Palestine. Roosevelt counted on Niles to advise him on which Jewish leaders he had to see and which ones he could put off without harming his political capital with Jewish constituents. As Niles explained to one of Truman's aides, "I envision my chief job to protect the President" from those who might "create some political damage."[23]

Urging Truman to retain his services, Matthew Connelly told Truman that if he lost Niles, "he would lose somebody who would be completely loyal to him," who had backed him for nomination as vice president in 1944, and who could be invaluable working for Truman in the same job he had had for FDR, as an adviser on minority groups. Niles, Connelly said, "was a very bright political analyst. He was quiet, he was receptive, he was never out in front." In FDR's days, the staff called him the "back stair boy at the White House."[24] Truman took Connelly's advice and put Niles in charge of information coming to the White House that had to do with Jewish issues, including Palestine.[25]

Born in 1890, Niles grew up in Boston's rough North End, where his father was a tailor. Always a hard worker, he earned a place at the elite Boston Latin School and upon graduation got a job at the information office of the Department of Labor during the First World War. Not being able to afford college, Niles engaged in a process of self-education. He gravitated toward the Ford Hall Forum, a Boston institution that was a center of lectures and discussions for the city's intellectual community, eventually becoming its associate director. In that capacity, he made important political connections. Eventually he came to Washington to work for the New Deal and became a personal assistant to Harry Hopkins, FDR's top aide.[26]

Somewhat of a loner with a passion for anonymity, Niles lived and breathed politics. A lifelong bachelor, he lived at a hotel close to the White House, where he followed the same routine every weekend. On Friday he would board a train at Union Station bound for New York City, where he would attend a Broadway play. The next morning, he would go back to his hometown of Boston, where he would visit his sister and preside over the regular Sunday-night meetings of the Ford Hall Forum.[27]

Like Rosenman, Niles had been a member of the anti-Zionist American Jewish Committee but Hitler's war against the Jews had challenged their faith in the protection that assimilation could offer the Jews of Europe. Rosenman summed up his own trajectory: "Before Hitler came to power I was an anti-Zionist, then after that I became a non-Zionist. At the end of the War I became a full believer in the idea of political Zionism."[28] Both men's sympathies lay with the so-called moderate Zionists represented by Nahum Goldmann and Chaim Weizmann, and they were close to Supreme Court Justice Felix Frankfurter, one of the most important American Zionist leaders. Niles also had a long-standing relationship with Rabbi Stephen Wise. Niles, the shy bachelor, had secretly been in love with Wise's daughter Justine, whom he had met at a Ford Hall Forum meeting while she was a student at Radcliffe. Even though nothing had come of that relationship, Niles had remained close to her father. Whenever Wise came to the capital to lobby for his cause, Niles welcomed the chance to see him.[29]

During his first days in office as president, Truman did his best both to gain support on the Hill and to persuade both the opposition Republicans and the majority Democrats to cooperate with him in carrying out FDR's legacy. Rosenman thought that Truman was "overawed, particularly at the beginning, about the responsibility, the authority, and most of all, the *lonesomeness* of the job." He was very conscious of the fact that he held the job only because of FDR's sudden passing. It was FDR's policies that had gained the overwhelming approval of the nation. Each time Truman took a step, Rosenman put it, "he would say to himself: 'I wonder what Roosevelt would have done?' " The Oval Office had a picture of FDR on the wall, and Truman would glance at it and say to Rosenman, "I'm trying to do what he would like." Rosenman thought Truman looked to him for advice because "he knew that I knew what Roosevelt would have liked."[30]

In the days and even months ahead, Truman continued to tell people that he did not want to be president under such terms. Senate Majority Leader Alben W. Barkley of Kentucky, exasperated, finally told the president it was not good to appear so self-deprecating. He let Truman know that he should appear to be confident, as did speaker of the House Sam Rayburn.[31]

Truman had reason to feel insecure. He was a modest man who was stepping into very big shoes. But because of his long tenure as a senator, one thing he did know was how to deal with Congress, as well as with the Democratic Party. He had always been a loyalist to the party of FDR and the New Deal. Some with a more conservative bent were predicting that Truman would move away from Roosevelt's New Deal legacy. They were disappointed when he presented his postwar program to Congress in September 1945, in a speech Rosenman helped him to write. He urged extension of unemployment compensation, an increased minimum wage, farm price supports, extension of TVA projects to other rivers, and creation of a permanent Fair Employment Practices Commission–one of the key demands of the NAACP and northern liberals. Truman also called for national health insurance and expansion of the Social Security program. It was not, as his biographer Alonzo Hamby has pointed out, a moderate program; it was clearly one that was strongly left of center.[32]

In foreign as well as domestic affairs, Truman was becoming his own man. As time passed, he would make major decisions that would affect the state of the world for decades to come, and he did so decisively. He alone decided to make the call to drop atomic bombs on both Hiroshima and Nagasaki, to proceed with the creation of the new United Nations, to respond to North Korea's aggression against South Korea in 1950 with military intervention, to dismiss the popular World War II General Douglas MacArthur during the Korean War for insubordination, to offer the Marshall Plan for reconstruction of Western Europe, and to create a new military alliance, the North Atlantic Treaty Organization (NATO), which would protect Europe's security against Soviet expansionism. It was, the most recent historian of the period has noted, "a new conceptual worldview of America's international role," one commensurate with the need to fight and win what came to be called the Cold War.[33]

There was one area of foreign policy that Truman felt confident to handle and that he felt strongly about: the terrible situation of the Jews in Europe and the promises made to them for a homeland in Palestine. Hitler's rise to power in Germany, and the resulting destruction of European Jewry, paralleled Truman's own rise to political power. Truman had been elected to the Senate at the age of fifty in 1935. His years representing Missouri coincided with the strong support many Democrats gave to the establishment of a national home for the Jewish people in Palestine.

Truman, who was a lifelong history buff, was fascinated by the biographies of great political and military leaders, especially of his hero Andrew Jackson. Perhaps because from the age of eight he wore thick eyeglasses, which precluded some of the rough-and-tumble activity of his friends, Harry was drawn to quieter pastimes. He became an excellent pianist and for a while entertained the thought that he might make it his career. Truman was also an avid reader and supplemented his home library by regularly using Independence's three-thousand-volume public library.[34]

Truman claimed his interest in Palestine went back to his childhood. Raised as a Baptist, he had read the Bible "at least a dozen times" before he was fifteen. He felt the biblical stories were stories about real people, and he felt he knew some of them better than the actual people in his life.[35] Truman had an almost fundamentalist belief in the Bible and as an adult looked to it for inspiration and guidance. As a child, he had been told that what was in the Bible was the truth and that "the fundamental basis of all government" could be read about in its pages and "started with Moses on the Mount."[36] He believed this was true of America and in his memoirs wrote, "what came about in Philadelphia in 1776 really had its beginning in Hebrew times."[37] Truman drew on the Bible as a source of knowledge of the history of ancient Palestine.[38] And in the Bible he read of the Jewish people's longing to return to their ancient homeland and God's desire for them to do so. His favorite psalm was number 137: "By the rivers of Babylon, there we sat down, yea, we wept, when we remembered Zion."[39] Sam Rosenman thought that Truman's religious training and familiarity with the Old Testament gave him, as one of his biographers put it, "a sense of appropriateness about the Jewish return to Palestine."[40]

Truman was not alone among American presidents in this belief. Christian Zionism had its roots in the earliest days of the republic, when John Adams had supported the idea of the Jews' returning to Judea as an independent nation, although he thought that they would "possibly in time become liberal Unitarian Christians."[41] Every president from Woodrow Wilson, who had given his approval to the Balfour Declaration, down to Truman, expressed sympathy for, as FDR put it, the "noble ideal" of the restoration of Palestine to the Jewish people as a homeland.[42]

Truman's first public support on behalf of Jewish settlement in Palestine came on May 25, 1939, eight days after the British announced the White Paper, which greatly limited Jewish immigration into and land purchases in Palestine. He asked that a newspaper article be printed as an appendix in the *Congressional Record,* along with remarks he had attached to it. The article, which appeared in *The Washington Post,* had the provocative title "British Surrender–A Munich for the Holy Land." The author, Barnet Nover, argued that the new British policy was a betrayal of the Balfour Declaration, in which the British had agreed to the creation of a Jewish homeland in Palestine. Truman added in his attachment that the British government "has made a scrap of paper out of Lord Balfour's promise to the Jews," which amounted to nothing less than another addition "to the long list of surrenders to the axis powers."[43]

Two years later, Senator Truman joined the American Palestine Committee, a pro-Zionist group. Rabbi Stephen Wise thanked him "on behalf of hundreds of thousands of organized Zionists" who wanted to express their "deep appreciation of your action in agreeing to join the American Palestine Committee, thereby lending your support to the Zionist cause."[44] The committee was organized in 1932 by Emanuel Neumann, with the support of Supreme Court Justice Louis D. Brandeis. The founders meant it to be a vehicle through which prominent individuals in American political life, most of them religious Christians, would be able to offer support for the Zionist cause. The committee languished when Neumann decided to go to Palestine. Upon his return to the United States in 1941, Neumann moved to reconstitute the dormant organization. Rabbi Stephen Wise had introduced Neumann to New York's senior senator, Robert F. Wagner, who immediately agreed to be the group's new chairman.

At a luncheon held at the Shoreham Hotel in April 1941, a cross section of leading American political, civil, religious, labor, and cultural leaders came to hear a keynote speech by Senator Alben Barkley (who would later become Truman's vice president). The majority leader told his distinguished audience that their membership reflected the overwhelming sentiment of the nation in favor of creating a Jewish national home in Palestine. The audience was composed of more than two hundred members of Congress, Cabinet officers, and religious leaders. The Jewish people, he told the assembled dignitaries, "gave the world the great moral laws and principles that have come down to us from the days of the Hebrew prophets." America looked forward to the day when "the Jewish people shall again come into its ancient inheritance." It was, he said, "their rightful heritage."[45]

It was left to Chaim Weizmann to present to the group the major Zionist goals. Accurately predicting that at the war's end the Jewish survivors would be "impoverished, dislocated, crushed and torn out of the economic fabric of which they had been a part," the Zionist leader emphasized that only Palestine could be the place to give them refuge and a new homeland. Others had advocated exotic and improbable areas as a substitute–the first of many being Uganda–but these places did not have "the ties which bind an ancient people to its ancestral home." Pausing to emphasize the importance of his remarks, Weizmann told his audience, "Fate, history, call it what you may, has linked the national destiny of the Jewish people with Palestine."[46]

Senator Harry S. Truman joined sixty-eight other senators known as "the Wagner group" in signing a statement that called for "every possible encouragement to the movement for the restoration of the Jews in Palestine" and seconded a 1922 congressional resolution stating that it was the goal of the United States to create a national home for the world's Jews in Palestine. Since Hitler had taken power in

1933, the statement read, 280,000 more refugees had come to Palestine. They were "streaming to its shores despite restrictive measures" enforced recently by Great Britain. To seek a home for them in Palestine was "in accordance with the spirit of Biblical prophecy."[47]

On the twenty-fifth anniversary of the Balfour Declaration, November 2, 1942, the American Palestine Committee issued a statement calling on the world's powers to enforce the promise made by the British "to open the gates of Palestine to homeless and harassed multitudes and to pave the way for the establishment of a Jewish Commonwealth." The survivors of Hitler's policy of extermination, the statement continued, must be able "to reconstruct their lives in Palestine." The day following FDR's sudden death, the Jewish Agency's Political Department issued a note to its members, pointing out that Harry S. Truman had signed this important declaration. It was cautiously optimistic that it had a friend in the White House.[48]

Truman's greatest expression of sympathy for the plight of Europe's Jews was revealed in the speech he gave on April 14, 1943, to a massive crowd gathered at the Chicago Stadium. The seriousness of the rally was evident in the name given the event by its sponsors: "United Rally to Demand Rescue of Doomed Jews." The event was sponsored by a number of Jewish organizations, including the American Jewish Committee, the American Jewish Congress, B'nai B'rith, the Jewish Labor Committee, the Emergency Committee for Zionist Affairs, and various Jewish federations, rabbinical bodies, and Jewish charities.[49]

Truman's speech reveals his strong feelings. U.S. troops had returned from the First World War with, he told the audience, "a consuming hatred for the forces of oppression." Comrades who had lost their lives could not be honored, Truman continued, "unless the liberty they fought for came into being." He continued:

> In conquered Europe we find a once free people enslaved, crushed and brutalized by the most depraved tyrants of all time. . . . The people of that ancient race, the Jews, are being herded like animals into the Ghettoes, the concentration

> camps, and the wastelands of Europe. The men, women and the children of this honored people are being . . . actually murdered by the fiendish Huns and Fascists.

"Today–not tomorrow," he continued, "–we must do all that is humanly possible to provide a haven and place of safety for all of those who can be grasped from the hands of the Nazi butchers." They had to have access to free lands and needed the United States to aid the oppressed and to show "national generosity." It was not a Jewish problem, he concluded, but "an American problem," and it had to be faced "squarely and honorably."[50]

Senator Truman's sympathy for the Jewish cause led him to move outside the ranks of the moderate mainstream Zionists. He joined Peter Bergson's organization, which sought to create a Jewish army that would fight in the European theater in conjunction with the Allied forces. The militant Bergson had a supporter in Congress approach Truman and ask him to join the effort. But this particular solution did not grab him. "I think the best thing for the Jews to do," Truman wrote Bergson, "is to go right into our Army as they did in the last war and make the same sort of good soldiers as they did before."[51] Bergson was persistent.

Before the end of 1942, Truman had changed his mind, and signed a petition supporting a Jewish army, which appeared first as a two-page ad in *The New York Times* and then in the *Chicago Daily News,* the *Los Angeles Times,* and *The Washington Post.*[52] Titled "A Proclamation on the Moral Rights of the Stateless and Palestinian Jews," the ad galvanized the public with its jarring and dramatic petition, highlighted by a drawing of a Jewish soldier–Star of David on his helmet–carrying a sick or wounded religious Jew who is holding a Torah in his hands. Its main point: "We proclaim our belief in the moral right of the disinherited, stateless Jews of Europe and of the stalwart young Jewish people of Palestine to fight–as they ask to fight–as fellow-soldiers in this war, standing forth in their own name and under their own banner, fighting as The Jewish Army." The signal, Bergson wrote, had to come from America: "To commiserate is not enough." Justice alone demanded that Jews be allowed to form their own fighting force. The world, said Bergson, needed to see "the army of the Fighting Jew."[53]

Those expecting a call for a Jewish homeland found no mention of it in the ad. Bergson's tactic was to focus on one issue that could garner the most support. More people would support a call for a Jewish army–since Hitler's war was one against the Jews–than would support a call for a Jewish homeland. The ad was signed by more than twenty-five senators, hundreds of representatives, the major labor leaders from both the AFL and CIO, prominent educators, newspapermen, and governors and mayors. Some of the best-known cultural figures signed including Cecil B. DeMille, Humphrey Bogart, Oscar Hammerstein, and Raymond Massey.

Bergson, however, soon overplayed his hand. The Roosevelt administration, under pressure to take steps about the Jewish refugee crisis, convened a twelve-day conference on the island of Bermuda in April 1943. The meeting was a formality that might as well never have taken place. Both Britain and the United States simply reiterated existing policy: they maintained tight control over immigration in their own homelands and refused to support free Jewish entry into Palestine. Little was accomplished. When it was over, the administration saw fit not even to release the recommendations to the press. The conference did not address the question of how to achieve any mass rescue of the Jews.[54] For Bergson, all the Bermuda Conference accomplished, as one sympathetic biographer wrote, was to succeed "in dampening the pressures for action by giving the appearance of planning steps to rescue the Jews."[55]

Bergson's answer was to create more pressure, once again through a major ad campaign in America's largest newspapers. First, on the day the conference began, Bergson took out an ad in *The Washington Post* demanding that the delegates act, not merely talk. Then, when it ended, the group again took out a full-page ad in *The New York Times* with the inflammatory heading "To 5,000,000 Jews in the Nazi Death-Trap Bermuda Was a 'Cruel Mockery.' " Bergson explained that to the trapped Jews of Europe, who lived in "Hitler's hell," the deliberations at Bermuda had meant to them nothing less than the last "ray of hope" for escaping from "torture, death, starvation and agony in slaughter-houses." As it turned out, they had cherished an illusion. Hitler had now learned that the world's powers had given him carte blanche to continue the process of extermination. Bergson concluded that stateless and Palestinian Jews should form commando squads that would go deep into territory controlled by Germany, "bringing their message of hope to Hitler's victims."[56]

The ad infuriated Senator Scott Lucas (D.–Illinois), who was a member of the U.S. delegation at Bermuda. Speaking to the Senate, he condemned the Bergson group as "aliens" who enjoyed America's hospitality more "than they can get at any other place under God's shining sun." Bergson and company, Lucas charged, were "taking advantage of the courtesy and kindness extended to them" by now attacking U.S. policy and decent American efforts. He was so peeved that he asked both the State Department and the FBI to see if Bergson was subject to the military draft, as well as exploring how he had raised money for his many ads.[57]

Bergson saw the new ad as a continuation of the call he had made earlier for creation of a Jewish army, which was again listed as the sponsor. He assumed, therefore, that he could simply reprint the names of the people who had endorsed the earlier call in this new ad. Alongside the text were the names of the same thirty-three senators who had signed off on behalf of the Jewish army. That list included the name of Senator Harry S. Truman, who considered his colleague Lucas to be a good friend. For Harry S. Truman, loyalty to one's friends stood above all else. Bending to Lucas's wish, he publicly resigned from the Committee for a Jewish Army. Some have used Truman's resignation as proof that he was not as favorable to the Jewish cause, as he claimed to be in his memoirs. But, as Bergson himself explained to his biographers, all Truman told him was "I wish to resign from the committee as of now." That meant, Bergson added, that "he didn't put any aspersions on the past." He did it for one reason. Truman told him personally, "I like my friends. Scott Lucas is a friend of mine." Bergson replied to Senator Truman, "Well, Senator, you know we are right and he is wrong." Truman agreed. "Yes," he told Bergson, "but he's a friend of mine, and I'm loyal to my friends, and I want to help Scott Lucas." He did it, Bergson thought, "as one politician to another."[58] From this point on, Truman informed Bergson, "greater caution [should be] exercised in publishing the names of Senators who favor *our cause*"[59] (our emphasis).

Truman sent a copy of his resignation letter to Rabbi Stephen Wise, who was glad to receive it. Wise, who had always seen Bergson as a competitor and a political enemy, wrote Truman that Bergson's work "has been a source of considerable embarrassment to the organized Zionist movement." They feared the "appalling plight of the Jews of Europe" might be obscured by the fracas over Bergson's tactics. But Wise had to agree with Bergson that he thought the Bermuda Conference had failed to tackle the issue in any effective way.[60]

"It is fellows like Mr. Bergson," Truman wrote Wise, "who go off half cocked . . . and that cause all the trouble." He now felt that the ad Bergson had placed in the newspapers might "be used to stir up trouble where our troops are fighting," and hence he concluded it was "outside my policy" to be part of such a group. By this time, it was apparent by Truman's tone that Lucas had given him a talking-to. Truman incorrectly argued–perhaps seeking to placate Wise–that the Bergson ad had been "used by all the Arabs in North Africa" and led them to "stab our fellows in the back."[61]

Although Truman said he maintained his commitment to the rescue of the Jews and his sympathy for the cause of a Jewish homeland, he now followed FDR's policy of deferring the issue until the war was won. In 1944, when the pro-Zionist forces succeeded in getting their friends in Congress to introduce a congressional resolution on behalf of a Jewish homeland and the abolishment of the White Paper, Truman disappointed many of his colleagues by supporting the president and opposing its passage. Some historians have argued that his position indicated that he was not really a supporter of Zionism and was only engaging in empty rhetoric.[62]

But Truman felt the administration's reasons for opposing the resolution made sense and that he was acting in the American national interest. As he explained in a letter he sent out, the resolution had to be "very circumspectly handled until we know just exactly where we are going and why." Of course, Truman noted, his sympathy "is with the Jewish people." But financing the war to its conclusion was the prime consideration. He did not want to take any step that might "upset the applecart," he wrote, *"although when the right time comes I am willing to help make the fight for a Jewish homeland in Palestine."*[63] Opposition to the resolution was a tactical issue. As Truman explained to his pro-Zionist colleague Senator Wagner, "we want to help the Jews but we cannot do it at the expense of our military maneuvers."[64]

The leaders of the Zionist movement were understandably disappointed by Truman's failure to support the congressional resolution. Compared to the enthusiastic support that the junior senator from Missouri, Bennett C. Clark, gave the resolution, Truman's vote must have dismayed the Jewish community living in Saint Louis and Kansas City, the largest Missouri cities. Clark, a member of the Foreign Relations Committee, gave a rousing speech to the Senate on March 28, 1944. He denounced the administration for opposing the resolution and for not acting to abrogate the British White Paper, which was soon to expire, ending the possibility of further Jewish immigration into Palestine. Unless the White Paper was repudiated or modified, he said, "the tragedy of the Jew in our time will be infinitely worse than the tragedy throughout the ages. Where, then, Mr. President, will be our vaunted Christian civilization? Where, then, will be the long-time policy of this government? Where then will be our national self-respect? The time has passed when mere words will halt continuation of this tragedy," he bellowed. "Action is needed." He told his colleagues in the Senate that for the past few nights he had been reading Dr. Lowdermilk's book *Palestine, Land of Promise,* and he urged them to do the same. In Palestine, he said, the soil was being reclaimed and "an ancient land is being returned to the fruitfulness which the Creator intended."[65] When Clark's Zionist speech is compared to Truman's hesitancy, one can see why the Zionists thought Truman was not completely on their side. From their point of view, Truman seemed unwilling to argue with the administration's overall foreign policy.

American Jews, as well as the entire nation, were still in mourning for the passing of Franklin Delano Roosevelt. A few days after FDR's death, Rabbi Stephen Wise presided over a Carnegie Hall memorial service, held under the auspices of his Reform congregation, the Free Synagogue. Three thousand, two hundred people filled every seat, and 2,500 more had to be turned away. Those unlucky mourners had to rush home and listen to the program on the New York radio station WMCA.

Roosevelt, Wise told the audience, had been "a warm and genuine supporter of the Zionist cause." Referring to his last visit with Roosevelt the previous March, Wise emphasized that FDR had told him how he was trying to get the Arab nations of the Near East to appreciate "the miracle which the Jewish rebuilders had wrought in Palestine" but had been candidly disappointed at his "partial sense of failure . . . as far as Jewish hopes were concerned." Wise explained that if FDR had not succeeded, it was only because of the influence of his "counselors in the State Department and the Colonial Office in England who exaggerated the importance and power of the . . . Near East rulers." Before he died, Wise claimed–again inaccurately–that Roosevelt had been preparing for yet another conference on the Palestine issue, "the solution of which was bound to be the establishment of a free and democratic Jewish Commonwealth in Palestine."[66]

The next day, Sumner Welles, FDR's former undersecretary of state, spoke before 1,500 women of the New York City chapter of Hadassah, the Zionist women's organization. Before he spoke, the audience rose to its feet in tribute of the deceased wartime leader. "A tower has fallen," Welles stated. No other American leader, he thought, "has done more for our country." Welles offered another solution for Palestine: the one favored by the State Department calling for the creation of an international trusteeship by the new United Nations, which would replace the British Mandate.[67]

That kind of talk made the committed Zionists quite nervous. How would the new president, who they knew had been sympathetic when he was a senator, now treat their movement and its goals? Would he stay true to the policies he had espoused while in the Senate? Many of the prominent Zionists, who were used to dealing with Roosevelt, did not know much about the new president. Truman, Emanuel Neumann wrote, "almost completely lacked [FDR's] finesse and sophistication." Yet they thought he was more of a straightforward midwesterner, who meant what he said. Moreover, he had "generous humanitarian impulses," which led them to hope that their movement might have an ally in the White House.[68]

When Eliahu Epstein asked the Zionist leader Louis Lipsky if he had any impressions about Truman, Lipsky pointed to Truman's decision to join the American Palestine Committee. Lipsky added, "Many Jewish friends speak well of his personal integrity, sense of justice, and feelings of friendship and sympathy toward the Jewish people."[69] Senator Wagner, who had asked Truman to join the American Palestine Committee years earlier, was equally favorable. Truman, he told Epstein, was "endowed with a keen sense of humanity and justice, and after what has happened to the Jews under Hitler, we will find the president an attentive listener to our just claims and appeals."[70] New York congressman Emanuel Celler said he believed Truman to be "straightforward and modest." Truman was not going to reverse FDR's overall foreign policy, and he thought the prospects were good that "we may be able to gain Truman's help in arriving at a satisfactory solution to the Palestine problem." As a senator, he noted, Truman had responded positively to every appeal made to him on behalf of Zionist issues. Almost as an aside, Celler added that he knew Truman would face pressure from those opposed to Zionism, and it was too early to know if he would "bear the burden of his office" and be able to stand up to them.[71]

Celler had good reason to be concerned. On April 18, the president received an important letter from Secretary of State Edward Stettinius, presenting the views of the Department of State. "It is very likely," the secretary warned, "that efforts will be made by some of the Zionist leaders" to obtain commitments from him in favor of "unlimited Jewish immigration into Palestine and the establishment there of a Jewish state." Declaring his and the public's sympathy for the situation of Europe's persecuted Jews, Stettinius warned that the Palestine question was a highly complex one that went beyond the plight of the Jews of Europe. He warned Truman that it would be best if he did not make any public statement on the issue without asking State for full and detailed information. Because the United States had vital interests in the region, the subject was one that should be handled with the greatest care.[72]

Two days later, the Zionist leaders had their first meeting with Truman when a small delegation from the American Zionist Emergency Council (AZEC) went to the White House. The group, led by Rabbi Wise, had expected to spend close to an hour with Truman, but as they walked in, Secretary of State Stettinius, Acting Secretary of State Joseph Grew, and the U.S. ambassador to the Soviet Union, Averell Harriman, were also waiting for an audience with Truman. It didn't look like an opportune time for a full meeting, but Wise and his group were pleased when Truman began the meeting by telling them that "he welcomed the opportunity to discuss the matter . . . because he was sympathetic to the Zionist cause." When he had more time, he said, he wished to hear more, but at the moment he wanted to assure them "that he would follow the policies laid down by the late President Roosevelt on the Palestine issue."[73]

Wise answered that he did not know if Truman understood the reasons why the Jewish people favored a national homeland in Palestine. Truman was polite but, as he said years later, "in those days nobody seemed to think I was *aware* of anything."[74] Two days earlier he had received a letter from Secretary of State Stettinius "telling me to watch my step, that I didn't *really* understand what was going on over there and that I ought to leave it to the *experts*."

To the contrary, Truman believed he was well versed on the situation and had prepared for the meeting by reviewing FDR's public statements and record regarding Palestine. He reread the Balfour Declaration and familiarized himself with the history of demands for a Jewish homeland and the Arab and British views about it. Remembering the visit in his memoir, Truman noted that he "was skeptical" as he looked over the record and came across "the views and attitudes assumed by the 'striped-pants boys' in the State Department." Truman at the time felt that State "didn't care enough about what happened to the thousands of displaced persons," and he thought that he could both "watch out for the long-range interests of our country while at the same time helping" the persecuted European Jews to find a home.[75] Most important, Truman told Wise that State would not be allowed to make Near Eastern policy while he was president. "I'd see to it that I made policy," Truman assured him, and State's job "was to carry it out."[76] As he recalled the meeting, he believed that he had made that very clear to Wise.

A few weeks later, Wise bumped into Jewish Agency representative Eliahu Epstein in San Francisco at the first meeting of the new United Nations. Epstein asked him if Truman was sincere in his sympathetic statements toward the Jews, or would he feel differently now that he was president? Wise told him that at the April 20 meeting, "he detected no difference whatsoever between Truman the senator and Truman the president, regarding his views on Zionist affairs, except now he is in a position to act rather than simply to speak or promise." Wise understood that the Jews' greatest problem was not Truman but overcoming "the opposition of the State Department." In fact, he added, Truman himself had "referred to this in no uncertain terms during their conversation."[77]

Truman had told the nation that he would continue the policies of the beloved former president. But if Truman thought he knew what Roosevelt's policy was in the Middle East, he was in for a surprise. Soon after, Acting Secretary of State Joseph Grew wrote to him that he thought it important to let him know "that although President Roosevelt at times gave expression to views sympathetic to certain Zionist aims, *he also gave certain assurances to the Arabs which they regard as definite commitments on our part*" (our emphasis). FDR had authorized the State Department to inform the heads of Arab governments that no decisions affecting them would be taken without consulting both Arabs and Jews. The Arabs, Grew stressed, "have made no secret of their hostility to Zionism," and it would be "impossible to restrain them from rallying with arms, in defense of what they consider to be an Arab country.[78] Truman should, Grew advised, renew FDR's assurances to the Arabs.[79] In a series of letters to the various Arab potentates submitted to him, Truman, without thinking too much about it, dutifully signed as Grew suggested.[80]

Later, Truman would be "embarrassed at having given such a pledge," Loy Henderson, the head of the Office of Near Eastern and African Affairs at State, recalled.[81] But at the time he was trying to honor Roosevelt's wishes. Speaking in October with Lord Halifax, the British ambassador to the United States, James F. Byrnes, who had replaced Stettinius as secretary of state, told him that Truman "is greatly disturbed about this" and if he were asked to make such a pledge today "I don't think it would be made." Moreover, Byrnes argued, FDR had made the statements when he "was an ill man," and he thought that had it not been for his ill health, he would not have made such promises to the Arab monarchs.[82]

Harry Truman had diligently tried to pursue the policies that he thought Roosevelt would have favored. In the case of Palestine, it quickly became clear that he would not be able to. FDR had had the war as an excuse not to take action on behalf of the Jews. He had told them that the best way he could help was to win the war, a policy that Senator Truman had accepted. But now the war would soon be over. The simple man from Missouri, who had never intended to be president, would find it impossible to live with FDR's obfuscations. He would be confronted with decisions FDR had avoided. Was the new president, so inexperienced in foreign affairs, up to the challenge?

THREE

FROM SAN FRANCISCO TO POTSDAM

Two weeks after Truman assumed office, the struggle over the future of Palestine shifted to San Francisco, where the delegates' main task was to prepare the charter of the new United Nations. The venue chosen had great symbolism. The conference was held at the War Memorial Opera House, a Beaux-Arts structure that had opened in 1932 and had been built as a tribute to those who had lost their lives in the First World War. A landmark in the city, its impressive columns and massive arched windows were similar to those of the Louvre in Paris. It was considered by many to be one of the most beautiful theaters in the nation, and those entering its grand hall could not help but be impressed. This time, the Opera House was not filled with maestros and divas. Instead, as one participant noted, its long, wide corridors were "crowded with men and women of every race, many dressed in national costume, including the delegates from Saudi Arabia, who wear traditional Arab robes and kaffiyehs."[1]

Although Palestine itself would not be on the agenda, crucial decisions affecting its final status could be made.[2] The Arab strategy was to propose new principles and guidelines for trusteeship by introducing a clause in the U.N. Charter recognizing the rights of only one people in a mandated territory. The fact that they made up the majority in Palestine would give them the right to decide the future of the country. This might sound like a reasonable, even democratic argument, but if the Arab proposals were adopted, wrote the Jewish Agency representative Eliahu Epstein, "the legal foundations of the Palestine Mandate and of the national home would be shaken." The agency's "main and urgent objective at this conference," wrote Epstein, "must be to protect the present international status of Palestine."[3]

The Arabs thought they had a good chance to execute their plan. Five Arab states–Egypt, Iraq, Saudi Arabia, Syria, and Lebanon–had founded the Arab League in Cairo on March 22, and now each state had official delegates at San Francisco. The Jews had none. Chaim Weizmann was especially angry that the Arab states had been admitted as full U.N. partners–despite their dismal war record and active support of the Axis–while the Jews, who had more than 1 million soldiers in the Allied armies and 35,000 Palestinian volunteers, were "relegated to the corridor."[4]

Before his death, Roosevelt had told Rabbi Stephen Wise that he wanted the Jewish Agency to be invited to the founding meeting in an official capacity. Wise conveyed this to Truman, but the president went along with State's recommendation to create a "consultant" capacity, in which interested nongovernmental parties could participate. The State Department chose two groups: the American Jewish Conference (coalition of Zionist groups) and the American Jewish Committee (non- or anti-Zionist) to have the official status of consultants to the U.S. delegation. In choosing these two organizations, State reinforced FDR's and afterward Truman's practice of giving both groups legitimacy as representatives of American Jewry. Showing that the Jewish community was divided over Zionism benefited State's anti-Zionist policy. But this time the groups united, agreeing that whatever their differences concerning the creation of a Jewish state, they would work together at the conference to defend Jewish rights under the British Mandate in Palestine.

The Jewish Agency, however, could not even get its representatives access to the conference floor. Writing from Palestine, David Ben-Gurion bitterly lamented, "We are a people without a state, and therefore a people without credentials, without recognition, without representation, and without the privileges of a nation, without the means of self-defense, and without any say in our fate."[5]

When Epstein arrived in San Francisco from Palestine as a representative of the Jewish Agency, he kept a detailed diary of his experiences. It was his first trip to the United States, and in his diary he recorded his impressions as he tried to figure out how to conduct himself and be effective. Born in Russia, Epstein had become a Zionist and fled to Palestine in 1925. At first, he had joined other young people working as a farmhand to build the Jewish homeland. He had begun to study at the Hebrew University and in 1930 attended the American University of Beirut. The Jewish Agency had recognized his gifts, and from 1934 to 1945 he served as director of the Middle East division of the Jewish Agency's Political Department. In that capacity, he had gone on many diplomatic missions to Arab capitals and created a wide circle of contacts. By the time he arrived in America, he could converse in Russian, Hebrew, Arabic, English, and French.

The warm, outgoing Epstein was the kind of person who could talk to almost anyone, even adversaries, in a cordial way. Therefore he tried to meet with as many people as he could in San Francisco: Zionist and anti-Zionist Jews, Arabs, the British, delegates from other countries, and American journalists and politicians. He was amazed that in America all occasions, even breakfast, were used for business, and he wasted no time in getting started. Epstein found many shortcomings in the American Zionist movement. Among all the Zionists he talked to, he complained in his diary, he had not "met a single person with the requisite level of either background knowledge or practical experience in Arab or Middle Eastern affairs."[6] Nor did American Zionists know what the Arabs were doing to promote their cause in the United States. There was little contact between Zionists in the United States and the State Department's Division of Near Eastern Affairs, he noted, advising his Jewish Agency colleague Nahum Goldmann that something should be done about it.

He also found the Jewish presence in San Francisco chaotic. More than twenty Jewish organizations had representatives in San Francisco, and they were all busy distributing their materials and proposals to the conference's delegates.[7] Among them were representatives of the anti-Zionist American Council on Judaism, which claimed that Judaism was only a religion and Zionism was a betrayal of their faith. The American Jewish Committee supported Jewish immigration and demanded an end to the British White Paper but opposed the creation of a Jewish state in Palestine. Peter Bergson and his followers were also there, presenting their demands for recognition of what they called the already existing "Hebrew nation" in Palestine. The Bergson group inflamed all of the other Zionists. AZEC sent out a press release that Bergson's group "represents no one and is responsible to no one."[8]

The Zionists, through AZEC, did set up an effective nine-man committee. They gave out more than forty separate press releases, translated into three languages, to the delegates and the reporters from the world's press corps. And the American Jewish Conference and AZEC together sponsored eighty-eight mass rallies throughout the United States, all meant to demonstrate strong public support for a Jewish homeland in Palestine.[9]

The largest rally, held in New York City–the home of the largest Jewish population in the world–drew more than sixty thousand people. Every seat in Lewisohn Stadium of the City College of New York was filled to capacity. The overflow crowd spilled out onto a playground packed with people who listened closely through speakers set up for the occasion. Senator Robert F. Wagner was loudly cheered as he proclaimed that British policy was nothing less than a revival of "the disease of appeasement." Addressing the British government directly, Wagner accused it of going back on its word to the Jews and said it was time that it redeemed the pledge made to it long ago in the Balfour Declaration. Rabbi Stephen Wise, joined onstage by Rabbi Abba Hillel Silver for the first time since they had had a public disagreement the previous December, demanded that "the Jewish case be placed upon the agenda of San Francisco."[10]

The first person Epstein wanted to meet with at the San Francisco Conference was Judge Joseph Proskauer, the head of the small but influential American Jewish Committee. He wanted to see if there was any common ground between them and the Zionist organizations and, if not, to win the judge over to his side. He was pleased to find that Proskauer's overriding concern was "the fate of the survivors of the Holocaust and his wish to find shelter for them." Now Epstein urged Proskauer to take the next step. It was not enough to advocate bringing Jewish immigrants to Palestine, Epstein told him; it was necessary "to see to their absorption and ensure their future in the land." That is why the Zionist dream "*had* to be realized." The Zionists, he told Proskauer, sought political independence "in order to preserve and develop the progressive, democratic character of the society we have built in Palestine. We are surrounded by feudal and reactionary neighbors who are hostile to Zionism and who see in the Zionist venture a threat to their oppressive regimes. We can ensure our future only when we become masters of our own destiny."[11]

Epstein lobbied any delegation willing to meet with him. The French delegates, he found, were the most supportive of Jewish claims for Palestine. On May 8, he met with Jan Masaryk, the foreign minister of the postwar Czech government. Masaryk recounted his father's prewar visit to Palestine and told Epstein how impressed he had been with the new Jewish agricultural settlements. That visit, Epstein wrote, had "deepened the elder Masaryk's Zionist sympathies and influenced his colleagues in the Czech government in the same direction."[12]

Epstein also met with various Latin American delegates, whose votes and position would become very important in the 1947 U.N. vote for partition. The Chilean delegate proved supportive, as did the Mexican. The Jews could rely, they told him, "on the understanding and assistance of Latin America. The Holocaust in Europe has shocked public opinion . . . and they understand our struggle for free Jewish immigration to Palestine and the achievement of political independence."[13] The delegates from South Africa, Holland, and New Zealand were also supportive. Much to his surprise, Mostafa Adle, the chairman of Iran's delegation, told him that the Iranians "are favorably inclined toward the Jewish people and the Yishuv" and "proud of the historic event of the Jewish people's return from Babylonian captivity to Zion by permission of their king, Cyrus." Their current official stand of neutrality was prompted only by tension on their borders with the Soviet Union, and Adle asked that Epstein not make their meeting or his comments public.[14] Finally, Epstein received encouragement from Australia's delegate, Richard H. Nash, a member of his country's Senate and a leader of its Labour Party. Nash knew from his fellow countrymen who had visited Palestine about the Zionists' achievements and their contributions to the labor movement there. They had come back to Australia devoted friends of the Zionists' cause. Jewish Palestine, they told Nash, was "an island of devotion and loyalty to the democracies" amid a sea of "espionage, subversion and deceit."[15]

The Jewish groups were concerned with countering the well-financed Arab lobby. Si Kenen, the PR representative of the American Jewish Conference, pointed out to Epstein that big oil, with concessions in Saudi Arabia, was spending a small fortune to finance Arab propaganda. Senior personnel of the companies, he reported, had been assigned to the Saudi Arabian delegation to help it and the other Arab delegations with public relations, especially in their anti-Zionist activities among delegates and journalists."[16] There were reports that the Saudis were giving journalists boat rides around San Francisco Bay. "The entertainment was lavish beyond description," Epstein wrote, and had been paid for by the big oil companies. "The seagoing merriment was certainly not in keeping with traditional Wahhabi Puritanism," he noted in his diary, "but this did not seem to bother the Arabs aboard, who toasted the journalists drink for drink."[17]

Epstein found the Arab representatives at the United Nations confident. Jews, the Arab delegates told him, should be content to live as a minority in Palestine under the reign of an Arab regime. Three Arab delegates–two from Iraq and one from Syria–told him that because of the Arab League, the Jews had no chance to win against the Arabs. Newly discovered rich oil deposits, they believed, meant that the Americans would now have to depend on Arab goodwill and cooperation, instead of the other way around. The Jews simply could not overcome "both Arab hostility and the enormous power and political influence of the international companies with interests in Arab countries." They should be practical and realistic and reach an accommodation with the Arabs of Palestine that would ensure their existence and future in a Palestinian Arab state.[18] Later during the conference, the Syrian prime minister, Faris al-Khouri, told Epstein that the creation of the Arab League meant that the Balfour Declaration had been annulled, along with any special privileges the old Mandate had given the Jews of Palestine.[19]

For all the Arabs' optimism, their plan to make changes to the trusteeship guidelines met with the opposition of the British, French, and American governments, whose representatives saw it as a threat to the peace and stability of countries that had diverse racial, tribal or national populations. The British feared internal disorders in Palestine and, with the French, wanted a policy that would uphold their status as colonial power. They would not support a proposal that would have given the Arab majority in Palestine the sole right to decide the future of the region. Any change from the terms of the Mandate, the British in particular feared, might open a Pandora's box that would affect their privileges as a mandatory power. By the time of a vote, only the Arab powers at the U.N. conference voted for their own proposals.[20]

To fill what he thought was a gap in the Zionists' knowledge, Epstein made a study of how strong the Arab propaganda apparatus was. He counted six major pro-Arab organizations and institutes working in the United States, all of which had a pro-Arab attitude to the Palestine question. Among them was the New York–based Near Eastern College Association, founded by American Presbyterians. The best-known educational institution connected to it was the esteemed American University of Beirut. At San Francisco, Epstein counted twenty-nine former AU students in the senior posts of the Arab delegations. He noted that many Americans on the university's staff had served in the American armed forces during the war and in diplomatic missions and intelligence services operating in the Middle East. Now some of them were "advisers on Middle Eastern affairs in different government departments in Washington, influential in the framing of United States policy in the region and in its implementation." Two important pro-

Arab educational centers in the United States, he found, were the Institute of Oriental Studies of the University of Chicago and a similar center at Princeton University, both of which were also training future diplomats and government specialists in the Middle East.

Their potential effectiveness led Epstein to worry. "I have been struck," he wrote, "by the fragmentary knowledge our people display on the activities of the Arabs and their friends in this country, who seek to increase Arab influence in the public and political life of the United States and to weaken our position." He feared that American interests in the oil-rich Arab states would actually make "the tasks of our opponents easier and ours more difficult." In the past year the Foreign Service Education Foundation had been organized to train U.S. diplomats. Its focus, it seemed to Epstein, was the Middle East, and much of its funding came from American oil companies.[21]

During the summer and fall of 1945, the Arab League organized its own PR campaign in the United States, allocating a budget of $2 million for the first year. Officially the campaign was designed to familiarize Americans with all aspects of Arab life, but its main goal was to counter Zionist arguments, particularly the claim that it would be in the strategic interest of both Britain and the United States to see the creation of a Jewish state. Offices were opened in Washington, London, and Cairo, with headquarters in Jerusalem. From its Washington office, the staff worked to get their message out to journalists, businessmen, universities, church groups, student groups, organized labor, and non-Zionist Jews.[22]

Despite their creation of a well-organized and -financed campaign, the Arab states and their supporters were at a disadvantage. Only 100,000 Arabs lived in the United States, while the Jewish population numbered 5 million. In a democratic country, votes counted. Moreover, the image most Americans had of Arabs was of a backward, feudal people whose life was far removed from that of the modern world. *The New York Times* featured a report written by C. L. Sulzberger that served to reinforce the prevailing view. "The Arabs," Sulzberger wrote, "have so far proved themselves an impractical people, unable in the long run to merge their local differences . . . to form themselves into fully independent self-governing units." They lived in "semi-servitude" that left them to fight one another in what Sulzberger called "regional, sectarian and dynastic wars." Moreover, it was the search for oil–it was thought that Saudi Arabia alone had more untapped oil than the entire amount in the United States–that was sucking the Western powers into "one of the world's traditional cockpits for trouble."[23]

The Zionists were beginning to win the public relations war. All the work they had done to gain American support during the war seemed to be paying off. The number of members of American Zionist organizations swelled from less than 250,000 in 1943 to 400,000 by 1945.[24] In April, David Niles received the results of a poll taken by Hadley Cantril, "Public Opinion Toward Creation of Jewish State in Palestine," taken the last week in March. The results had surprised the pollster, who found that about half of Americans knew about the idea of establishing a Jewish state in Palestine. Sentiment, among those polled who had an opinion on the issue, ran about three to one in favor, and a great majority of those thought the U.S. government should use its influence to establish it.[25]

An appraisal of the Zionists' success was offered by British Ambassador Lord Halifax in his report to the British Foreign Office assessing the nature of the American relationship to the Jews, Zionism, and the Arabs. "The Jews have made Palestine and the experiment of establishing a national home there for Jews, interesting to the world at large. This interest has been greatly stimulated by the toll of suffering and persecution to which European Jewry has been subjected." Half of the world's Jews, he reminded the British diplomatic office, lived in the United States. Moreover, many of them held important positions in the Truman administration, and because so many were concentrated in New York City, they could possibly influence the outcome of elections. They could therefore "exert considerable pressure on the Administration, in Congress and on public opinion."

The American public, Halifax thought, had no strong opinion on the issues concerning Jews, aside from favoring loosening of the restriction on Jews who wanted to immigrate to Palestine. And they blamed the sad plight of the Jews on the British government, since there was a feeling in liberal quarters that more Jews could have been saved from Hitler if Britain had admitted more to Palestine. Halifax saw this feeling in part as a result of anti-Semitism. "The average citizen," he thought, "does not want them in the US and salves his conscience by advocating their admission to Palestine," with the result that "on this issue the Jews can therefore carry with them both liberal humanitarians and many anti-Jews."[26]

While the delegates met at San Francisco, the world community was about to receive a shock. On April 18, readers of *The New York Times* learned about the extent of Nazi Germany's war against the Jews. A delayed dispatch dated April 16 informed its readers of what the reporter called "the horror, brutality and human indecency" of the Dachau and Buchenwald concentration camps. German neighbors who had feigned ignorance of the camp's existence were confronted with the reality when they were taken to see the remaining twenty thousand Jewish prisoners, "many of them barely living." They learned for the first time of the torture rooms, execution chambers, gas chambers, and medical facilities, where Nazi doctors had performed fiendish experiments on living human beings. It was a scene that might have been a picture of hell: "The stench, filth and misery here," the reporter wrote, "defied description."[27]

Yet the press tried. The May 7 issue of *Time,* the nation's major newsmagazine, confirmed the daily papers' reports. Here at Dachau, the reporter wrote, "was concentrated the flower of Nazi sadists whose business was torture and death." Accompanying the U.S. Seventh Army as it liberated the camp and saw the remaining prisoners–*Time* put the figure at 32,000–they found the following grotesque scene: "Outside one building, half covered by a brown tarpaulin, was a stack of about five feet high and about 20 feet wide of naked dead bodies, all of them emaciated." Readers learned about the gas chambers and the mass death of Jews that had taken place there, leaving only a few "pitiful, happy, starved, hysterical men," who crowded the soldiers and reporters asking for help, food, and cigarettes–they were people who had "lived in a super-hell of horrors" and were now "driven half crazy by the liberation they have prayed so hopelessly for."[28]

"The World Must Not Forget" was the title of yet another magazine report about the Nazi atrocities. Harold Denny called it "cruelty such as the world has never known before." Words and photos would have to be used, but, Denny assured his readers, "they were less horrible than the reality." He and others saw their task as a simple and necessary one: "The world must know and it must not forget." It was clear that the Allies would now have to deal with the Germans, who had become a "mentally and morally sick people." Those liberating the camps had been forced to walk into "nightmares of human savagery and depravity."[29]

Soon the reporter's observations would be confirmed by members of Congress. When, in mid-April, General Dwight D. Eisenhower, shocked by what he saw as he toured the camps, asked American newspaper reporters and editors to fly in from the United States to see what he had, American government officials went with them. The group included four senators and six congressmen. "The barbarous treatment these people received in the German concentration camps," Eisenhower told them, "is almost unbelievable." He made available to the press all the resources of his command, so the American people would learn why their men had been fighting.[30]

The representative from Connecticut Clare Boothe Luce was one of the members of Congress who accepted Eisenhower's offer. Luce, a playwright, social activist, and mainstay of the society pages, was married to the publisher of *Time* and *Life,* Henry Luce. Thus her opinion carried great weight. Luce said she hoped the people of the United States would see the newsreel footage that the Army photographers were taking as she visited the camp sites. One reporter informed readers of how "she visited the basement crematorium where, in a white walled room, thousands had been hanged from iron hooks. Her prisoner guide told her how the executioner had used clubs shaped like potato mashers to kill victims who did not die quickly enough in the noose. She saw the elevator which carried the victims up to the furnaces. Outside the crematorium she saw a wagon stacked high with shriveled bodies. She did not remain long, saying: 'It's just too horrible.' "[31] On April 22, eight congressmen toured Buchenwald, one of the most infamous German camps. All said they "were shocked beyond belief" at what they saw and were told by the former prisoners. " 'This is the most horrible thing that anyone could conceive,' said Rep. Carter Manasco, a Democrat from Alabama."[32]

Harry Truman read these gruesome reports along with the rest of America, and he saw the harrowing newsreel footage that made the reports real for many. The evidence of what had taken place literally gave him nightmares. "It was a horrible thing," he told CBS News in 1964. "I saw and I dream about it even to this day."[33]

The question now was, what would happen to the survivors? On Truman's very first day in office, Eliezer Kaplan, the treasurer of the Jewish Agency in Palestine, reported to the American people that the situation of the Jewish people in Europe was unbearable. "They are afraid to stay in Hungary, Romania and Poland," he told the press. Even though some European nations had quickly passed new laws guaranteeing equality, European Jews knew that such legislation often meant nothing. "A whole generation has been raised on anti-Semitism," Kaplan explained, and the survivors of the Holocaust "did not want to stay in the countries where their mothers, brothers and sisters were killed." They knew that non-Jews had stood by and done nothing and even participated in the acts of mass killing. Hence, Kaplan argued, the new United Nations had "the responsibility to enable all parentless Jewish children to gain entry into Palestine."[34]

In London, Weizmann had delivered a memo to Churchill describing the desperate straits of the Holocaust survivors, as well as that of Jews who were living in Arab lands. Joseph Linton, the Jewish Agency officer in London, was demanding that the agency assume full powers to decide Palestinian immigration policy. In the meantime, he asked the British high commissioner to grant 100,000 entry permits to Palestine immediately.[35]

The Americans were trying to alleviate the grim refugee situation in Europe created by the war. From May through September 1945, the military repatriated some 6 million refugees, but another million and a half refused to or could not go back to their former homes. Many of them were Jews. The military at first did not distinguish between civilian refugees who had been displaced as a result of bombing or fighting and the persecuted Jews who had survived the death camps. Instead, they were treated like any other group of displaced persons (DPs), perhaps even worse. Sometimes they found themselves put into the same barracks as Germans who had been displaced by the war and who had supported Hitler during the years of the Third Reich. German Jews released from Polish and Czech camps found that they were not given ration cards, on the grounds that they were Germans and hence part of the enemy. Some Jewish refugees found that they were put up with their former camp guards, who now also claimed to be refugees. And a worse insult, as the historian Leonard Dinnerstein has written, was that "the Allies put Germans–ex Nazis–in charge of the DP camps while the inmates were held as virtual prisoners with armed guards surrounding the centers."[36]

For many Jews the situation had not improved since the arrival of U.S. troops. In May and June 1945, 18,000 Jews died of starvation and disease in Bergen-Belsen; at Dachau–the same camp the press had reported on–sixty to one hundred survivors of the Holocaust were dying each day.[37] Not all the mistreatment of the former Nazi prisoners was due to the inability of the troops to care for so many refugees. In one notorious case, that of the camps administered by General George S. Patton, Jr., mistreatment of the Jewish refugees was due entirely to Patton's anti-Semitism. Patton's Third Army controlled the southern zone, where most Jewish DPs were to be found. The camps Patton was responsible for had barbed wire and were patrolled by armed guards. Patton wrote in his diary on September 15 that others "believe that the Displaced Person is a human being, which he is not, and this applies particularly to the Jews who are lower than animals." The Jews had to be kept under armed guard, Patton explained; otherwise they would flee, "spread over the country like locusts," and then be rounded up and some of them shot, only after they had "murdered and pillaged" innocent Germans.[38]

The Jewish victims struggled to find an identity other than "displaced persons," which is how the world referred to them. The Nazis had sought to dehumanize them, leaving them with numbers tattooed on their arms, but they called themselves "the survivors" and adopted their own symbol. The graphic symbol that they displayed when possible was a tree felled at its base that was dried and lifeless. Out of the stump sprang a new shoot, alive and well. Though aware of their losses, one observer reported, they "insisted that their survival meant that they, and not Hitler, had triumphed." In time the tree's new shoot would grow to maturity, and the "continuity of the life of the tree would not be broken."[39]

Reports about the shameful treatment of Europe's remaining Jewry led to a new round of intensive activity on the part of America's Jewish organizations. Jewish members of Congress made an official protest to the Department of War. In the executive branch, Secretary of the Treasury Henry Morgenthau, Jr., urged by Chaim Weizmann, suggested to Truman that he create a new cabinet-level position to look into it. Truman rejected the proposal. Morgenthau then went to the State Department, told it of the stories he was hearing about the terrible conditions of the Jews, and asked it to mount an immediate investigation. Acting Secretary of State Joseph Grew approved the plan once he was convinced that the study would concern itself only with the plight of the Jewish refugees, for which something had to be done, and not with the Zionists' goals of having the Jews enter Palestine and set up a Jewish state.

The protests of the Jewish groups, combined with Morgenthau's efforts, proved successful. State agreed to send Earl G. Harrison, the former U.S. commissioner of immigration and dean of the University of Pennsylvania Law School, to make his own inquiry into the situation.[40] He was charged with investigating the needs of the Jewish refugees and determining where they wanted to go. Truman approved the Harrison Commission on June 22. When Epstein heard the news that Harrison would be sent to report on the refugees, he was elated. The Jewish Agency office in San Francisco, he noted in his diary, "was in a state of great excitement." Zionist leaders in Philadelphia had phoned to say that they thought "Harrison was deeply interested in the question . . . and can be relied to discharge his task with skill and a sense of responsibility." As for their own Zionist plans, Epstein made it clear that they would use the Harrison trip "as a lever to put strong pressure on world opinion to support our urgent demand that Palestine be opened to mass Jewish immigration."[41]

Nahum Goldmann, the senior Jewish Agency representative in America, decided to take Epstein's advice and meet with the State Department. On June 20, he went to the Division of Near Eastern Affairs and reported to Assistant Chief Evan Wilson on the bleak situation of European Jewry. "The mood of the Jewish people was running to one of desperation. They had seen millions of their fellow Jews ruthlessly murdered, their homes destroyed, and their culture completely stamped out." They comforted themselves with the realization that at the war's end, they would see their aspirations in Palestine realized. During the war, Goldmann stressed, he and Weizmann had urged that they "follow a policy of moderation and not to expect a solution . . . along Zionist lines before the end of the war in Europe."

However, he warned that if something weren't done soon, more militant leaders would arise to replace the moderate ones. Behind the scenes, in Palestine, he hinted, militant elements were working, which might not be able to be contained. The State Department knew that there was truth to his claim. Active in the background in Palestine were members of the militant group of Zionists organized into the Irgun. Unlike Palestine's mainstream Zionists, the Irgun advocated the use of terror against both the British and the Arabs and was responsible for the assassination in 1944 of Churchill's friend Lord Moyne, the British minister of state for the Middle East and owner of the Guinness beverage company. "Anything might happen," Goldman warned. Sixty thousand young men had been trained and were ready to fight for their homeland. He himself, he continued, "had been branded a Quisling" while visiting Palestine, and at any moment a moderate such as Weizmann might be replaced by a militant such as Abba Hillel Silver. Unless something favorable was done for the Jews, control of the Zionist movement "might pass to those not averse to violence," and "there might even be actual bloodshed in Palestine."

Attached to Evan Wilson's report of the meeting with Goldmann was a cover memo from Loy Henderson, who added that some Jewish youths were being driven into the arms of Moscow and that it was quite possible that violence would soon erupt in Palestine. Alluding to the possibility that Silver could replace the moderate Rabbi Wise in the Zionist leadership in America, Henderson wrote, "we have reason to believe that there is considerable truth in the claim that the extreme Zionists are gaining support among Jews here and abroad."[42]

If the State Department was worried, its fears were confirmed when David Ben-Gurion, then the chairman of the Jewish Agency executive in Jerusalem, arrived in the United States a few days later. Ben-Gurion, who would become Israel's first prime minister, made a strong impression among those who met him for the first time, such as the American lawyer Bartley Crum. Crum, who would soon become deeply involved with the Palestinian issue, was, like others, mesmerized by Ben-Gurion. A short man, but "stockily built," he wrote, "with a halo of white hair, a determined jaw set as in stone, with piercing blue eyes under heavy white shaggy brows, was an extremely forceful personality."[43]

Ben-Gurion was the epitome of the early Jewish settlers in Palestine. He had come from Poland as a seventeen-year-old in 1906, the scion of an old Zionist family. During the First World War, he had volunteered for the Jewish Legion and gone to fight under General Edmund Allenby as part of the Jewish Battalion of British troops. Returning to Palestine in 1920, he had quickly risen to the top of the Yishuv leadership, founding the trade union body the Histadrut. In 1935, he had been elected chairman of the Jewish Agency executive, and presided over Jewish immigration to Palestine from Hitler's Germany.

After arriving in the United States, Ben-Gurion, accompanied by Goldmann and Eliezer Kaplan of the American branch of the Jewish Agency, paid another visit to the State Department. After tracing the situation from Balfour to the White Paper, Ben-Gurion told Evan Wilson and Loy Henderson that "this intolerable [British] regime, had to be dealt with." If it were not, "the Jews could not continue indefinitely to put up with the breach by the administration of its obligations to the Jewish people." The Yishuv had to be allowed to make its own decisions and not have to take into consideration the demands of Arabs in Egypt, Iraq, or elsewhere. He understood that Arabs in Palestine were a different matter, since they had a legitimate interest in the future of Palestine. Perhaps naively, he reflected that "Jews and Arabs had lived there in amity for many years and there was no reason why they should not continue to do so."

Again he stressed that the Jews had supported the Allied military effort and expected that the Allies would honor their promises to tend to the Yishuv's needs now that the war had ended. "The world," he emphasized, "must not underestimate the strength of the Jews' feeling on this point." They did not seek a fight with the British government but would, however, fight if necessary to defend their rights, and the "consequences would be on Great Britain's head if the Jews were provoked into some action." What if the British listened to them, Henderson asked. How would they deal with the Arabs? I "know the Arabs well," Ben-Gurion responded, and "they would not really put up any kind of a fight." The leaders of the Arab states would have little luck in rallying their people on behalf of the official Arab position.

Is not your immediate objective, Henderson then asked, to increase Jewish immigration into Palestine? Of course, Ben-Gurion replied, that was imperative. But it was not enough. The time had come to grant this and the Jews' other demands, including the immediate establishment of a Jewish state.[44]

Ben-Gurion and Goldmann mistook Henderson's response to them. "There is reason to believe," they reported, "that Mr. Henderson takes a more sympathetic approach to our problem than his predecessor, Mr. Wallace Murray, whose antagonism was notorious."[45] A seasoned State Department expert on the Middle East, Henderson actually shared his colleagues' pro-Arab orientation. He cordially listened to Ben-Gurion and the other Zionists, as giving them such an audience was part of his job.[46]

Henderson was convinced that Palestine would not be able to absorb the European Jewish refugees who wanted to go there, that any U.S. support for the Zionist program would harm U.S. interests in the Islamic and Arab world, that a favorable policy toward Zionism might interfere with the flow of oil, and that the United States had to counter Soviet policy, which was tending toward support of the Arab states and seeking to make inroads in the Middle East.[47] Soviet machinations concerned him a great deal. In the early years of his career, Henderson had worked on Soviet and Eastern European affairs. He had lived in the Baltic states and in Moscow, where he had witnessed the brutality of Stalin's regime and developed "an intense dislike of the Soviet Communists and their sympathizers abroad."[48]

His opposition to Zionism began when he was U.S. ambassador to Iraq from 1943 to 1945. The leaders of Iraq told him that if the United States backed the Zionist state in Palestine, "the whole Arab world would begin to feel that the United States had become an enemy of the Arabs." Pro-Western Arabs would quickly be deposed, and the entire Middle East would become anti-American. Before he came back to assume his duties as the new head of the Office of Near Eastern and African Affairs at the State Department, he traveled to Saudi Arabia, Lebanon, and Egypt. There he found the same attitude as that held by the Iraqis.

When he returned to Washington to assume his new position in the State Department a few days after FDR's death, he knew that "the problem of Palestine would be one of the crosses which I would have to bear." As for the refugees, Henderson's view was that they should be sent to other "civilized" Western nations. In Palestine, he later wrote, "they would not find the happy, quiet Jewish National Home which they were looking for." It was "the Zionist juggernaut," as he called it, that was responsible for the idea that Jews had no place but Palestine to go to. But where else, he could not say. The Zionists believed that such an opportunity might not come "in another thousand years"; hence now was "the time for the Jews to be assembled again in their old homeland."[49]

Ben-Gurion did not wait long to act. On July 1, he convened an extraordinary meeting in New York City, at the apartment of Rudolf G. Sonneborn. Sonneborn, the secretary-treasurer of a multimillion-dollar corporation, was the son of a German Jewish Baltimore family whose wealth came from his family's oil and chemical business. The two had met in 1919 at the Versailles Peace Conference, where Sonneborn had been secretary to a Zionist delegation, and although they came from disparate backgrounds–one a man of wealth, the other a rough-and-ready man of the Yishuv–they had become friends.

New York City was sweltering that day, and the hot, fetid air rose to unbearable levels as the massive concrete buildings picked up the heat and made outdoor activity forbidding. Sonneborn, whose penthouse duplex did not have air-conditioning, opened all the windows and told his butler to prepare lots of cool drinks and ice. They would be needed, especially after the assembled guests heard what Ben-Gurion had to say.[50]

Ben-Gurion was worried. This meeting, he feared, was his last hope. The previous month, he had traveled through the United States, and what he had heard made him discouraged. Jewish leaders favored help for the Jewish refugees in Europe, but he found only platitudes of support and a lackadaisical response to his requests for real monetary support and total commitment to whatever had to be done to achieve success for his movement. He sought action, and he intended to get it. He had asked Henry Montor, the director of the United Jewish Appeal, to help him find ten to thirty prominent men who were interested in Palestine and who would come to a meeting in New York–on quick notice and without any public announcement

The two were pleased to find that all seventeen men they had invited had come–some traveling from Los Angeles, Miami, and other American cities and even one from Toronto. There were one rabbi, five lawyers, and eleven businessmen. Ben-Gurion minced no words. He told them that 5 million to 6 million Jews had been exterminated by the Nazis. "The great centers of Jewish population in Eastern Europe . . . were no more. The Jewish communities of Western Europe were decimated. After years of hiding or living in concentration camps . . . those Jews, who had survived the Nazi massacres were demoralized and displaced, not merely from their homes but from all recognition as self-sustaining human beings." The survivors wanted to go to Palestine, the Yishuv needed them, and no other country wanted them.

Then he made some startling predictions. He had just returned from London, he told them, and was convinced that the Labour Party would win the elections. His audience assumed this would be a good thing, since Labour was on record as supporting the end to the White Paper and supporting the Jewish National Home. Ben-Gurion quickly disabused them of their illusions. He believed that once Labour was in power, the new government would quickly rescind its pro-Zionist pledges and would soon give up its Mandate. The Foreign Office saw stability only with a pro-Arab Palestine and would convince the new government to follow its line. Moreover, Ben-Gurion now told his audience what he really believed–that with Britain out of Palestine, one thing was inevitable: the Arab nations, despite their differences, would unite to invade the land, with the intent of destroying the Yishuv. This time, however, the Jews would fight to defend themselves.*

The Holocaust had proven an observation about the Jewish situation he had made in Palestine in 1933 to be all too true. He had written then, "We have sinned in this land, in all other lands, we have sinned for two thousand years, the sin of weakness. We are weak–that is our crime."[51] Ben-Gurion did not want to make this mistake again. Now there could be but one solution: the survivors had to be brought to Palestine by the Zionist movement, using its own ships. Once there, the sick would be cared for and the young men and women would be trained to be ready to fight the invading Arabs, should the time come.

Ben-Gurion did not ask his guests to make any promises, and no one took notes on the meeting. Leonard Slater, who interviewed many of the participants, noted that "what was remembered best by each . . . was the uncertainty he felt at the end about what was expected of him."[52] In hindsight, Rudolf Sonneborn noted that they were being asked to "form ourselves into . . . an American arm of the under-

* This statement countered the argument used by American Zionists. Rabbi Stephen Wise, for example had argued that the Arabs were totally dependent on the goodwill of the Great Powers. If the Great Powers acted in concert, Wise wrote, "the whole Arab 'bogey' will be found to be the myth it is." See Stephen Wise and Herman Shulman, "Memorandum on Palestine," July 3, 1945, in President's Official File, no. 771, HSTL.

ground Haganah. We were given no clue as to what we might be called upon to accomplish, when the call might come, or who would call us. We were simply asked to be prepared and to mobilize like-minded Americans. We were asked to keep the meeting confidential."[53] Later, they would learn that what the Yishuv leader wanted from them was funds that could be used to purchase arms. Ben-Gurion would later write in his personal journal, "That was the best Zionist meeting I have ever had in the United States."[54]

While the U.N. delegates were organizing in San Francisco, Harry Truman's major concern was the negotiations with Winston Churchill (and later Clement Attlee) and Joseph Stalin, scheduled to begin at Potsdam on July 17–the conference they hoped would settle the remaining issues left over from Yalta. Here the world's victorious leaders would discuss how to handle the administration and political order of Germany, which had surrendered unconditionally on May 8. Truman was already concerned about the Soviet Union's many challenges to agreements made at Yalta, especially to ensure democracy in Poland, and was intent upon his promises to honor the course chartered by the late president.

With other pressing questions on the heavy agenda, such as what steps they should take to win the war against Japan, the situation of the Jews and the demand of the Zionists for a Jewish state were not uppermost in the president's mind; nevertheless, they weighed on him. As Truman sailed to Potsdam, he brought with him several reports on the subject that he had pledged to study and consider: a petition from senators and congressmen, a report from the official Zionist coalition regarding Palestine and the Jewish refugees, a letter from the American Jewish Committee, and a brief from the State Department. All gave the president conflicting advice. Truman promised only one thing: he would study them all carefully and then chart his own course.

The petition Truman had with him had been circulated in the Senate under Robert F. Wagner, Jr.'s, name. The son of German immigrants, Wagner had grown up in a small Manhattan basement apartment. Entering Democratic politics in 1904, he was first elected to the Senate in 1926 and reelected four times after that. Known mostly for his commitment to the New Deal and his sponsorship of the Wagner Labor Relations Act, Wagner had the overwhelming support of New York's large Jewish community.[55]

Not only was Wagner responsive to the needs of New York's striving Jewish working-class community, many only one generation away from czarist oppression, but also, as the chairman of the American Christian Palestine Committee, he was an ardent Zionist. He had gained the respect and support of the Zionists during the war years, when he had stood in the forefront of the effort to ease American immigration for Europe's Jews. Wagner had proved his mettle when he joined with Republican Senator Robert A. Taft to sponsor the proposed congressional resolutions of 1944 in support of a Jewish homeland. He was proud of his efforts in those years to aid the victims of Nazi persecution and do what he could to save them from certain death. "In his lifetime," Wagner told Epstein, "he had certainly committed many sins, but still hoped to get to heaven because of the staunch advocacy of Jews whom he had saved and who would testify before the Almighty in his favor." He also told him that as a good Christian, he felt it was his duty to do all he could to assist the return of the Jews to their homeland.[56]

Now, in the postwar era, Wagner was adamant that Jewish goals finally be realized. Palestine, he believed, was not only indispensable to the rehabilitation of the few European Jews who had survived but equally essential "to the health of Europe." The Christian world that had allowed the Holocaust to happen "has Palestine on its conscience," he told Epstein, and "if it would regain its moral self-respect, it must promptly do justice to Palestine."[57]

Representative Helen Gahagan Douglas, the national secretary of the American Christian Palestine Committee, sent out the letter to members of Congress, asking them to sign the petition. A former actress and the wife of the actor Melvyn Douglas was a longtime liberal activist who worked in the film community. Like so many others, Douglas was horrified about what she called the Nazi regime's "monstrous campaign of systematic murder and torture that has almost annihilated the Jewish communities of Europe." Noting that the "civilized world" had been shocked by the revelations about the death camps, she urged that all those she was contacting on a list of distinguished Americans help her "do all we can to insure justice to the remnant of European Jewry." The letter would be addressed to the president and would stress that both political parties favored unrestricted Jewish immigration to Palestine, with the eventual outcome that it would become "a free and democratic Jewish Commonwealth."[58]

AZEC helped gather signatures and tried to make sure that the petition would include majorities from both the House of Representatives and the Senate. To attain that goal, the group's strategies included getting influential members of each community in America to write their representatives, urging them to sign.[59]

After a two-month campaign, the petition was completed. The organizers wanted to release the letter before the president went to Potsdam, in the hope that it would lead him to deal with Palestine at the meeting. At first they prepared for a major press conference, to be attended by both the majority and minority leaders of both the Senate and House, but when Wagner met with the president, he requested that a public demonstration of that sort not take place. Instead, Truman asked that Wagner present it to him personally at the White House and said he would take it with him to Potsdam and consider it carefully. Expressing sympathy with Zionist goals, Truman promised Wagner that he would "do everything he could at the forthcoming conference" but that publicity would harm their current dealings with Near East problems.[60] Following Truman's request, Wagner presented Truman with the petition in person. It was signed by 54 Senators and 250 members of the House, favoring establishment of Palestine as a Jewish state.[61]

A week later, an AZEC official, Herman Shulman, reported to his executive committee that he had spoken with one of Truman's close advisers (probably Niles), requesting a meeting with the president and Rabbi Wise before Truman left for Potsdam. The unnamed aide had reported back that Truman had said he would be happy if they would prepare a comprehensive brief dealing with the Palestine issue in all its aspects and promised that he would give it "his very careful attention."

The president wanted to remind them, however, that he had pledged to follow FDR's policy and that eventually he planned to meet with them.

Shulman also reported that he and Wise had next met with Robert Hannegan, the chairman of the Democratic National Committee and a friend of Truman from Missouri. Both men were impressed with Hannegan's apparent sympathy for their arguments. It was he, Hannegan told them, who had personally told Truman at the Chicago Democratic National Convention to include the Palestine plank in the party platform. Hannegan also asked for the brief, which he promised to personally deliver to Truman and others who were on their way to Berlin. After receiving the twelve-page document, Hannegan told them he had given it to Truman as well as to Secretary of State James F. Byrnes. Byrnes read it immediately and told Hannegan that he was "sympathetic to the Zionist aspirations."[62]

In their report, Wise and Shulman argued that the White Paper of 1939 directly contradicted the terms of both the Balfour Declaration and the Mandate, which had been endorsed by every American president since Woodrow Wilson. It had been issued by the Neville Chamberlain government in response to Arab terrorism and was "in conformity with its general program of appeasement." Moreover, they emphasized, the Grand Mufti of Jerusalem, the leader of the Palestinian Arab community, was openly pro-Axis and had fled to Germany during the war. Unlike the Arabs, the Jews had fought the Axis with the other Allies, putting off their own demands to the war's end. "Their dead lie scattered on the battlefields of Greece, Syria, and North Africa, and their economic and military contribution was all out of proportion to their numbers." The Jews had waited patiently, but now, with the war over, they could no longer tolerate the White Paper.

Only Palestine, the report continued, could be the destination of Europe's Jewish remnant and that of the one million Oriental Jews, who were subject to persecution in Arab lands. The Yishuv was ready to take its first million new residents. Noting that Palestine had four and a half million unoccupied acres that were regarded as uncultivable wasteland, they reminded Truman that most of the flourishing settlements had been built on similar land. In just twenty-five years, as the Jewish population of Palestine had grown, the Jews had "established some three hundred Jewish agricultural settlements, and have made Palestine the industrial center of the whole Middle East, with thousands of factories and small workshops."

Wise and Shulman then noted that the Arabs had benefited from the Jews' efforts to develop Palestine. As evidence, they pointed to a statement made to the British House of Commons in 1938 by the British secretary of state for the colonies, Malcolm MacDonald. MacDonald said that the Arabs could not claim that the Jews were driving them out of their country. In fact, there had been a dramatic growth in the Arab population that could only be attributed to "the Jews who have come to Palestine bringing modern health services and other advantages."

Finally, they argued that if the United States and the other Allied powers came out for a Jewish state and free immigration to Palestine, the Arab nations would be forced to accept it as "an accomplished fact." If the Great Powers made it clear that they would be ready to use force if necessary, the need for actually carrying through would not exist and the Arab threats would disappear. And if the United States did not carry through on pledges made beforehand for a Jewish state, it would be nothing less than a "gross betrayal of the principles of international law and good faith."[63]

Truman received quite different advice from the American Jewish Committee's leader, Judge Joseph Proskauer, whose recommendations he also had with him. Declaring their opposition to a Jewish state, Proskauer wrote that his group distinguished between favoring Palestine as homeland and place of refuge and that of statehood. FDR himself, he informed Truman, had told them shortly before his death that "he had come to this belief and that he saw in the extreme Zionist agitation grave danger for the world and for Palestine itself." But, he added, the AJC supported the consensus of all Jewish groups for "liberalization of Jewish immigration into Palestine."[64]

The brief prepared for Truman by the State Department warned him of a British trap to ensnare the United States in the Palestine problem. It recommended that if Truman spoke to Churchill, he do so with the understanding that the United States realized that Palestine was primarily a British problem, since Britain held the Mandate. It predicted that the British would make some concessions on immigration and would seek U.S. backing and support for it, but since the British themselves were divided on the subject, they could then turn around and "blame US pressure for policies that the Arabs may not like." Therefore, the president shouldn't commit himself to anything. Moreover, State warned him, the Zionist Jews were "growing restless" and were "determined to force a decision on Palestine this summer." Moderates might not be able to restrain them, and the "thoroughly aroused" Arabs would oppose any changes to the White Paper by force of arms. He might be criticized by American Jews if he did not take their wishes into account, but if he took a pro-Zionist stance, the United States could draw the hatred of the entire Arab world.[65]

With this advice under his belt, Truman wrote to Winston Churchill on July 24. "There is great interest in America in the Palestine problem," he told the wartime prime minister. British policy continued to "provoke passionate protest from Americans"–a fact of which Churchill was more than aware. Truman summarized their feelings in words that indicated his sympathy: "They fervently urge the lifting of those restrictions which deny to Jews, who have been so cruelly uprooted by ruthless Nazi persecution, entrance into the land which represents for so many of them their only hope of survival." Truman informed Churchill of his hope that the British government would lift the White Paper restrictions. Understanding that they would not be discussing this at Potsdam–as Jewish groups had hoped–Truman urged Churchill to act and not let the issue be subject to "prolonged delay."[66]

Two days after he sent the memo to Churchill, the British electorate voted overwhelmingly for the Labour Party, throwing the wartime unity coalition government and its leader out of office. Churchill was to be replaced by a Labour prime minister, Clement Attlee. Zionist leaders were overjoyed; the Labour Party had a long history of commitment to Zionism.[67] In Tel Aviv, "there was rejoicing and dancing in the streets," since Palestine's Jews knew that British Labour "appeared to be definitely and strongly committed to the Zionist program."[68]

As opposition leader, Attlee had been at Potsdam with Churchill, and now that he was prime minister he found Truman's memo to Churchill waiting for him on his desk. Attlee's formal response was terse: he would not give the president his views until his government had time to consider the issue, to which he promised careful consideration.[69] Soon the Zionist movement received disappointing news vividly recalled by David Horowitz, one of the pioneer settlers and an economist for the Jewish Agency: "Then came the great shock. It all set out as a faint and feeble whisper, which quickly assumed substance and spread like wildfire. Within no time at all the incredible truth had come out: the British Government intended to maintain the hated White Paper policy in all the articles of its repression."[70]

Truman tried to take it easy on his way home from Potsdam. He played poker with Secretary Byrnes and started each day aboard the *Augusta* with a slow walk around the deck. The Big Three meeting was over, and the president reflected on the successes and potential pitfalls of Potsdam. He had arranged the terms for the management of postwar Germany. Yet, as it later would become glaringly obvious, Truman had acquiesced to Stalin's insistence that the newly liberated Eastern European nations fall under the domination of the Soviet Union. In particular, Truman agreed to the consignment of German territory in the east to Poland, although the new Polish government was already under the control of the Soviet Union. Truman and Byrnes had negotiated a "spheres-of-influence" peace, which both men hoped would ensure the continued cooperation of America and Russia in the postwar era. Their concessions were necessary, they thought, in order to gain needed agreements with the Soviets. Of special concern was Truman's hope that Stalin would agree to enter the war against Japan, relieving the United States of bearing the entire burden of the scheduled military campaigns in the Pacific. On this he was successful, as Stalin agreed to move Soviet forces from Europe to the Pacific in time to enter the war in mid-August.[71]

As Truman was eating lunch in the sailors' mess hall, he received a message that the *Enola Gay* had dropped an atomic bomb on the city of Hiroshima, which he hoped would save American lives and end the war with Japan more quickly. A major invasion of Japanese islands would now not have to take place, although the Japanese leaders did not at first accept defeat and pledged to continue fighting. It was not until a second atomic bomb was dropped on Nagasaki that the emperor announced he would accept peace and agreed to an unconditional surrender to the United States. The president announced the news to an anxious nation on August 14. For the next two days, the entire country celebrated, from gatherings in the smallest of town squares to the massive outpouring of people in New York City's Times Square.[72]

"Things have been in such a dizzy whirl here," Truman wrote to his mother. Everyone had been operating at a breakneck pace, and he wished that they would calm down so he could turn his attention to the country's domestic problems and his reform agenda. "It is going to be political maneuvers," he thought, "that I have to watch."[73] On August 16, at Truman's first press conference after his return from Potsdam, he was asked a question that he did not expect, since the press briefing was to deal only with European affairs: Had the question of a Jewish state in Palestine been raised? a member of the press asked. Had it been mentioned during the conversations held with Churchill and Josef Stalin? No to the first question, Truman replied, but he had discussed it with Churchill and later Attlee. The president paused, wanting to move on to other issues, when he was asked a more direct query: "What is the American position on Palestine?" The president's answer was straightforward. "The American view of Palestine is that we want to let as many of the Jews into Palestine as is possible to let into that country. Then the matter will have to be worked out diplomatically with the British and the Arabs so that if a State can be set up there, they may be able to set it up on a peaceful basis. I have no desire to send 500,000 American soldiers there to make peace in Palestine."[74]

Truman was on his way to developing his own Palestine policy. Though he wanted Britain to abrogate the White Paper and let the

Jewish refugees go to Palestine, he was not going to take responsibility for creating a Jewish state. That would be the responsibility of the British and the Arabs. The State Department warnings had obviously alarmed him about the likelihood of having to send American troops to keep the peace if the Zionist project progressed. It would be better, Truman thought, if he kept his hands off of it.

It wasn't even clear how strong Truman was on the immigration issue. American press reports indicated that Congressman Adolph J. Sabbath of Illinois had met with the president and that Truman had told him that although he favored the admission of Jews into Palestine, "he was afraid that Arab opposition would be too great" and preferred concentrating more on "Jewish rights in Europe" and less on Palestine.

Rabbi Silver felt that Truman's statement was hardly a victory for the Zionists and could even be seen as a step backward. The best that could be said was that Truman was at least discussing the issue with the British government and that he appeared to have "good will with regard to Jewish immigration into Palestine." On the other hand, it might encourage the Arabs to take actions that would necessitate sending in U.S. troops. If it came down to such a choice, Silver feared that Truman was "prepared to give up the idea of a Jewish state."[75]

The Arabs were also puzzled over the meaning of the president's statement. The counselor to the Egyptian Legation told Loy Henderson they feared Truman's words "might indicate a change in policy of this Government which would give rise to great unrest in the Arab world." Any sudden move that proved prejudicial to Arab interest, he threatened, "might well set the Arab world in motion and result in violence on a wide scale."[76] The Syrian minister to the United States, Nazdem al-Koudsi, on the other hand, saw Truman's statement as positive. He emphasized the president's assurance that nothing would be done without consultation with the Arabs and that the United States did not contemplate sending armed forces to Palestine. He feared that other Arabs might focus only on the part of the statement favoring free immigration of Jews to Palestine.[77]

Loy Henderson had reason to worry. Addressing the secretary of state, he reported that the president's press conference had been "bewildering to the Arabs," since they believed that FDR had promised Ibn Saud that the United States would not support the Jews in Palestine. Henderson attached the text of a top secret memorandum of the conversation between Roosevelt and the Saudi monarch held on board the *Quincy* in February 1945. The translation by William Eddy had been approved by both sides and served as "an agreed minute of the meeting." The secret memorandum reported on Roosevelt's promise that he would "do nothing to assist the Jews against the Arabs" and "make no move hostile to the Arab people." Noting that FDR had reiterated these points in a letter to Ibn Saud, he pointed out that all the Arab leaders seemed to know about this. "As this pledge has never been divulged by us to any other parties," Henderson told Byrnes, "it would appear that it was Ibn Saud who made it known to the Arab League."[78]

Alarmed that the United States might be changing its policy, on August 24 Abdul Rahman Azzam, the secretary-general of the Arab League and undersecretary for foreign affairs in the Egyptian government, announced publicly that FDR had given a pledge to Ibn Saud "that he would not support any move to hand over Palestine to Jews." Ibn Saud, he added, had also told Roosevelt that if Palestine were given to the Jews, "he would start a war against the Zionists and all who supported them" and would "never rest until I and all my sons have been killed in the defense of Palestine." Then Ibn Saud had stood and, placing his hand in Roosevelt's hand had said, "Swear that you will never support the Zionists' fight for Palestine against the Arabs." According to Azzam Bey, FDR then "shook Ibn Saud's hand and pledged he would not support the Jews against the Arabs."[79]

President Truman was very puzzled by the assertion that FDR had made such a pledge to Ibn Saud. He sent Admiral Leahy, who was at the meeting between FDR and Ibn Saud, the official summary of the conversation. Was there any contradiction between what Assam Bey asserted and the official transcript prepared by William Eddy? he asked.

The purpose of FDR's meeting, Leahy responded, "was to endeavor to get the King of Saudi Arabia to agree to a compromise that would permit some of the displaced European Jews to find homes in Palestine." Roosevelt had made an "excellent presentation of the Jewish Palestine difficulty as seen from his point of view," he continued. "With great dignity, courtesy, and smiling, the King said that the Jews and Arabs now living in Palestine have learned to live together in peace, but that if Jews from outside Palestine with their foreign financial backing and higher standards of living are imported, they will make trouble for the Arab inhabitants. When this happens, as a good Arab and a 'True Believer,' he will have to take the Arab side against the Jews and he intends to do so." President Roosevelt replied, Leahy wrote, "that he had no intention of getting involved in hostilities between Arabs and Jews." Leahy said that based on other conversations with FDR he accepted Saud's statement to mean that he would go to war in defense of the Arabs in Palestine if more Jews immigrated there. "I do not believe that President Roosevelt at any time said that he would not support a plan to establish a Jewish colony in Palestine," Leahy concluded. "He plainly did not, however, intend to go to war with the Arabs for such a purpose." Finally, Leahy assured Truman that Eddy's translation was accurate.[80]

To assure himself of Leahy's accuracy, Truman then sent Leahy's answer to Sam Rosenman. What did he think? the president asked. Agreeing with Leahy, Rosenman wrote Truman, "there is nothing inconsistent between this statement of what President Roosevelt said and your statement to the press conference the other day [August 16]." "Furthermore," Rosenman added, "I do not think that opening the doors to Palestine is in any sense an act which is a 'move hostile to the Arab people.' Nor does it in my opinion contravene the conversation of President Roosevelt."[81]

While Truman was pondering the implications of FDR's meeting with Ibn Saud and what it meant for U.S. policy, he received the results of Earl Harrison's investigation into the situation of the Jewish refugees. Harrison's report and recommendations would challenge Truman to find a solution for the refugees and overcome the difficulties of sending them to the place they most longed to go: Palestine.

FOUR

THE PLIGHT OF THE JEWISH DPs: THE HARRISON REPORT

On the morning of August 24, Truman sat down for his regular Saturday breakfast meeting with his staff and some close friends, including his press secretary, Charlie Ross, Matthew Connelly, Sam Rosenman, Admiral James K. Vardaman, General Harry Vaughan, and his assistant press secretary, Eben Ayers. The night before, he had read Earl Harrison's report, and he told the group that it made him sick. "The situation at many of the camps," Truman said, "especially with respect to the Jews, was practically as bad as it was under the Germans."[1]

Harrison's report troubled him, as it would anyone who read it. When Harrison arrived at Dachau on July 22, Rabbi Abraham Klausner, a chaplain in the U.S. Army, introduced himself. Klausner, who had arrived at Dachau right after it had been liberated, had been writing reports to the major Jewish organizations about the poor treatment of the Jewish survivors. By the end of August, his efforts had helped to establish an Office of the Advisor on Jewish Affairs to the General Command in the European theater. Klausner looked over the itinerary that the Army had drawn up for Harrison and told him he thought it reflected only what it wanted him to see, mainly the official collection centers. Harrison asked Klausner to alter it so that he could get an accurate picture of the refugee situation. The rabbi took him and

Joseph J. Schwartz, the European director of the American Jewish Joint Distribution Committee (usually referred to as "the Joint"), to see the partially bombed buildings where hundreds of refugees were housed because there was nowhere else to go. Then he took him to see the overcrowded camps.[2]

Harrison's report told an abysmal story. In Germany and Austria, he had found many of the DPs living under guard behind barbed-wire fences, sometimes in the most notorious concentration camps "amidst crowded, frequently unsanitary and generally grim conditions, with nothing to do, and with a serious lack of needed medical supplies." The DPs subsisted on bread and coffee, and the observers had seen many "pathetic malnutrition cases." A high death rate among the former prisoners continued after liberation. At Bergen-Belsen, some twenty-three thousand people had died, most of them Jews. Not only did the prisoners live in unsanitary conditions and have little food, some of them "had no clothing other than their concentration camp garb–a rather hideous striped pajama . . . while others . . . were obliged to wear German S.S. uniforms." And it was distressing that there was no organized effort to help them find out what had happened to or to locate their loved ones.[3]

"Beyond knowing that they are no longer in danger of the gas chambers, torture and other forms of violent death," Harrison continued, "they see–and there is–little change." An essential point, he emphasized, was that the Jewish DPs had to be viewed as Jews, not just as simple refugees, undistinguished from the others. If they were not, one would be "closing one's eyes to their former and more barbaric persecution." Most of the Jews, Harrison reported, wanted to leave Germany and Austria as soon as possible and go to Palestine. Many had relatives in Palestine, he wrote, while others "having experienced intolerance and persecution in their homelands for years, feel that only in Palestine will they be welcomed and find peace and quiet and be given an opportunity to live and work."

The situation was so bad that Harrison put it in the strongest possible terms: "As matters now stand, we appear to be treating the Jews as the Nazis treated them except we do not exterminate them. They are in concentration camps in large numbers under our military guard instead of S.S. troops. One is led to wonder whether the German people, seeing this, are not supposing that we are following or at least condoning Nazi policy."

Pending their emigration to Palestine, Harrison suggested setting up separate quarters for the Jews, so they could receive special attention. It was a matter of "raising to a more normal level" the status of a people who had "been depressed to the lowest depths conceivable by years of organized and inhuman oppression." It was a matter of "justice and humanity." To really end the inhumanity, the Jewish DPs should be allowed into Palestine as quickly as possible. "The civilized world," Harrison concluded, "owes it to this handful of survivors to provide them with a home where they can again settle down and begin to live as human beings." Harrison concluded by seconding the request of the Jewish Agency that Britain immediately allow emigration of 100,000 of the European Jewish DPs.

Truman's first reaction upon reading the Harrison report, Niles's assistant, Phileo Nash, recalled, was shock. Particularly upsetting to him was to find that when asked, most of the Jewish DPs told authorities that they did not want to come to the United States, they were afraid to, and that they would feel safe only in a country where they had sovereignty, which to them meant Palestine. Their fear was that what had happened to them in Germany could happen elsewhere just as easily.[4]

Harrison's report galvanized Truman to do something about the plight of the Jewish DPs in Europe. He wanted "a rapid and constructive settlement" since he considered their dire situation a problem of the "highest humanitarian importance and urgency." If the majority wanted to go to Palestine, Truman thought they should be allowed to do so. He agreed with Harrison that the 100,000 refugees would be able to easily be absorbed into Palestine as well as get jobs, given the severe shortage of labor there. The Jewish community in Palestine was ready to receive them and help them rebuild their lives.[5]

Truman's strong feelings on the subject were indicated by the letter he wrote to the new Labour prime minister, Clement Attlee. Sending him a copy of Harrison's report, Truman urged Attlee to lift the quota and immediately allow 100,000 Jewish refugees to enter Palestine. "No single matter," Truman wrote Attlee, "is so important for those who have known the horrors of concentration camps for over a decade as is the future of immigration possibilities into Palestine." To be effective, Truman emphasized, "such action should not be long delayed."[6] Truman felt the issue was so important that he wrote the letter himself, without consulting any speechwriters or informing the State Department. His hope was that he would get faster results by a personal appeal to the new prime minister rather than by going through normal diplomatic channels.[7]

Though Truman leaned heavily on Byrnes and other advisers for his European policies, he wanted more control over issues relating to Palestine and the Jews. This was noted by the new undersecretary of state, Dean Acheson. "By the time I took up my duties . . . in September 1945," Acheson recalled, "it was clear that the President himself was directing policy on Palestine." At this time, according to Acheson, Truman was separating his short-range goals from the long-range solution, which he felt the United Nations would address. Acheson, like others at State, did not share the president's views. He believed Palestine did not have the capacity to absorb any more Jews "without creating a grave political problem" and imperiling all Western interests in the Near East. He had learned from Felix Frankfurter, who was a close friend, and Justice Louis Brandeis to "understand but not to share the mystical emotion of the Jews to return to Palestine and end the Diaspora." Despite his disagreements, Acheson attempted to carry out the policies desired by the president and not to undermine him.[8]

Acheson was right about one thing: the State Department and Truman were at loggerheads. The same day Truman wrote his letter to Clement Attlee, the Division of Near Eastern Affairs prepared a memo for Secretary Byrnes to read before going to London, where he was scheduled to attend a meeting of the Council of Foreign Ministers. Despite the American Zionists' increasing demand for free immigration into Palestine, State argued, the Jews simply were not in a "position to have the knowledge and information" necessary to make a rational judgment. They would have no places to live, be subject to outbreaks of diseases such as typhoid and bubonic plague, and be subject to unemployment once Palestine industry moved from a wartime to a peacetime production schedule. Most important, it warned, any large increase in such immigration would precipitate considerable violence and possibly civil war in Palestine. To deal with this, the United States would have to allocate both troops and military supplies to the region, which would be necessary to maintain security. Its solution: "The United States Government should not favor mass or unrestricted Jewish immigration into Palestine," since it would commit the United States "to a definite policy in favor of the establishment of a Jewish State" without consultation with the Arabs. It would also have the most harmful effect on "American interests and prestige." The United States should definitely not do anything, State recommended, until it consulted with the British government.[9]

While the president was trying personal intervention with Clement Attlee, he was blindsided by a leak to the press from former Democratic Senator Guy Gillette of Iowa. In August, Gillette had taken the job as chief political adviser to the main Bergson group, the Hebrew Committee of National Liberation.[10] In that capacity, he met with Truman, accompanied by Senator Owen Brewster of Maine and Senator Warren Magnuson of Washington. They informed him that they were planning to travel to London to pressure the British government. Truman requested that the group postpone their meeting until after Byrnes's negotiations concerning Palestine and other matters were over. According to the three, Truman told them that it was the United States' position that the doors of Palestine be immediately opened to the homeless European Jews, and he mentioned that he had written a letter to Attlee to that effect.[11]

Truman was furious when he learned that the two senators and Gillette had gone to the press and leaked his letter to Attlee without his permission and before he had received an answer. When Silver and Wise heard about the letter, they were delighted but cautioned that Truman's request to Attlee for admission of the 100,000 Jewish refugees into Palestine would not solve the Jewish problem. Only the establishment of a Jewish state in Palestine, the two Zionists argued, would achieve that. Advising that Zionists not be too enthusiastic about the president's proposal, Wise and Silver argued that even if it were fulfilled, "the governments concerned may feel that we are satisfied with this action and may be inclined to let it go at that."[12] Their point was not without merit. Had the British government accepted the admission of the 100,000, the Jewish Agency worried, it might have lessened its argument for the necessity of a Jewish State. "It could be argued," the historian William Roger Louis points out, "that British rigidity on the question of the 100,000 proved to be the most serious tactical error in the controversy with the Americans."[13]

While Wise and Silver were pondering Truman's actions, in London Chaim Weizmann met with Secretary Byrnes, who told him that *"the whole matter [of Palestine] was handled to a considerable degree by Mr. Truman himself"* and therefore Byrnes had to be *"careful not to make statements to which Mr. Truman might raise objection"* (our emphasis). Byrnes asked Weizmann how long it would take, if they got the 100,000 certificates, to convey the refugees to Palestine. Weizmann thought about eighteen months and added that the climate in Palestine was more temperate than Europe's, so that the people "could live temporarily under canvas and in barracks," but whatever the conditions, they would be much better than the ones in which they were presently living. Byrnes, who, Weizmann thought, was impressed by this, said he would inform Truman.

Byrnes also wanted to know why the Jewish Agency had rejected Britain's offer to use the 3,003 certificates still available from the White Paper's quota. Weizmann replied that since they considered the White Paper to be both morally and legally invalid, it did not want to appear to be honoring its terms. Byrnes disagreed with this tactic and said the Agency should have taken the certificates and asked for more. Weizmann then mentioned that some of the Agency's "British friends" were constantly telling him that they could do something if only the Americans could take some responsibility. Weizmann wasn't sure what that could be. He doubted that the British would want the United States to participate in the administration of Palestine, but maybe sending a token force to Palestine or helping fund a development project for the country such as Lowdermilk's scheme would help. Byrnes said that while he couldn't speak for the president, he thought the Americans would try to help.[14]

Weizmann then wrote Byrnes a follow-up letter reiterating that the Agency's immediate concern was to see the White Paper abrogated. This would not only eliminate immigration quotas but repeal the current land regulations, which discriminated against Jews by forbidding them to purchase land in most of Palestine. He assured Byrnes that the Jewish community would bear its full share of the financial cost for the refugees but that the United States and Britain would have to help transport them. Supporting the refugees in Europe, Weizmann knew, was becoming a financial burden for America, and he pointed out that the funds and supplies used to maintain them in the camps, where they had nothing to do, could better be used by "initiating them into productive careers in Palestine."[15]

It was hard for Great Britain to grasp that it was no longer going to be one of the Great Powers of the world. Its new postwar leadership was the first that had to operate in the context of the country's diminished role in the world. The war had cost Britain one quarter of its national wealth. The country could not pay its international debts without borrowing more from the United States. No longer able to carry out its responsibilities in the Middle East, it was now in the humiliating position of being a debtor nation to the Arab regimes, instead of the creditor it had been before the war. The new foreign secretary, Ernest Bevin, also had to deal with Stalin's apparent intention to create Soviet-controlled regimes in Eastern Europe and to replace British influence in the Middle East with Soviet influence, as well as contend with Mohandas Gandhi's independence movement in India.[16]

Visitors to London who had known the capital city before the war were depressed by what they saw. Postwar London, the Jewish Agency economist David Horowitz wrote, was "grey, dreary and immeasurably fatigued. The gaps in the long streets, the empty spaces, and the piles of debris were testimony to the ferocity of the aerial bombardments. . . . The monotony of the diet and incessant toil had left their mark on people's faces, bodies and spirit." London, he found, especially at night, "was dismal and joyless."[17]

Dominating the policies of the British government were Ernest Bevin and the prime minister, Clement Attlee. Attlee was the complete opposite in personality from Winston Churchill, who had inspired the nation during the war. Churchill had given eloquent and heroic speeches, meant to keep his countrymen looking ahead and buoying their spirits at the worst of times. Attlee, from all accounts, gave boring speeches and was a man lacking in charisma. The differences between Attlee and Bevin were also pronounced. Bevin, his biographer writes, was "a heavyweight in personality as well as physique, temperamental, passionate and egocentric," while Attlee was "spare, dry and undemonstrative."[18]

Bevin was "a large, powerfully built man," attorney Bartley Crum remembered, "with heavy shell-rimmed spectacles and a way of holding his hands at his sides, fists clenched." Bevin is today thought by many to have been an anti-Semite. But in 1945, he was regarded by his colleagues as a man who had been supportive of Zionist goals before the war. Whether or not he was anti-Semitic, his most recent biographer makes it clear that he did not comprehend the depths of despair world Jewry felt after the Holocaust. Nor did he understand that the Jews of Palestine were intent upon achieving a Jewish state and were ready to fight for it under the banner of their underground army, the Haganah.[19] Bevin assumed that the Zionist agenda would be set by the moderates he knew in London, led by Chaim Weizmann, who preferred using gradualism and diplomacy to reach their goals.[20]

Both Bevin and Attlee gave great weight to the views of their Foreign Office advisers, who, like their State Department counterparts, overestimated the strength and fighting power of the Arab states. These men advised the new government that Britain had to support the Arabs, given the country's dependence on Middle Eastern oil and its need to keep control of the Suez Canal, which was a gateway to British possessions in Asia.[21]

On September 10, Secretary Byrnes personally delivered Truman's letter of August 31 to Clement Attlee. Two days earlier, he and Bevin had received recommendations from the Labour cabinet's Palestine Committee. It was not good news for the Zionists. There could not be Jewish mass immigration into Palestine, the committee concluded. The White Paper could not be abrogated, and any continuance of a limited immigration of Jews had to depend on the Arabs' agreement.[22]

On September 14, Attlee finally replied to Truman. Promising a more thorough letter a bit later, Attlee said he had been concerned when Byrnes had told him about Truman's plans to issue a statement about Palestine that would disclose Harrison's conclusions. Such a statement, he informed Truman, "could not fail to do grievous harm to relations between our two countries." Moreover, Attlee complained, the Palestinian Jews were refusing to accept the allotted immigration certificates and were demanding instead immediate granting of the entire 100,000 they and Truman had asked for. Attlee warned that should Truman actually issue his statement, it would precipitate "a grave crisis" that would greatly interfere with the work of reconstruction after the war.[23]

As promised, Attlee telegraphed the president his longer and considered response two days later. First, he strongly disagreed with the view that Jews in the DP camps should be treated differently from the other victims of Hitler. To do so, he argued, would provoke "violent reactions on the part of other people who had been confined to these concentration camps." He claimed that there was little difference in the way non-Jewish prisoners had been tortured and their general treatment. Attlee told Truman that Jews could not be put in a special racial category that would put them "at the head of the queue." His suggested solution was to send 30,000 immediately to a camp that was available for use in Philippeville, Algeria, and another in Fedala, in Morocco, that could hold five thousand more. Finally, he reasserted that Britain had to take into consideration the view of the Arabs, whom both FDR and Churchill had promised to consult on the issue. To "break these solemn pledges," Attlee stressed, would "set aflame the whole Middle East." And it was Britain alone that would have to restore order.[24]

Truman was annoyed by Attlee's dismissal and his insensitivity.

He wrote a perfunctory response saying only that he understood the complications "from your point of view" and that he would take no further action until Secretary Byrnes returned to the capital from Europe.[25] Almost immediately, the president called David Niles into his office. The president, Niles recalled, had been familiar with the pro-Zionist statements made over the years by Labour Party leaders, and he assumed that these reflected the official position of the party. In his report, Harrison had quoted the words of the Labour leader Hugh Dalton, who had said in May 1945 that it "is morally wrong and politically indefensible to impose obstacles to the entry into Palestine now of any Jews who desire to go there." At Truman's request, Niles gathered other resolutions and statements adopted by Labour over the years, including pro-Zionist statements made in previous years by both Attlee and Bevin. Reading these, Truman asked Niles, "How can we trust the Labour people in London when they do not respect their pledges? Today they are cheating the Jews, and where is the assurance that they won't cheat *us* tomorrow?"[26]

Truman was disturbed that Attlee's response was not only a retreat from Labour's long-held position on Palestine but seemed devoid of "all human and moral considerations." He decided that he would no longer deal personally with Attlee on the subject of Palestine and the Jews. Instead, he instructed Byrnes to formally raise the matter again when meeting with Bevin the next time he was in London. Byrnes, the most skilled of negotiators, would find that Attlee and Bevin refused to meet him halfway. It seemed that an unbridgeable gap was growing between America and Britain on the issue. Niles believed that the British intransigence on any kind of compromise on immigration helped Truman come to the conclusion that the Jews, without sovereignty, were helpless to control their destiny and to rectify all that had been taken away from them.[27]

On September 23, the British government announced its final decision. The Palestine issue, it proposed, should be referred to the new United Nations along with a statement that all the Allied powers had a responsibility to deal with the issue in common. Since the United States rejected any proposal that it share the administration of Palestine along with the British, no other path was open other than continuing with the stopgap measure of admitting the final allotted certificates under the White Paper and wait for resolution by the United Nations.[28] Responding to this announcement two days later, Chaim Weizmann, speaking at an emergency Zionist conference in Britain, castigated the Labour Party for reneging on its many previous pledges. The Jewish Agency, he announced, had formally rejected the British plea to accept the certificates and again declared the illegality of the White Paper.[29]

Truman seemed tense when he met with Silver and Wise on the September 29. "The war is far from over," he told them. The United States had failed to reach agreement with the Soviets at the London meetings of the Council of Foreign Ministers. Urging the Zionists to be patient, Truman complained of ethnic pressure coming at him from different groups, especially the Italians, the Poles, and the Jews. He would not "be confronted by past commitment," he told them. He would "work in his own way."[30] Later that day, Truman met with the two leaders of the American Jewish Committee, Jacob Blaustein and Judge Proskauer. According to their recollection, Truman told them that he was irritated with both Silver and Wise, who were "insisting as they do constantly for a Jewish State," which he assured them was "not in the cards now . . . and would cause a Third World War."[31] Truman assured them, however, that he was seeking a prompt and substantial increase in the number of certificates for Jewish immigration into Palestine. Political questions had to be put aside, Truman said, and the "humanitarian factor placed foremost."[32]

The Zionist movement's first public response to Truman's call for the 100,000 Jewish refugees to be allowed to go to Palestine was to keep up the pressure through mass rallies and demonstrations, which they hoped would stir public opinion and reinforce Truman to continue his efforts. On September 30, 20,000 people packed Madison Square Garden and almost 40,000 more stood outside in cold weather for three hours to hear speeches by loudspeaker. Republican Governor Thomas E. Dewey of New York was the featured speaker, followed by

Rabbi Abba Hillel Silver, who, Lord Halifax reported to London, used his "usual eloquence" to give a "moving and effective" presentation.[33]

The next day, the Senate chambers were filled with senators arguing in favor of immigration for the 100,000. The tone of their speeches, Halifax reported, "was vehement and occasionally bitter."[34] In general, observed Halifax, "the tempo of agitation over Palestine is rising here," and the Zionist movement was flooding Congress and the White House with letters and postcards. It was, he thought, "taking advantage of the fact that the elections of 1946 are approaching and that both parties are anxious to capture the Jewish vote particularly in the key state of New York."[35]

Dewey's headlining the event was particularly vexing to the administration. With midterm elections coming up, the Democrats argued that the humanitarian cause of the European refugees should not be politicized. The New York governor had unabashedly called for what Truman was hedging on, support for a Jewish home in Palestine. Five months after the war's end, Dewey told the crowd, the European Jews "remain victims." They had been promised the Jewish homeland a quarter of a century earlier, Dewey said, and "there was no legitimate reason for its continued denial." The New York rally was the first in a continuing series of events in major American cities. Two weeks later, at a packed Chicago Stadium, the crowd heard speeches by Senator Alben Barkley of Kentucky, who asked Truman "to stand firmly" in his desire to see the Jews get their homeland in Palestine. Thousands of people signed postcards to be sent to Truman, demanding that he use U.S. power to get Britain to fulfill its promises made in the Balfour Declaration.[36]

The jubilation that the Jews in Palestine had felt in July at Labour's victory was now giving way to anger and disillusionment. They felt betrayed by the direction Bevin and Attlee were taking on immigration despite their party's recent promises. The quiet diplomacy of Weizmann was not working, and Ben-Gurion was ready to adopt more militant tactics. Preparing to use direct action against the British authorities, Ben-Gurion moved to fold the Irgun and Stern militants, whose violent actions the Haganah had previously opposed, into one new unified command together with the Haganah. The units would now be called the Jewish Resistance Movement (JRM). Bevin was not happy when he learned of their new alliance. When Weizmann met with Bevin on October 5, he found him in a belligerent mood. The Jewish Agency had turned down the last certificates left over from the White Paper. Bevin greeted Weizmann with "What do you mean by refusing certificates? Are you trying to force my hand? If you want a fight you can have it!"[37]

When the two continued their discussion five days later, Bevin was in a better frame of mind. Weizmann did his best to impress on him the seriousness of the Yishuv's growing militancy and the effect it could have on the British if something weren't done for the refugees. Britain, he told Bevin, could soon find itself pushed out of the Middle East. The granting of 100,000 entry permits, Bevin replied, would hardly solve the political problem. He was so certain of this that he impetuously told Weizmann "he would stake his political career" on its solution.[38]

On October 10, the Palmach, the elite group of the Haganah military, freed two hundred illegal Jewish immigrants who had been rounded up by British troops and confined to a camp near Haifa. The raid was followed by a half-day general strike that included violent demonstrations. This new use of armed resistance against British Mandate authorities further convinced Bevin and his associates that they were correct to stand firm against any efforts to allow increased immigration. The British government was convinced that, despite Weizmann's peaceful orientation, the entire Yishuv was now supportive of terrorism. If anything, Bevin's opposition to the Zionists was hardened rather than moderated.[39]

The adoption of "guided terrorism" (aimed mostly at physical targets, not people) led to an irretrievable split between Chaim Weizmann in London and the Yishuv leaders in Palestine. The split had first come to a head at the World Zionist Conference held in London that August. Rabbi Silver, Moshe Sneh, the commander in chief of the

Haganah, and Ben-Gurion openly called for using forceful resistance to complement diplomacy. Weizmann called this irresponsible talk and offered the delegates his hope that the Labour government would fulfill Britain's historic promise to the Jews. The delegates honored Weizmann but overwhelmingly rejected his proposed course of action. The subsequent announcement of the Labour government that it would continue the White Paper policy vindicated the militants' position and isolated Weizmann.[40]

On the day of the first Palmach raid, two months after the conference, Ben-Gurion wrote to Weizmann, telling him he would not meet with him since Weizmann had only a "fictitious responsibility" as a leader of the movement. Weizmann, quite disturbed, replied curtly that he had received the letter, "and its contents distress me very deeply."[41] Weizmann might have lost his faith in the British government as well as the Yishuv militant leadership, but he was adamant about the necessity to avoid the use of force in the fight to gain a Jewish state. Disassociating himself from the JRM, Weizmann sent a message to the Yishuv, condemning all violence and pleading for restraint. It was to no avail. Weizmann had to acknowledge that the majority favored expanding the military campaigns against the British authorities. His statement marked the end of his leadership of the Jewish Agency.[42]

In the midst of these problems, Truman was facing a potentially embarrassing development. At his September 26 press conference, he had denied "that President Roosevelt had made any commitment to King Ibn Saud not to support Jewish claims if and when they arise." Moreover, Truman told the press that "there was no record of any conference between the King of Arabia and President Roosevelt" in which FDR had made any such statement. He added that *"he had looked through the records of the foreign conferences very carefully and had found no such commitment"* (our emphasis). That meant that a statement made by Abdul Rahman Azzam, the secretary-general of the Arab League—that FDR had shaken hands with Ibn Saud and pledged that the United States would "never support the Zionists' fight for Palestine against the Arabs"–was incorrect. Truman noted that even if Roosevelt had done that, "he would not feel bound by any such understanding."[43]

Truman's statement distressed the Arab League, and it immediately let State know it. On October 9, Loy Henderson sent Byrnes a memo regarding what he called "urgent problems relating to Palestine." The ministers of four Arab states were requesting a meeting with Byrnes to discuss what they thought was a firm commitment from the United States that it would not sanction any basic change in Palestine without consulting both Arabs and Jews. King Saud, Henderson reported, had sent President Truman a message indicating that he wished to make public the text of conversations held between himself and FDR after the Yalta Conference. Most problematic was that he also wished to release the text of letters FDR had sent shortly before he died, which gave him assurances regarding American policy.[44]

Henderson told Byrnes that in his opinion the United States had "no adequate basis for refusing King Ibn Saud's request to publish President Roosevelt's letter to him of April 6, 1945." Truman should immediately choose a date to release it and make sure that this decision was made known to the Saudi Legation at Jidda.[45] Clearly, the United States had to preempt any release of the correspondence between FDR and Ibn Saud.

Truman turned to Sam Rosenman for advice on how to handle the issue. After speaking with Byrnes about it, Rosenman told the president that he found it very embarrassing to be at odds with the State Department. Nevertheless, he told Truman that he strongly disagreed with the advice that Byrnes had given him. Truman himself had not taken any position that would have supported FDR's April 5 letter to Ibn Saud. Truman, Rosenman advised, should take the position that admitting 100,000 Jews to Palestine did not mean there was a "change in the basic situation." As he interpreted the Ibn Saud–FDR correspondence, Roosevelt had not made any promise beyond saying he would first consult with both Arab and Jewish leaders but "there was no intention on his part that he would have to obtain their consent before he took action." Thus Rosenman advised Truman to continue to pressure Attlee for the admission of 100,000 Jews to Palestine. Next he should call a conference of both Jewish and Arab leaders and formally consult with them, thereby fulfilling FDR's promise. After that, the judge told Truman, "you can take whatever action you wish." Finally, Rosenman thought that publication of the FDR–Ibn Saud correspondence should be postponed until mid-November–after the elections–and as for the present, he did "not see why we should publish it ourselves at this time."[46]

Truman quickly read Rosenman's memorandum and phoned him at home to tell him "he thought he agreed with it" but would wait to make a decision after a meeting with him and Press Secretary Charlie Ross the next morning. At the meeting, Rosenman first reiterated the points of his memo, then added that he had told Byrnes he did not think his suggestions "were repudiating what President Roosevelt had said *because President Roosevelt had been on both sides of the fence*" (our emphasis). It was correct to view FDR's April 5 letter to Ibn Saud as "very bad and . . . pro-Arab," but at other times Roosevelt had stated that he favored Jewish immigration to Palestine and even the establishment of a Jewish commonwealth there. Truman had to take into account that FDR's letter was written a week before he died, and Rosenman was sure he did not fully understand it.

The president finally decided that he would simply let the Arabs publish the correspondence, which they were going to do anyway.[47] Rosenman was disappointed with the timing. Alluding to its political impact, he told Truman that Postmaster General Robert Hannegan had informed him that "the repercussions over it in New York are terrific."[48]

The letters exchanged by Roosevelt and Ibn Saud were published on October 19: the monarch's letter dated March 10 and FDR's answer of April 5. The Arabs alone, Ibn Saud wrote, "had a natural right to Palestine," a fact that, he stated, "needs no explanation." He then proceeded to give the explanation he said was not necessary. Arabs had been in Palestine since 3000 B.C., while the "Jews were merely aliens" who had come there in intervals and been turned out in 2000 B.C. Any historical claim by the Jews was nothing but "a fallacy." To allow the Jews entry, Saud argued, would be to allow them to enter a land "already occupied" and would then "do away with the original inhabitants," which would be "an act unparalleled in human history." The Jews' ambition was not only to occupy Palestine but to "take hostile action against neighboring Arab countries." To allow a Jewish state would mean "a deadly blow to the Arabs," who for generations would "defend themselves . . . against this aggression." The Allies, he told FDR, had to "fully realize the rights of the Arabs" and "prevent the Jews from going ahead" in any matter that would threaten all the Arab nations.

Roosevelt had chosen not to deal with or answer any of Ibn Saud's specific arguments about the Jews, history, Palestine, and the situation of the Arab states. Rather, he had issued a brief reply. He assured Ibn Saud he had given his letter "most careful attention," and he alluded to their conversation at the Great Bitter Lake, during which he had obtained "so vivid an impression" of his views. It was then, FDR assured him, that he would "take no action . . . which might prove hostile to the Arab people." The policy of the United States remained unchanged.[49]

Accompanying the release was a statement issued by Byrnes reiterating that Roosevelt's stated policy was that the United States would not adopt any proposals that would change the basic situation in Palestine without "full consideration Jewish and Arab leaders." President Truman, he said, adhered to this policy. At the same time, Byrnes announced that the United States would "continue to explore every possible means of relieving the situation of the displaced Jews of Europe."[50]

The letters shocked the Zionists and the entire Jewish community. The AZEC leaders scheduled a plenary meeting, to be held the night after they were released. Silver's first reaction was that the letters were being issued under what he called "diplomatic coercion" and could not be viewed as the definition of U.S. policy. The worst aspect of FDR's words, however, "was his failure to reject the false and slanderous utterances" made by Ibn Saud regarding Jewish history in Palestine.

What bothered him about Secretary Byrnes's statement accompanying the release of the letters was that he gave no indication of the U.S. government's current position on Palestine. "After all we thought that we had accomplished in securing the commitments of Mr. Roosevelt and Mr. Truman . . . in enlisting pro-Zionist expression on the part of the American people," he told the other AZEC leaders, ". . . we now find that everything is brushed aside as though it did not exist." "We are confronted by a government," he concluded, "which feels that it has absolutely no obligation towards us."[51]

After their meeting, the leaders composed a nine-page memorandum, to be delivered by Silver and Wise on the twenty-third to Secretary Byrnes at a scheduled meeting and sent to the president two days later. U.S. policy was "clear and unmistakable," they argued, and was in favor of a Jewish commonwealth in Palestine. Both FDR's letter and Byrnes's statement upon its release took no cognizance of the past announced policies whatsoever. It was "deeply disturbing," they wrote to Byrnes, "that it should not have been found necessary to make affirmatively clear that American policy on Palestine has already been established by the public pronouncements of the Presidents of the United States . . . a policy predicated upon the right of the Jewish people to rebuild their National Home." If the real policy was not made clear, they predicted, "serious doubts and misunderstandings" would arise on the part of the American people.[52]

Most upsetting to them was Roosevelt's failure to answer Ibn Saud's "vilifications of the Jewish people." These slanders, they noted, should not "have been allowed to stand unchallenged by one who knew how false those statements are." Obviously not counting on Byrnes or the State Department to correct Ibn Saud, they proceeded to answer his arguments in detail. AZEC's leaders rejected the claim that the Arabs had legal or moral title to sovereignty over Palestine. Though they conceded that the Arabs had conquered it more than 1,300 years before, their rule was intermittent. Palestine was ruled by Christians during the Crusades and conquered by the Turks in 1518. It had remained a neglected backwater of the Ottoman Empire for three hundred years. "In the eroded, poverty-stricken and disease-ridden country which within the last few decades the Jewish people set out to reclaim," the report continued, "it was difficult to recognize the land of milk and honey described in the Bible. In the twenty years between the two World Wars the Jews have done much to repair the ravages of the previous 1300."

As to ethnic claims, about 75 percent of the Arabic-speaking people now in Palestine were recent immigrants or the descendents of people who came in "comparatively recent times." "If Palestine exists as a separate concept," they concluded, "it is because of its immemorial association with the Jews and Jewish History. At no time was there a Palestine Arab State. . . . The Pan-Arab claim to Palestine is an attempt to add yet another to the immense, but for the most part thinly populated and underdeveloped territories of the independent Arab states." They added that while the Jews had "conquered deserts and swamps, revived agriculture and industry and established in Palestine a sturdy, self-reliant community," the "great mass of the people in the various Arab states are kept down in ignorance and fanaticism, in dirt and wretchedness by a ruling class which shows little or no interest in the improvement of their miserable lot."

The statement ended with a fact sheet on the history of U.S. policy on Palestine, detailing support for a Jewish commonwealth from presidents, Congress, governors, labor leaders, and educators. To listen to Arab threats, they concluded, would be nothing less than an act of "encouragement to terrorism." If the United States were firm, the Arab states would accommodate to reality and accept a Jewish state. This was simply another example, they concluded, of "Roosevelt charm without any firm commitment of any kind."[53]

The Zionists and their friends might have been justifiably angry, but Felix Frankfurter cautioned them "not to make tactical mistakes which might harm the cause with President Truman." They must not forget that "he is the man who is to decide what shall be done."[54]

The British were growing tired of the American criticism of British policy when it was British soldiers who were trying to keep the peace and who were under attack. Ambassador Halifax put it this way in July:

> As it is the United States are in the completely illogical but, for them exceedingly comfortable position that they cannot be ignored in the Palestine problem . . . and yet they do not have to bear any share of the responsibility. For the Americans to be able thus to criticize and influence without responsibility is the most favourable and agreeable situation for them and, I must suppose, the exact converse for us.[55]

Now the British press was demanding that if the United States kept up its criticism, it was bound to either help find a solution or keep quiet. Attlee and Bevin decided that the only way to resolve the standoff was to involve the Americans in helping to find a solution. On October 19, the same day the Roosevelt–Ibn Saud letters were made public in the United States, Great Britain came up with a new proposal. It had decided to put aside for the moment its plan to bring the Palestine situation before the United Nations and would instead try to work out a joint Anglo-American solution. Would the United States, the British government asked, like to join it in the creation of a new Anglo-American Committee of Inquiry to study the issue and hopefully bring it to a satisfactory conclusion?

FIVE

THE SEARCH FOR CONSENSUS: THE ANGLO-AMERICAN COMMITTEE

The Harrison Report's impact on American public opinion was of great concern to Attlee and Bevin. Pressure from the United States, the country from which they were seeking a hefty loan to rebuild their economy, meant that Britain could not simply reject or ignore the report's main proposal–the immediate admission of 100,000 more Jews into Palestine. However, diplomatic experts in its Foreign Office thought that in order to shore up Britain's weakening position in the Middle East, it was essential to keep Arab goodwill. This meant maintaining the White Paper with its restrictions on Jewish immigration. Bevin, in particular, was wary of Stalin's expansionist aims in the Middle East and was afraid that the Arabs might turn to the Soviets.

Attlee and Bevin were also quite aware of the long record of Labour's pro-Zionist statements, since they had made some of them. The best defense being an offense, the British government now proposed a new Anglo-American effort to study the refugee problem and to come up with a joint solution. Convening yet another committee to study policy options would buy the British more time before they had to make tough decisions concerning the Arabs and the Jews. If all went well, such a joint committee might lead the United States closer to the British position, to share some of the responsibility and perhaps even to enter a partnership with Britain in the administration of Palestine.[1]

On October 19, Britain's ambassador to the United States, Lord Halifax, formally proposed to Secretary of State Byrnes the creation of a joint Anglo-American Committee of Inquiry. As for specifics, he submitted suggestions for the committee's terms of reference. The committee would be given the task of visiting British- and American-occupied Europe and assessing the refugee crisis.

Challenging the Harrison Report's conclusions, Halifax told Byrnes that the British government refused to accept the view that the present living conditions of the Jews were any worse than those of the other victims of Nazi persecution. Rather than allow them to leave Europe, it was their hope and policy to enable the Jews in Europe "to play an active part in building up the life of the countries from which they came."[2] Hence, the British proposed that the new committee would investigate the position of the Jews in Europe and the possibility of their emigration into other countries outside Europe, including the United States.[3]

Halifax then explained the logic of London's position against admitting 100,000 Jews to Palestine. First and foremost, he made it clear, Britain did not want to "fly in the face of the Arabs." Of lesser concern–but nevertheless most important–was that the United States' stance was both embarrassing to Britain and hurting the two countries' relations. Britain could not accept the view that most of the Jews should leave Germany and the rest of Europe, because to have them leave "would be to accept Hitler's thesis." Moreover, the British claimed that the Zionists were using intimidation to stop Jews in Palestine from moving *back* to Europe.[4]

While Byrnes was negotiating with Halifax over the terms of reference, the president again sought Sam Rosenman's advice. "I think it is a complete run-out on the mandate," Rosenman told Truman. "I certainly do not think that you ought to agree to it." The terms of reference made no connection between the refugees and Palestine (this, of course, was what the British wanted to avoid). As he saw the British position, it was one of "temporizing, appeasing and seeking to delay the settlement of the issue." Most of the proposed committee members would carry out unnecessary work and be charged with obtaining data that could be found out in a few short days. The only valid purpose for such an investigation, he believed, would be to "determine just how many people could be absorbed into Palestine per month."

Taking Rosenman's advice into account, Byrnes told Halifax that from America's viewpoint, the terms of reference proposed by the British government needed to mention Palestine as one possible destination for the refugees, which they did not. Indeed, Byrnes complained, it seemed meant to "divert the mind of the committee from Palestine" to that of finding other countries that might take the European Jewish remnant. If it did not discuss Palestine as part of the solution to the homeless refugees, he predicted, "the Jews are going to say this is just another trick and nothing will be done." Therefore, Byrnes demanded the committee address how many Jews would actually be able to be absorbed into Palestine.[5]

The British were anxious to announce the new committee, but Rosenman raised the political issue, telling Truman he did not understand why there had to be any statement issued right before the New York mayoral elections that might hurt the chances of the Democratic candidate confronted by voters who might very well not view the new committee as favorably as he did.[6]

The president decided to give Byrnes the go-ahead to accept U.S. participation in the new committee but took some of Rosenman's advice. Rosenman had suggested a time limit of thirty days, "which would take the sting out of the charge of stalling and delay," but Truman decided to give the committee 120 days to complete its work.[7] The terms of reference included an important qualifying caveat: the committee would first look at "conditions in Palestine as they bear upon the problem of Jewish immigration," as well as carrying out estimates "of those [Jews] who wish, or who will be impelled by their conditions, to migrate to Palestine."* The announcement of the committee would take place after the November 6 election, which would prevent it from becoming a political football. Finally, Byrnes stressed that the president's agreement did not mean he had changed his mind on the immediate immigration to Palestine of 100,000 Jewish refugees.[8]

Bevin, whose outsized ego was legendary, was willing to bend to Truman's demands, as he assumed that eventually the committee would reach the conclusions he favored: that Palestine would not be the answer for the European Jewish refugees because of Arab opposition and because it was too small a land area to absorb tens of thousands of new refugees.

As Rosenman predicted, in New York there was an immediate negative reaction when Truman announced the creation of the committee on November 13. Representative Emanuel Celler called the committee "just another British dodge and stall." British policies, he threatened, might call for reevaluation of any U.S. economic aid and sharing of atomic energy information with Britain that Attlee sought. "I am surprised," Celler said, "that President Truman has fallen into the British trap."[9] Rabbi Baruch Korff of New York, who had led a parade of 1,000 rabbis to the capital a day earlier to demand the immediate admission of the 100,000 to Palestine, told the press that the president's acceptance of the committee was nothing less than "a death sentence to European Jewry." Ironically, Korff said, the idea of the demonstration had come from "a member of the White House staff," a man who had suggested it take place "before it is too late."[10] Although the march had been "spontaneous" and no organization had orchestrated it, he affirmed that it had been "inspired" by this White House staff member on a train ride from Washington to Boston.[11]

On November 18, Israel Goldstein, president of the Zionist Organization of America, spoke at the group's convention, held in Atlantic

---

* The last sentence of the terms of reference finally read, "The establishment of this committee will make possible a prompt review of the unfortunate plight of the Jews in those countries in Europe where they have been subjected to persecution, and a prompt examination of questions related to the rate of current immigration into Palestine and the absorptive capacity of the country."

City, New Jersey. Goldstein denounced Truman in the harshest of terms and demanded that the United States have nothing to do with the proposed British committee. "No committee is needed," he told the assembled delegates, "to study whether Palestine has room for the 100,000 immediately." In fact, Palestine needed Jewish laborers and farmers. Why was President Truman so willing to defer his first request for the 100,000? Goldstein warned Truman that "the Zionists are done with illusions" and were "impatient with delay, and were determined to do all that the situation demands."[12]

The highlight of the convention was Chaim Weizmann's speech. "A tired old man in a wrinkled gray suit," one news article reported, "stood on the auditorium's platform, and as he has done for scores of years, pleaded the case of his people before the world." His presentation was, as usual, eloquent and powerful. "He spoke slowly without oratorical gesture, without raising his voice, seemingly without bitterness against Britain, as he explained why the Jewish people must be granted a state of their own in Palestine."[13]

Weizmann expressed his profound disagreement with the new Anglo-American Committee. By now he had expected the White Paper to be history, he told his audience. Instead, he sadly noted, "another document [has been] added to those which seek to repudiate the solemn covenant of 1917 between Great Britain and the Jewish people."[14] When Bevin announced the formation of the Anglo-American Committee to Parliament on November 13, he warned that the Jews shouldn't ask for any special treatment. He said, "If the Jews want to get too much at the head of the queue they face the danger of another anti-Semitic reaction." At the convention, Weizmann reminded Bevin that not too long ago the "Jews had the highest priority in the queues which led to the crematoria of Auschwitz and Treblinka."[15] What Bevin called a queue, Weizmann answered, was in fact a "simple request for survival." Was it, Weizmann asked, "getting too much at the head of the queue if after the slaughter of six million Jews, the remnant of a million and a half implore the shelter of the Jewish homeland?"

Challenging Bevin's assertion that Jews should remain in Europe,

Weizmann asked that no Jews be forced to live in nations "where they saw their wives mutilated and burned, their sons and daughters buried alive, their parents turned into white ash." The return to the countries of their birth was, for many, not an option. Moreover, Weizmann saw the Anglo-American Committee as an insult. During the war, the democracies had said there was no way to save the suffering Jews of Europe. "Now we are told that the survivors must wait until another inquiry will establish the exact measure of help they will require."

If Weizmann was respectful of Truman, Rabbi Silver was his usual militant self. Originally, Silver admitted, he had been "heartened by President Truman's request of Prime Minister Attlee that 100,000 Jews, principally from the concentration camps in Europe, be permitted immediately to go to Palestine." Sadly, he now had to say, "we had overestimated the determination of the President." Instead of getting action, the president had accepted "the shabby substitute of an investigating committee, that . . . transparent device for delay and circumvention, against his own better judgment." Once again, America had given in to Arab chieftains and British policy makers at the expense of the Jews. Chastising Truman for not demanding admission of the 100,000 "with all the prestige and authority of his office," he suggested that the president withold economic help desired by Britain in return for their granting free immigration into Palestine.[16]

The entire Zionist movement, from moderate to militant, appeared to be united in its condemnation of the new committee and of Harry Truman's decision to have the United States participate. These objections were seconded by the former first lady Eleanor Roosevelt, whose advice Truman valued. She had supported Truman's call for admission of the 100,000 Jews to Palestine, even though at this time she opposed the creation of a Jewish state. "I am very much distressed," she now wrote Truman, "that Great Britain has made us take a share in another investigation of the few Jews remaining in Europe. . . . Great Britain is always anxious to have someone else pull her chestnuts out of the fire."[17] "I'm very hopeful," Truman responded, "that we really shall be able to work out something in Palestine which will be of lasting benefit. *At the same time we expect to continue to do what we can to get as many Jews as possible into Palestine as quickly as possible, pending any final settlement*"[18] (our emphasis).

Though Truman was committed to helping the refugees rebuild their lives in Palestine, he was adamant that he was not about to let Zionist pressure force him to commit to send American troops to Palestine to enforce a Jewish state. "I told the Jews," he wrote an old colleague, Senator Joseph H. Ball, "that if they were willing to furnish me with five hundred thousand men to carry on a war with the Arabs, we could do what they are suggesting . . . otherwise we will have to negotiate awhile." Agreeing that it was an "explosive situation," he did not think "that you, or any of the other Senators, would be inclined to send a half dozen Divisions to Palestine to maintain a Jewish State." He told Ball that he was trying to "make the whole world safe for the Jews," but, he explained, "I don't feel like going to war for Palestine." On second thought, he wrote "Do not send" across the letter.[19]

In fact, Truman was not convinced of either the need or the wisdom of creating a Jewish state. He told the publisher J. David Stern, who owned *The Philadelphia Record,* that although he favored the creation of a democratic state in Palestine, he did not favor one based on religion, race, or creed. Palestine, he thought, had to be "thrown open" to Jews, Arabs, and Christians alike. It should aspire to be a pluralistic society like that of the United States.[20] Truman had clearly been influenced by the arguments of non-Zionist and anti-Zionist Jewish groups, the American Jewish Committee, and the American Council for Judaism (ACJ). The president had just received a visit from Lessing J. Rosenwald, the head of the ACJ, who had presented him with a seven-point program. Rosenwald's proposal would ensure that Palestine would not be a Jewish state but a country open equally to people of all faiths. Therefore, Rosenwald argued, the issue of the European Jewish DPs should be treated separately from that of Palestine. The refugees would be settled abroad according to their preferences, as administered by the United Nations.[21]

The same day Truman saw Rosenwald, Ambassador Halifax brought Chaim Weizmann, who was a British citizen, to the White

House. Thanking Truman for his position on the 100,000 European Jews, Weizmann gave him his candid views on the Anglo-American Committee. "The whole ground," Weizmann argued, "had been repeatedly worked over during the last twenty years." Interrupting Weizmann; Truman said he thought the Jewish problem should not be viewed simply through the prism of Palestine, and he then deprecated use of the term "Jewish state," telling the Zionist leader he favored the term "Palestine state." Truman argued that "there were many Jews in America, representatives of whom he had been receiving before he had seen Weizmann, who were not at all keen on the Palestine [i.e., Zionist] solution." Weizmann was obviously disappointed. Truman closed the meeting by telling him, "The United States wants a solution and we shall have to see whether we cannot work it out."[22]

Weizmann was so disturbed by his discussion with Truman that he wrote him a nine-page single-spaced typed letter.[23] It was perhaps the most thorough attack on the positions of the anti-Zionist Jewish groups that the president would ever receive. Unlike the American Zionist leaders, who often appealed to their constituencies first and the administration second, Weizmann knew how to address an American president.

Weizmann began by thanking Truman for his insistence on the 100,000. He was only seeking with this letter to "place on record" his views of everything the president had touched upon during the brief meeting. The new Anglo-American Committee, Weizmann wrote the president, "cannot bring to light any new facts, as there have been a whole procession of committees on Palestine and every aspect of the problem has been investigated" since 1937.

Next Weizmann took up the issue of the absorptive capacity of Palestine. The one useful question the committee could look at was how to develop Palestine through irrigation, since studies had proved that large stretches of land that had never been touched could indeed be cultivated, which would provide room for hundreds of thousands of families who would be employed in agriculture and industry. If this were accomplished, he informed Truman, Palestine would be able to absorb three to four million additional people. The truth, however, was that "only Jews are capable of initiating and of executing such development schemes." The Arabs and the British would not, since for them it was not a matter of sheer existence. The Jews would have to do it if they were to be able to bring in hundreds of thousands who would become citizens of the Jewish homeland. They would also have to make room for the oppressed Jews from the Orient and Arab states. "We are convinced," he told Truman, that the Jews could be settled in Palestine provided the land had been properly developed.

Challenging Ernest Bevin's claim that the Balfour Declaration did not promise a Jewish state in Palestine, Weizmann noted that its intent had been to create a home not for individual Jews "but a home for the Jewish People." As for Jews who argued that they were only a religious community–a clear reference to Lessing Rosenwald's position–Weizmann countered that the Zionists had never intended for the new national home to "become a 'religious' or theocratic state." Palestine would be a "modern and progressive" nation, with "no stress on the religion of the individuals who would form the majority of its inhabitants." He assured Truman that it would be "a secular state based on sound democratic foundations," with a system similar to those of the United States and Western Europe.

Already, he emphasized, the Jews living in Palestine had in essence set up a government with state functions, but without recognition and without executive powers to enforce its decisions. All the work of generations had not been undertaken to establish another Jewish minority society–a ghetto, as Weizmann called it. As for the Arabs, Weizmann reassured Truman, the Jews and the Zionists "desire nothing better than to live on the most peaceful terms with the Arabs" and that the latter's rights in a Jewish homeland would be scrupulously safeguarded, with equal opportunity and without regard to race or creed. Like many other Zionists of the day, Weizmann wanted to believe that the Arabs, despite their protests, would accept a Jewish state and their opposition would not "turn into active armed conflict."

Finally, Weizmann proposed that the president consider giving a speech in which he would tell the Arabs that the Allied armies had saved them from Fascist enslavement, and that the Allies had the right to ask them not to hinder the settlement of the homeless Jewish people in Palestine, which was but "a small notch in the vast under populated Arab peninsula." The Jews, he assured them, knew how to treat minorities with dignity, and they would also be assured guarantees of civil and religious rights by the United Nations.

Senators Robert F. Wagner and Robert A. Taft were also concerned about rumors circulating on Capitol Hill that the Jewish homeland would be a theocratic state. They were also angry and confused about Truman's seeming about-face on their latest resolution (introduced on October 26, 1945) supporting the refugees' immigration to Palestine and supporting a homeland for the Jewish people. Just a short while before, Secretary Byrnes had encouraged them to reintroduce their failed 1944 resolution in the Senate. He had told them that the administration did not object and that they should "go ahead."[24] The 1944 resolution had spoken of creating a "Jewish commonwealth" in Palestine; the new version promised that the doors of Palestine would be open to all Jews, so that they could create a "free and democratic Commonwealth." The word "Jewish" before "Commonwealth" was eliminated, thereby possibly obtaining a few more votes from Senate and House members who otherwise might have been reluctant to sign on.[25]

The senators wrote to Truman that since the resolution had been introduced, a campaign had taken place against its basic preposition, "that the Jews shall have the right of free entry into Palestine so that they may reconstitute it as a democratic commonwealth." Particularly objectionable was the charge that they favored a theocratic state in Palestine based on religious and racial discrimination. Calling it an "insidious campaign," the senators set out to explain their position to Truman in order to dispel such misconceptions.

Tracing the long history of the concept of a Jewish commonwealth from the days of the Versailles Peace Conference after World War I to the present Democratic Party platform, they assured Truman that what was meant was that in their ancestral land of Palestine, "the Jews should be free to grow into a majority and not be kept down artificially to the position of a minority in which they find themselves in every other country in the world." The senators' own resolution was only giving renewed expression to a long-standing U.S. and Allied policy. Noting the agreement with its emphasis on a democratic state open to citizenship by all citizens of Palestine, whose rights would be protected, they explained to Truman that anyone who was well informed could not be claiming that the Zionists wanted a theocratic state in Palestine.

Jews were not only a religion but a people with distinct cultural features, like Czechs, Greeks, or Irish, argued the senators. Like those peoples, Jews were entitled to their own homeland. Thus the Mandate created at Versailles in 1919 spoke of the historical connection of the Jews to Palestine and their need to reconstitute a national home there. The present campaign was meant by opponents of a Jewish homeland to "confuse the public, to deprive the Jewish people of their established rights, and to assist the British government in evading its obligations."[26]

At the end of the first week of December, the White House announced the composition of the Anglo-American Committee of Inquiry. The American group was to be chaired by the conservative Judge Joseph Hutcheson, whom Acheson had described as a "fiery Texan and friend of the President."[27] The rest of the group was composed of Frank W. Buxton, the editor of *The Boston Herald* and a friend of Felix Frankfurter; James G. McDonald, the former League of Nations high commissioner for refugees; Bartley C. Crum, a left-leaning San Francisco lawyer; William Phillips, a former ambassador to both India and Italy; and Frank Aydelotte, the director of the Institute for Advanced Studies at Princeton University. The British members were led by its chairman, Sir John Singleton, a High Court justice. The British members were Wilfrid P. Crick, an economic adviser to the Midland Bank; Richard H. S. Crossman, a Labour MP; Lord Morrison (Robert Craigmyle, Baron Morrison); Sir Frederick Leggett, a deputy in the Labour cabinet; and Major Reginald E. Manningham-Buller, a Conservative MP.[28]

The premise of the British members, Crossman later wrote, started with the assumption that "the whole idea of a Jewish national home is a *dead end* out of which Britain must be extricated" and that the legitimate demands of the Arabs had to be satisfied. The British members worked under the assumption that they had to explore how to find homes for the dispossessed Jews in Europe and that they had to destroy the Zionist case for immigration into Palestine.[29]

The American group had quite a different perspective. James McDonald, in his previous posts, had been an ardent supporter of Europe's Jewry and, as high commissioner for refugees, offered them his support and tried to get as many of them as possible out of Nazi Germany. Having met Hitler in 1933, he was an early witness to the Führer's intentions. As he told audiences upon his return to America, Hitler had told him, "I will do the thing that the rest of the world would like to do. The world does not know how to get rid of the Jews. I shall show them."[30] McDonald had thought he should be taken at his word.

The British, rightfully, suspected that McDonald was privately pro-Zionist. Joining them in this suspicion was Loy Henderson. Henderson claimed he had not been aware of it at the time. Niles had put McDonald's name on the list of potential committee members. Henderson later concluded that "McDonald had been campaigning actively for the establishment of a Zionist state in Palestine" and while high commissioner for refugees "had been supporting Zionist propaganda being carried on in the refugee camps."[31] In fact, as stated earlier, by 1944 McDonald had publicly recognized what he called a "transcendent role for Palestine" and had concluded that the remnant of Jewry still alive in Europe could not return to their old homes in the countries in which they had lived. "Only in Palestine," he said, "will most of them feel that they have returned home." Since his views were well-known to all, we must assume that Truman and his staff knew it when his name was added.

Bartley Crum, who was registered as a Republican, was nonetheless a civil libertarian and a man friendly to and supportive of left-wing causes, most especially the militant CIO. Yet his own firm had many big business clients, as well as Jews known to be anti-Zionist. He had started out supporting the candidacy for president of Wendell Willkie but had swung over to become a major supporter of FDR. Crum was first suggested as a member by Truman's adviser David Niles, who had gotten to know Crum in 1944, when he had chaired the Republicans for Roosevelt group.[32] Later, Crum asserted that Truman had told him that on three different occasions, the State Department had rejected his name when it was sent to them for approval. Crum believed State was wary of him because of his endorsement of left/liberal causes such as support for refugees after Francisco Franco had won the Spanish Civil War. It was only because of Truman's insistence that he serve that State relented.[33] Crum's suspicions were correct. The State Department security officer concluded that Crum should not receive clearance since he was a member of several "united front" (i.e., pro-Communist) groups and was noted "for his demagogic speeches." Henderson himself had initialed a memo opposing the issue of a security clearance for Crum.[34]

Before leaving with the committee, Crum met with Leo Rabinowitz, an American Zionist, to discuss how the committee would handle its work. "He confirmed my impression," Rabinowitz reported to Epstein, "that he is most sympathetic to our point of view, or at least receptive, and certainly devoid of prejudices." Because of Crum's background, Rabinowitz thought that he was "in a position to ask much from the present administration."[35] McDonald was impressed with Crum. He was "an amazingly energetic and keen student," McDonald wrote in his daily diary. "With his political connections closer than any of the rest of us, he should prove invaluable."[36]

Judge Hutcheson provided balance. He was shrewd, honest, and determined to find a "just solution," McDonald thought. But the judge was strongly opposed to Zionism and most likely sympathetic to the State Department and British arguments. "He has very strong feelings against any form of Jewish state," McDonald confided to his diary, "and is quite unsympathetic to anything which smacks of Jewish nationalism."[37]

From the beginning, there was a British-American split in the committee that would increasingly grow wider. The Zionists expected that the British members would be unsympathetic to their cause. A possible exception was Richard Crossman. He was at age thirty-nine a Labour member of Parliament. Previous to that, he had been a Foreign Office intelligence officer, an Oxford don, and an assistant editor of the liberal *New Statesman* and *Nation,* the flagship pro-Labour/Left magazine. Epstein viewed him as a moderate supporter of Labour's traditional pro-Jewish position. "Dick," Eliahu Epstein reported, would be "a practical but difficult member. . . . He is impatient, obstinate, dogmatic, but has a good brain."[38]

Epstein did not know, however, that even Crossman started out with his countrymen's assumptions. Crossman believed that the White Paper had to be maintained and that the Zionist position was wrong. He agreed with Ernest Bevin that to view Jews as a nation was an anti-Semitic reflex. The survivors had to be liberated from the separateness Hitler had forced them into and become assimilated Europeans with full rights and duties wherever they settled. That meant rejecting Zionism, which he thought only strengthened the walls of a spiritual concentration camp. To advocate their cause would mean that one would be joining the anti-Semites who wanted to take Jews from Europe and put them all in Palestine. And no single place was worse, thought Crossman, "for a persecuted people than this strategic key point in which the whole Arab world is against them."[39]

Before the hearings began in Washington on January 7, 1946, the American delegation was summoned to the White House, where Harry S. Truman met with them. No problem concerned him more deeply than the fate of the DPs, he told the assembled members. It was the obligation of the "democratic world to give these people who had wronged no one a chance to rebuild their lives," and he pledged that he and the American government would do all in its power to find a solution.[40]

The committee's schedule was intense. It would begin with hearings in Washington and London and then split up, its members going to DP camps in Germany, Austria, Poland, and Czechoslovakia. At the end of February, they would proceed to Cairo, Jerusalem, Damascus, Beirut, Baghdad, Riyadh, and Amman. Much of what the members would hear would be repetitive, since they had been given basic background reading and already knew the essence of the Zionist and Arab cases. By the end of the Washington hearings, Crossman wrote, "we generally knew what a witness was going to say before he said it."[41]

When Crossman arrived in Washington, he tried to analyze why the American and British perceptions of the situation were so different. He had thought that support for Zionism in the United States was limited to American Jews but was surprised to find that it had widespread support. America was still a pioneering nation, he observed, with a frontier mentality, and the idea that Jews who had been oppressed and confined to ghettos in Europe for decades could improve their lot by setting sail for a new country was part of the American grain. Palestine as a refuge, therefore, "comes naturally to an American."

To the Americans, Europe's Jews who sought to rebuild their lives in Palestine were the equivalent of the American settlers who had developed the West. The analogy carried to the Arabs, who were the equivalent of America's Native Americans. They were "the aboriginal who must go down before the march of progress," wrote Crossman. America had gained its independence by fighting a war with King George III; the Jews of Palestine, he thought, might conflict with his successors in the British Mandate. If that occurred, Crossman predicted, the Jews were "bound to win an instinctive American sympathy." That was not necessarily beneficial. He summed up what he saw as the American view in these words:

> The American knows that if an imperial power had espoused the cause of the Red Indians, maintaining that no settlement could be allowed which was damaging to their rights, and that development of the west could only be permitted according to the absorptive capacity of the country, half of the U.S.A. would still be virgin forest to-day. Because a nation's history conditions its political thinking, Americans . . . will always give their sympathy to the pioneer and suspect an empire which thwarts the white settler in the name of native rights.

The British, on the other hand, in Crossman's view, were made up of people who had never left and went back for generations. What they feared most was an invasion by a foreign conqueror. They also resented the idea that Europe was foundering "and that a million Jews must be rescued from the sinking ship." To them, Zionism was unnatural and was nothing but "the product of high-powered American propaganda." Thus the average Englishman sided with the Arab, whom he saw as "defending his thousand-year tenure of his country against the alien invaders."

When the hearings began in Washington and the witnesses appeared, Crossman experienced their testimony as "a monumental indictment of Great Britain." At times, he confided, he felt more like "a prisoner in the dock than a member of a committee of enquiry." The committee heard from the usual suspects: Rabbi Stephen Wise and the Zionist leader Emanuel Neumann, whom Crossman called "the organizing brain behind the indictment." Moving about continually in the back of the hearing room was Peter Bergson, a man who "looked more like a Russian university student than a terrorist," who was there as the Irgun's sole spokesman, as well as being the "skeleton which the orthodox Zionists obviously thought should have remained in the cupboard." All of them collectively made the British members feel that they "were personally responsible for the death of six million Jews."[42]

Crossman was irritated that in Washington there was almost a "complete disregard of the Arab case." He thought, "Why should these people from a safe position across the Atlantic lambaste my country for its failure to go to war with the Arabs on behalf of the Jews?" His mood improved when he met David Horowitz, who was with a friend of his from Tel Aviv. The three talked for more than three hours. Crossman was impressed with the case they made for the Jewish community in Palestine. The Palestinian Jews "belonged to a different world from that of the American Zionists. Palestine for them was not a cause which they had taken up, a gigantic piece of organized philanthropy, or a stick with which to beat the British. Palestine was their native country." "Perhaps," thought Crossman, "while the rest of the world was arguing, the Jewish nation had been born."[43]

The hearings had begun with a presentation by Earl Harrison, who restated the conclusions he had come to in his report on the status of Jewish refugees. Following him was Dr. Joseph Schwartz, the European director of the Joint Distribution Committee, who had accompanied Harrison in Europe and shared his views. The legal case for Zionism was presented by Emanuel Neumann, who argued the illegality of the White Paper and urged that the DP camps be closed down and the Jews living in them be sent to Palestine. The Arab case was presented by Dr. Philip Hitti, a professor of Semitic literature at Princeton University. A Christian Arab, Hitti argued that Palestine was part of Syria and had been home to the Arabs since time immemorial. A Zionist state would be an imposition on the Arabs and could be created only by force.

The biggest stir was the testimony on Friday the eleventh of Albert Einstein. His appearance was regarded by the Zionist leadership as their pièce de résistance, but they had to pull out all the stops to get him to testify. Meyer Weisgal, Chaim Weizmann's man in America, traveled to Princeton to convince him to participate. Weisgal walked up and down the streets with him for hours, imploring him to come. Finally agreeing, Einstein made the kind of demands associated with prima donnas. "I had to go with him on the train," Weisgal wrote. "I had to be with him at the sessions, I had to take care of him all the time."[44]

With his "great mane of flowing white hair reaching almost to his shoulders," Crum would write, "he looked like a patriarch stepping out of a Biblical tale."[45] He approached the witness stand "with adoring women gazing up at him like Gandhi–flashlights, movie cameras and so on."[46] Bearing what McDonald called "a sweet smile and a very gentle manner," the eminent scientist "proceeded to throw bombs in three distinct directions, blasting, as it were . . . the British, the extreme Zionists, and the Committee." The British were following a policy of divide and rule, he said, keeping Arabs and Jews apart the better to serve imperialism. He continued to make it clear that he also repudiated Zionism. And as for the committee itself, Einstein said in a "beatific tone" that it was "futile and a smoke-screen for the two governments."[47] Crum put it another way. Einstein, he said, argued that the British were responsible for all the turmoil. Its Colonial Office sponsored Arab-Jewish clashes to prevent the two peoples uniting and finding they did not need British rule. His own answer was an independent Palestine under the United Nations.[48] If the Zionists had high hopes for his testimony, they were sorely disappointed. A Jewish majority in Palestine was "unimportant," Einstein concluded, and he thoroughly disapproved of any nationalism. Hearing this, the "audience nearly jumped out of their seats."[49]

Next was London. The group traveled on the *Queen Elizabeth*, finding time to further study the reading material they had been given and to compare notes. Evan Wilson, the American secretary for the committee from State's Division of Near Eastern Affairs, accompanied them on the journey. Expressing the State Department's point of view, Wilson counseled the American members that if they reached a decision that could be conceived as too favorable to the Jews, "an aroused Arab world might turn to the Soviet Union for support."[50]

Tall and rangy, Harold Beeley, the secretary for the British and an expert on the political history of the Middle East, held views similar to Wilson's. In the ship's lounge he told Crum and Buxton that one could understand Palestine only in the context of the new Cold War with the Soviets. Since Stalin sought to move into the Middle East, he argued, the United States had to join with Britain "in establishing a *cordon sanitaire* of Arab states," which would become a strong link in the anti-Soviet chain if Palestine were made an Arab state.[51]

The American delegates were shocked when, on the third day aboard ship, they were given the State Department's confidential communications on Palestine, probably meant to sway them to the department's views. They had the opposite effect. After reading the outline of State's handling of Palestine, the Americans concluded that "each time a promise was made to American Jewry regarding Palestine," State immediately sent messages to the Arab leaders that those promises could be ignored, and that policy would not change. Crum saw this as "double-dealing," as well as sabotage of Truman's Palestine policy. He felt betrayed, thought the committee would do no good, and wanted to return home. Informing the British members of what the Americans had learned, Sir John Singleton responded, "It appears that Great Britain is not the only power who promises the same thing to two different groups."[52]

The British hearings took place from January 25 until February 1. According to Crossman, at a farewell luncheon for the committee in Dorchester, Bevin thanked them "for removing the responsibility from his shoulders for at least 120 days." He then slowly and deliberately announced that if they achieved a unanimous report, he would do everything in his power to see that it was implemented. The speech made a big impact on the committee members, especially the Americans. Maybe the committee wasn't a stalling device after all.[53] But in reality, Bevin believed that when the committee members saw the situation face-to-face, they would dismiss Truman's call to send 100,000 immigrants to Palestine and adopt his own view.[54]

In London, the committee heard more of the Arab case, including that of Prince Faisal, the second of the forty sons of Ibn Saud. As expected, Faisal was firm in his insistence that not one more Jew be allowed into Palestine. The Syrian delegate, Faris al-Khouri, argued that a Jewish state in Palestine would become a "vast imperialist power threatening the security of the entire Arab world." Turning to Judge Hutcheson, he bellowed, "Why don't you give the Jews part of Texas?"[55] "The Arabs," McDonald confided to his diary, "made an impression of such unyieldiness that it would be impossible to win them by any sort of compromise."[56]

While in London, James McDonald took time out to have a three-hour working dinner with an individual he described only as "an important American personality" and later as "one of the most important Americans in London." The individual, judging from what he said, was representing the White House, rather than the State Department. The committee, he told McDonald, had to "create the opportunities for enlarged immigration of Jews." It had to reach a conclusion that appealed to the conscience of the world but also had to be one that made it clear the United States would provide no military support for Palestine. Most significant, his dinner partner took a decided swipe at the State Department. "The men in the Near East division would be," he stressed, ". . . inclined to take the view of the Arab states." Therefore, when the American members had to communicate with the U.S. government later on, he gave McDonald the impression that it would be preferable to move directly through the president or the secretary of state.[57]

The British round of hearings having concluded, the committee split up and toured the DP camps, assessing the status and desires of the Jewish refugees. Sir Frederick Legget, Crossman, and Crum went to the American zone of Germany. There they found the DPs living in similar conditions to those that Harrison had seen, now only slightly better due to Truman's and Eisenhower's efforts. Abstract arguments now felt remote. Referring to Attlee's comment, which Bevin later repeated, that Jews should not push to the head of the queue, Crossman wrote, "That might go down in Britain; in Belsen it sounded like the mouthing of a sadistic anti-Semite."[58]

Speaking to refugees, they heard unimaginable horror stories. "What are you to say," wrote Crum, "when a man like yourself carefully extracts a snapshot . . . showing a pleasant-faced young woman with an infant in her arms and a little boy playing nearby with a pail in the sand? 'This is my wife and children,' he says. And he adds: 'They killed the baby with a bayonet and she and the child were burned in the crematorium.' "[59]

Near Frankfurt, the group was given a result of a poll taken of the 18,311 Jewish DPs there. Only thirteen wanted to stay in Europe. The rest all opted to go to Palestine. As they were reading the results, they heard the faint sound of marching feet outside the window. The DPs were marching three and four abreast, carrying a flag with the Star of David and a banner reading "Open the Gates of Palestine," some wearing the same concentration camp uniforms given them by the Nazis. They stood for an hour in the rain looking into the window at the commissioners, delivering their powerful message.[60]

Many of the DPs came from Poland and were the sole survivors of what once had been large families. Some had walked hundreds of miles to their old towns, only to have to turn around and trudge back to the camps. There was nothing left for them there. Some who returned never made it back, having been murdered by non-Jewish residents. The British Embassy in Warsaw reported that three hundred Jews had been killed in Poland alone between the end of the war and the end of 1945.[61]

In Munich they met some Jews who had recently arrived from Poland, through the illegal smuggling route established by B'riha, an organization originating in Palestine before the war. Resistance fighters during the war had kept up both spy networks and escape routes and were now using them again to smuggle Jews out of Eastern Europe. Supported by both the Jewish Agency and the Joint, they led parties of Polish Jews into the American zone and camps or to Palestine itself. By 1946, some 40,000 Jewish "infiltrees," as the Americans called them, had greatly enlarged the population in the American zone. Between 1945 and 1948, B'riha is estimated to have successfully moved 250,000 Jews into Palestine.[62]

Later, the committee met with General Clark, who told them that U.S. policy, unlike that of the British, was to keep the borders open: "We want to give the Jews trying to get out of Poland a chance to save their lives." They were told that the policy had been originated by General Eisenhower and was backed by Truman. But this was very difficult because of "transportation troubles, absence of adequate food supplies, and opposition from British sources." The British in Austria wanted to "compel the Jews to rehabilitate themselves in Poland" and was based on Bevin's assumption that if the Jews left Europe, it would mean Hitler had won the war.[63]

The committee found that the DPs' morale was highest where the people had some vision for the future. For the majority, it seemed, that vision was Palestine. Crum observed that amid all the despair, the young people who were preparing for life on communes in Palestine seemed hopeful and happy. But other members of the committee were not convinced of the refugees' desire to go to Palestine. Judge Hutcheson for one thought the demand for Palestine was only the result of "artful indoctrination by Zionist agents." Buxton, who had gone with him to Poland and Austria, disagreed. "The feeling we found," he responded, "was too deep, too passionate, too widespread to be accounted for in that manner." In a report he wrote for the other committee members, Buxton spoke about the underground railway established to smuggle Jews out of Europe. The passion they held could "not be checked by official steps of any kind, whether by "disappearance of Zionist propaganda, the elimination of the Jewish Agency in Palestine, or any other measure."[64]

During this same period, James McDonald and other committee members went to Paris, Bierbach, Constance, Austria, Bern, Zurich, and Lugano and then headed back to meet the rest of the group to continue their journey to the Middle East. Like those who had gone to Germany and Poland, McDonald met with a unanimous desire on the part of Jewish refugees to gain entry to Palestine. In a boys' camp in Lugano, McDonald wrote, "nearly everyone of them had lost all his relatives in concentration camps and had known little but terror and death. They had built a new world for themselves of dreams and hopes and would tolerate no questioning of their realization," which was to reach Palestine. "Their earnestness," McDonald confided, "tempted one to weep." A day later, visiting both boys and girls at a villa where they cleaned and worked in the gardens, McDonald observed, "To them, . . . the world and the whole future centers in Palestine. . . . They, as the others I had seen . . . were absolutely confident that there was no future for Jews anywhere in Central or Eastern Europe."[65]

Crum and Sir Frederick went on to Nuremberg, where they attended some of the war trials. The Army allowed them to view secret documents about Nazi policy and showed them unexpurgated films taken by the Nazis of their murder of Jews (which made Crum so ill he had to walk out). Crum also read extensive evidence implicating the grand mufti in the crimes against the Jews.[66] Crum convened a press conference. If the settlements were not cleaned out, he told the American newsmen, "sooner or later we will have a wave of mass suicides or they will fight their way to Palestine."[67]

All this made Crum become an advocate for the Jews and no longer a mere observer. Either he would get the committee to write an interim report (which it had been instructed it could write but didn't have to) calling for a clearing out of all the camps, or, he pledged, he would present what he found to the American people via the media. "The facts," he said, had to be revealed."[68] The other Americans also wanted an interim report written addressing the plight of the refugees but were opposed by the British members.

Finally, Judge Hutcheson asked Crum to stop his campaign. The judge had been informed by the president and the State Department that an interim report might call the committee's impartiality into question. Instead, when it filed its final report, it should make short-term and long-term recommendations.[69] Crum became even more agitated by this intervention, suspecting an anti-Zionist conspiracy. He threatened to resign, which would have been disastrous for the committee. Niles tried to calm him down, writing Crum that he had talked to Truman and the president wanted to assure Crum "that he has every confidence in you and that he hopes you will do nothing rash." Calling Crum a "good sport," Niles added that no one was suggesting how the committee should conduct itself. Crum calmed down, and the break was averted.[70]

In March, the group moved on to the Middle East, where they heard directly from Arab spokesmen of their opposition to both Jewish immigration and a Jewish state. The Jews, and all of Western civilization, they argued, had "no right to impose the solution of the Jewish problem . . . on the Arab world." Jewish colonial settlements in Palestine were simply a new variant of Western imperialism.[71]

In Egypt, Abdul Rahman Azzam, the secretary-general of the Arab League, made an elegant twenty-minute speech about his cousins the Jews, who had left the region and come back as Europeans and as imperialists. "We are not going to allow ourselves to be controlled either by great nations or small nations or dispersed nations," he told the committee. They also heard from Syria's spokesman, Jamil Mardam Bey, as well as others. In public, all of them had the same opinion, McDonald noted: "not one more Jewish immigrant, not one more dunem of Jewish land." They called only for "united and unlimited effort to block Zionism," and considered themselves to be in a state of war.[72]

Then the group moved on to Jerusalem, where they stayed at the King David Hotel and held hearings at the YMCA across the street. Weizmann was the first to give his testimony. He told the committee, "Here is a people who have lost all the attributes of a nation, but still have maintained their existence as a ghost nation, stalking the arena of history, maintained it for thousands of years. It is our belief in a mystical force, our conviction of a return to the land of Israel, which has kept us alive. . . . We are an ancient people. We have contributed to the world. We have suffered. We have a right to live–a right to survive under normal conditions. We are as good as anyone else, and as bad as anyone else. . . . I stand before young Jews today as a leader who failed to achieve anything by peaceful means." Despite all the promises of British and American statesmen, he continued, "Jews are able to enter Palestine only as illegal immigrants and have no freedom of movement in the land." McDonald thought it was "one of the most impressive and moving statements" he had ever heard.[73]

To Richard Crossman, Weizmann looked like a "weary and more human version of Lenin." He seemed moderate and reasonable, but Crossman thought that he was too old, ill, and pro-British "to control the extremists" among the Zionists. But Crossman was impressed by Weizmann's candor, as were other committee members, when he admitted that "the issue is not between right and wrong but between the greater and less injustice." Since it was unavoidable, he candidly told them, they would have to decide "whether it is better to be unjust to the Arabs of Palestine or the Jews."[74]

Two days later, Crossman and Crum drove out to Weizmann's home in Rehovot. The modernist building of white stone designed by Erich Mendelsohn, a noted German architect, was an anomaly in Mandate Palestine. Political leaders called it the White House, while children saw it as a fairy-tale castle. "The interior," Crossman commented, "is a show piece, polished stone floors, thick carpets, and a number of beautiful works of art, including the two Utrillos and the most beautiful T'ang horse which I have ever seen." Crossman felt that all of this was really his wife, Vera's, doing, since Weizmann himself "scarcely notices it." Taking them into a smaller room to talk privately, Weizmann told them that he thought the British would not accept any extremist solution, either Arab or Jewish, which was why he favored partition. He had been in favor of it in 1937 when a British commission had recommended it, and was still in favor of it because it would give the Jews national sovereignty. "I regard it as a practical solution," he told them. Moshe Shertok (Sharett) entered the room and told them that he could get his people to go along if Bevin said a Jewish state could be created in a divided Palestine. It would be seen as a "practical possibility," provided that the area for the Jews included Galilee and the Negev, as well as a separate Jewish flag, army, and representation in the United Nations.[75]

Traveling around Palestine, the group could not help but be impressed with the achievements of the Jewish community, especially its efforts at reclaiming Palestine's barren wastelands. Crossman visited Mishmar Ha'emek, a large Jewish collective run by Marxist Zionists (Hashomer Hatzair), which believed in cooperation with the Arabs and a binational state. He found it a lovely place "with turfed gardens, fountains, beautifully kept flower beds and 700 acres of plain land and vineyards, and forests on the hills." He felt that he had never met a nicer community anywhere. Yet he was disturbed by the inhabitants' naiveté. It was impossible to make them realize, he thought, that the beautiful place they had created "with the green turf and the fountain and the gold-fish and the magnificent memorial to the dead children of Europe, and the cooperative spirit, are all set inside a huge barbed-wire stockade in a hostile territory."[76]

Crum and Crossman were not alone in their admiration for what the Jews had accomplished. When they met Frank Buxton in the café at the King David Hotel, where he was relaxing after visiting a nearby kibbutz, they saw that "his eyes [were] welling up with tears." "I feel like getting down on my knees before these people," Buxton told him, "I've always been proud of my own ancestors who made farms out of the virgin forest. But these people are raising crops out of rock!"[77]

The poverty and lack of educational opportunities among the

Arabs of Palestine, on the other hand, disturbed Crossman and the other members of the committee. Though the majority of Arabs in Palestine appeared to be better off than their counterparts in most of the other Arab countries (their population was growing due to immigration, a higher birthrate, and a lower death rate), the discrepancies in living standards between the two communities could only cause continued friction. Crossman believed that "all the Arab hatred for the Jew is based on a resentment at the arrogance, wealth and superiority of the invaders." Only two hundred yards away from the lovely collective, he wrote, was "the stenchiest Arab village I have ever seen," where he was entertained by a sheikh and "seated on the floor of a filthy hovel drinking tea."[78]

Later in March, David Ben-Gurion gave his formal testimony. Crossman's description of Ben-Gurion was more playful; "a tiny, thickset little man with white hair–a Pickwickian cherub."[79] The militant Labor Zionist did not seem disturbed by the thought that the Jews would not get the whole of Palestine, observed Crossman. He and the other Eastern European Zionist leaders envisioned creating Israel as the first successful democratic socialist society. "I am a Socialist," Ben-Gurion explained, and the socialist commonwealth would stimulate "similar movements throughout the Arab world." He talked of inviting young Egyptians to come and train in the kibbutzim, with the result that the Zionist project would "capture the Middle East," as well as becoming the West's and Britain's bulwark against the Soviet Union.[80]

The second day, the entire committee heard David Ben-Gurion for the whole morning. Starting with a two-hour speech in which he explained the tenets of Zionism, he gave "no apology and no indication of the least doubt about ultimate success." McDonald concluded that there was no doubt that, if necessary, there would be "resistance to any move to liquidate or seriously weaken the Jewish position in this country."[81]

After Ben-Gurion, they heard a presentation from David Horowitz, the Jewish Agency's economist, who came ready with charts, statistics, and graphs. McDonald noted that he made "a stunningly clear and, to my mind, convincing presentation of the Zionist case, to the effect that Jewish development in Palestine had substantially benefited the Arabs."

Both McDonald and Crossman were impressed by Ben-Gurion's commitment and passion. Driving to his home on the March 26, they wanted to make their formal good-byes before leaving Palestine. Crossman felt an affinity for Ben-Gurion and others he had met during the trip, including Golda Meyerson (Meir), the chairwoman of the Histadrut Jewish Confederation of Labor, whom he regarded as socialist comrades. To his eyes, meeting with them was as if he were at a meeting of the Socialist International with Western members. They all acted as if they were "fanatically building what they believe will be the only free socialist society in the world," he recorded in his diary. Moreover, they acted as if their fate rested in the hands of the Anglo-American Committee. To his mind, he could not help but think of how Vienna's Socialists had been betrayed by the West when they thought they would be rescued and Hitler prevented from taking Austria in the 1930s.

They found Ben-Gurion very concerned that the committee would recommend disarming the Haganah and abolishing the Jewish Agency. He gave them a final warning: "Don't make the mistake of thinking of us as Jews like the Jews you have in London," he admonished them. "Imagine that we're Englishmen fighting for our national existence." Calling out as they left, he added, "Don't underrate our intelligence." The committee's bodyguard was concerned about their safety as they drove back to their hotel at night. "It's O.K.," Ben-Gurion joked. "I've telephoned the terrorists all along the route."[82]

The committee members listened to the representatives of the Arab Palestinians. In Jerusalem, Jamal Husseini, a representative of the Arab Higher Committee, presented the Arab case. Husseini was a relative of the grand mufti of Jerusalem, Haj Amin al-Husseini, the titular leader of the Palestinian Arabs and a well-known collaborator and partisan of Adolf Hitler and the Nazis during the war. The American members wanted to know what would happen if, as they wished, Palestine became an entirely Arab state. Would the Grand Mufti be proclaimed head of a new Arab Palestine government? If they succeeded, Jamal Husseini argued, at least 30 percent of Jews who had come to Palestine would leave, and the rest would accept living in Palestine under Arab rule.

As for the Mufti, he might well become head of the new Palestinian Arab state. He only had the interests of the Arabs at heart. As for his collaboration with Hitler, Jamal Husseini argued, the Mufti had only sought to "get something out of them in case they were victorious." Listening to his testimony, Bartley Crum thought to himself that al-Husseini was "Gerald L. K. Smith in a fez," referring to the notorious American anti-Semite.[83] "He was brilliant and fluent," McDonald commented, "but, from the point of view of the Americans . . . his open defense of the Mufti was a major strategic mistake."[84]

The final Arab statement was delivered in writing to the committee and later published in America by the Arab office as *The Problem of Palestine*. Its opening statement set its tone: "The whole Arab people is unalterably opposed to the attempt to impose Jewish immigration and settlement upon it, and ultimately to establish a Jewish State in Palestine."[85]

For many of the delegates, the most memorable session was that of Martin Buber and Dr. Judah Magnes, the two most prominent intellectuals in Jewish Palestine. Here the committee members heard their eloquent but perhaps irrelevant pleas to create a new binational (Arab-Jewish) state in the former British Mandate. Magnes, then the president of Hebrew University, gave what McDonald called an "eloquent" and "deeply moving" speech reflecting "a moral courage of the very highest kind." His solution, however, McDonald thought, reflected a poor sense of statesmanship as well as being completely impractical.[86] Crossman reached much the same conclusion. Acknowledging that the British viewed Magnes as a moderate who wanted conciliation with the Arabs, Crossman felt that his concept of a binational state "represented nothing real in Palestine politics" and that once it might have been possible but was now too late.[87]

Then the committee received a memorandum from the Jewish Resistance Movement, the armed unit that, after the war, had combined the fighters of the Haganah with those of the Irgun. As it functioned underground, its leaders could not testify, and the press was not given the memo. "We consider it our duty to warn you, against any attempt to impose an anti-Zionist political solution and mask it with a token increase of immigration permits." No minority status for Jews in Palestine would be acceptable, they informed the committee, nor would any "symbolic independence in a Lilliputian State." The committee, they admonished, had to be courageous and, most important, "decisive."[88]

Finally, the weary group went to Lausanne, Switzerland, in early April to hammer out their findings and, they hoped, to develop a consensus document. Switzerland, neutral during the war, was nevertheless a tense site for the members. Security was very tight, and the American members feared that they were under secret British surveillance and that their phone calls were bugged. McDonald thought his mail was being opened and complained that he was not allowed to go anywhere without telling the hotel concierge precisely where he was headed. When President Truman cabled the U.S. delegates' chairman, Judge Joseph Hutcheson, to let him know that he hoped they would arrive at a unanimous conclusion, the cable was first delivered to the British consul in Geneva. A second cable from Truman was also opened by the British before it was given to Hutcheson.[89]

Truman's request was hard to fulfill. Differences, particularly between the American members and the British group, threatened to tear them apart and make a satisfactory end next to impossible. Back home, the congressman from Brooklyn, Representative Emanuel Celler, wrote the president expressing his fears: "It is clearly evident from the rift that has arisen between the American and British members . . . that the British are determined to control completely this inquiry." He had heard rumors that the British had actually written a final report in London, without waiting for the delegates even to write their own. The British report, he was certain, would be one "pre-formed by the foreign policy of Great Britain."[90]

All of the committee members had been extremely moved by their experiences in Europe. The majority had concluded that anti-Semitism was still very strong; that little less than death would destroy the wish of the Jewish displaced persons to go to Palestine; and that the military wanted to close down the camps not only for the sake of the refugees but to help normalize life in Europe. Sir Frederick Leggett, who had been hostile to the Jewish position when the committee started, "became emotionally exhausted by the trip and resolved to do something to help those who had survived. He even started greeting everyone with 'Shalom.' " He told his colleagues, "Unless we can do something and do it soon, we shall be guilty of having finished the job Hitler started; the spiritual and moral destruction of the tiny remnant of European Jewry."[91]

But what was to be done? Sir Reginald Manningham-Buller argued that the Balfour Declaration had already been fulfilled in creating a homeland for the Jews and that Britain should be formally released from its promise to support an actual Jewish state in Palestine. The other British members, save Crossman, agreed with his perspective. Moreover, he favored disbanding the Haganah and altering the power of the Jewish Agency. The Americans answered back. Frank Buxton argued that the principle of "eminent domain," used to explain the American conquest of Mexico and the movement of American Indians into a modern society, applied to the situation in Palestine. Crossman emphasized the different culture of the Arabs and Jews and recommended partition as the best solution, as did Crick. They could not, McDonald summed up, avoid dealing with "the very wide and fundamental differences disclosed" during their meeting.[92]

The committee was still split on the issue of a binational state, the number of immigrants to be allowed in, the role of the Haganah and the Jewish Agency, and the issue of partition. The animosity was fierce. McDonald, Buxton, Crossman, and Crum favored immediate admission of 100,000 Jews: Phillips and Aydelotte leaned to the British view. It all depended on Hutcheson, who alone could persuade the other Americans. The discussions continued, and eventually Harry Truman's request for admission of the 100,000 seemed reasonable.

"The primary way to resolve the whole unsettled condition," Crum put it, ". . . was by doing precisely what the President of the United States had suggested." The committee called in military officers who ran the DP camps in Austria and Germany, and they agreed that all of those living there could be shipped to Palestine within the year.[93]

The final recommendations were meant to be tentative. Crossman explained that the members of the committee were not experts; they disagreed about much, and all they could really contribute was a set of guiding principles for Anglo-American policy. It was foolish, he thought, for them to work out a concrete plan that would probably eventually end up at the United Nations. They should make interim proposals "which could ward off the danger of war and give to the British government at least six months' breathing space in which to formulate its policy." As for settling the Jewish DPs in other countries, "The fact had to be admitted," he put it, "–shameful as it was–that Palestine was the only country where 100,000 Jews could be absorbed in the immediate future."[94]

Crossman was particularly astute in framing issues in a way that could gain support from the British members. In a long memo, he argued that they had to take a position that would isolate Zionist extremists and reinforce moderate leaders such as Weizmann, who were losing ground as the British continued to stand by the White Paper. Allowing Jewish immigrants into Palestine would satisfy Jewish demands and at the same time allow Britain to deal ruthlessly with uncompromising Jewish terrorists. Anything else would only turn all the Palestinian Jews "into a fanatical support of the extremists," as well as possibly leading to a war against British troops by the entire Jewish population. They had to accept, he told his colleagues, "a more humane and juster course."[95]

"In the end," Crum wrote, "it was the leadership of Judge Hutcheson which kept us all together." The judge worked around the clock, putting in twelve- to sixteen-hour days, shuttling between the two groups as he tried to achieve reconciliation.[96] On April 1, he convened a meeting in which he shared his recommendations with the committee. He thought that Palestine should be neither an Arab nor a Jewish state. Yet he insisted the Jews' achievements there had to be accepted, and nothing should occur that would disrupt the development of the Jewish homeland. Like Judah Magnes, Hutcheson suggested the creation of a binational state. As for partition, which some Zionist leaders had said they would accept, Hutcheson saw it as "a solution of despair" that would "satisfy neither the genius of the Jews nor that of the Arabs." In the short term, he argued, the committee should ask for the immediate "largest possible immigration" of Jewish DPs up to the 100,000, if necessary. The bottom line: there should be "substantial continuing immigration under the Jewish Agency" but not the creation of a Jewish state.[97]

The committee finally came to a unanimous decision along the lines suggested by Hutcheson: it recommended the immediate issuance of 100,000 certificates by Britain to allow Europe's Jewish DPs to go to Palestine and the revocation of the land and immigration regulations of the White Paper. For the long run, they rejected both an Arab state and a Jewish state in Palestine and called for a "country in which the legitimate national aspirations of both Jews and Arabs can be reconciled, without either side fearing the ascendancy of the other." The details would be worked out by the United Nations.[98]

The members signed it, feeling exhausted but relieved and proud of their achievement. Agreement had for days seemed impossible, yet they had put aside their differences and established a precedent for Anglo-American cooperation on this issue. Palestine was on the verge of war, they thought, and they hoped that their deliberations and recommendations might prevent it from occurring. Dissatisfied as they might have been, observed Crossman, each one of them felt the conclusions they had reached were better than what they could have achieved on their own.[99]

The committee had done its job. Its members had traveled and lived together under harsh and often difficult conditions. They had quarreled over differences, yet, at the end, they had come to a consensus. The committee knew that both sides would find many of their recommendations offensive, as well as wrongheaded. In a sense, they left it to others to decide on a final outcome and to reconcile the differences before making a new policy. But on one point they stood together: the 100,000 Jewish DPs had to receive immigration visas for Palestine immediately. Whether or not the British government would go along remained to be seen.

SIX

IMPASSE: THE REPORT THAT KICKS UP A STORM

As soon as the Anglo-American Committee finished writing up its recommendations, the interested parties dissected its implications and tried to figure out how to respond. Most British politicians thought the entire report, particularly the committee's recommendation for the immediate immigration of the 100,000 Jewish DPs into Palestine, was a sell-out to the Americans. Harry Truman, on the other hand, was encouraged that the report called for an end to the White Paper and the quick transfer of the 100,000 DPs. He thought the committee had made significant progress.

Meeting in Paris, Jewish Agency members David Horowitz, David Ben-Gurion, Moshe Shertok (Sharett), Berl Locker, Nahum Goldmann, and Arthur Lourie poured over the report. The most hostile reaction came from the Jewish Agency leader, David Ben-Gurion. The Anglo-American Committee's report, he declared, was nothing but "a disguised new edition of the White Paper, though more cleverly compiled." The report was framed to evade any decision on issues which the committee members disagreed. It only had one "saving grace," the demand for immediate immigration of the 100,000 DPs. Several of the others, however, saw the report as a "springboard for renewed political activity." Outright rejection could be dangerous, they argued, allowing the British to maintain the status quo, alienate American and British public opinion, and do nothing to help the refugees. They decided not to take a position on it.[1]

Predicting Ben-Gurion's hostile response, James McDonald rushed to head it off. In a hand-delivered note, McDonald told Ben-Gurion that whether or not the report was able to make a constructive contribution would depend upon the reception it got. He was certain that portions of the report would seem to both the Zionist leaders and the Arabs to be unwarranted and unjust, but he hoped that they would be able to utilize the opportunities recommended in the report and then "strain every effort to translate them into reality." To do that would require heroic efforts in Palestine, in Europe, and in America. Therefore, he pleaded that Ben-Gurion "take the lead in saying that ideolalogical [*sic*] considerations must, for the next few months at least, take a secondary place." Heated controversy, he warned, would only have an unfortunate effect on the U.S. government.[2]

Ben-Gurion was not impressed with McDonald's argument. If anything, it infuriated him. "Crum and McDonald," he wrote in his diary, "think they have achieved a brilliant triumph and done an historical service to the Jewish people! Now Crum demands the price–expression of thanks to Truman."[3] In a telegram he sent the same day to the American Zionists, Ben-Gurion argued that the committee was proposing a "British colonial-military state, which was no longer to be a homeland for the Jewish people, and which would never become a Jewish State." He urged American Zionists to tell Truman not to endorse its conclusions.[4]

At the end of April, McDonald and Crum met with Silver and tried to convince him not to publicly criticize the report. Would Silver really want to turn Truman away from the Zionist cause? they asked. If not, they suggested a compromise: Why not urge Truman to endorse immediate action on the 100,000 DPs and postpone a decision on the rest of the report? Silver reluctantly agreed. Crum then proposed his suggestion to Truman. The president, eager to have neither Silver nor the Zionist movement oppose him, magnanimously allowed Silver and Neumann to draft the statement he would make to the public announcing the committee's conclusions. The president wanted two points included: a sentence ensuring the protection of the holy sites and something about the Arabs' rights. Silver wrote a handwritten draft, and the administration and the AZEC council changed some of the text the next day.[5]

When the committee's report came out on May 1, Truman made his statement public:

> I am very happy that the request which I made for the admission of 100,000 Jews into Palestine has been unanimously endorsed by the Anglo-American Committee of Inquiry. The transport of these unfortunate people should now be accomplished with the greatest dispatch. The protection and safeguarding of the Holy places in Palestine sacred to Moslem, Christian and Jew is adequately provided in the report. One of the significant features in the report is that it aims to insure complete protection to the Arab population of Palestine by guaranteeing their civil and religious rights, and by recommending measures for the constant improvement in their cultural, educational and economic position.
>
> I am also pleased that the Committee recommends in effect the abrogation of the White Paper of 1939 including existing restrictions on immigration and land acquisition to permit the further development of the Jewish National Home. It is also gratifying that the report envisages the carrying out of large scale economic development projects in Palestine which would facilitate further immigration and be of use to the entire population.
>
> In addition to these immediate objectives the report deals with many other questions of long range political policies and questions of international law which require careful study and which I will take under advisement.[6]

Truman's announcement had an immediate impact. The president's statement, the journalist James Reston wrote, was more important than the entire Anglo-American Committee Report. The committee "merely had the power of recommendation," observed Reston, "while the President's comment is regarded as a statement of United States policy."[7] Loy Henderson understood this and informed British Ambassador Lord Halifax that he deeply regretted the president's statement; that the State Department had done all it could to prevent the president from issuing it; and that up to the last minute it had put all possible pressure on the White House not to do it. Henderson confided that there were forces in the White House that the State Department was not able to control.[8]

Bevin wasn't comforted by this explanation. He was tired of hearing one thing from the State Department and an altogether different one from the White House. He had sent Truman a message asking that no action be taken on the report prior to the British and American governments' having a chance to consult. But the president had not even given him a "courtesy consultation." Bevin's friend and first biographer, Francis Williams, wrote that reading Truman's statement "threw Bevin into one of the blackest rages I ever saw him in." He immediately prepared to dash off a letter venting his anger about British soldiers being killed by those he called "illegal Jewish terrorists."[9]

Bevin thought his request was quite reasonable. A few days before Truman made his statement, he had met with Secretary Byrnes in Paris and had let him know the British government would be prepared to permit immigration of the 100,000 DPs if they did not all go to Palestine. What worried Bevin most, he told Byrnes, was that the Jews were acquiring a great deal of arms, "most of them with money furnished by American Jews and are in a very aggressive frame of mind." Jewish immigrants, he said, were in fact being picked out by the Jewish Agency for their potential as soldiers. Britain's four divisions in Palestine were not sufficient to deal with any new population explosion, unless the United States shared in the responsibility, which meant sending American soldiers to Palestine.[10]

Bevin's sour mood was also due to the failure of a gamble he had made. He had agreed to the creation of the Anglo-American Committee because he assumed that if he picked reasonable men, who were neutral on the subject, they would end up recommending the policy he and the British government favored. The White Paper, in Bevin's view, had to stand until such time as the United Nations worked out a substitute. He had also assumed that they would come to the same conclusion as he had: that Palestine was not the answer to the plight of the Jewish refugees. They should instead be sent back to their original European homes and be assimilated. He had compounded his mistake by proclaiming that if the report were unanimous, he would do his best to implement it. But perhaps the most reckless statement he had uttered was that he would stake his political career on finding a solution to the Palestine problem.

Clement Attlee gave his own report to a meeting of the prime ministers of the Commonwealth (Australia, New Zealand, and South Africa). On the bright side, he told the ministers, the report did give Britain a "new opportunity for enlisting American cooperation." But unfortunately, he thought, the recommendation that Palestine would be neither Jewish nor Arab would be unacceptable to both groups. Most important, Attlee said, the Jewish armed forces, which he continually called "illegal," had been tolerated far too long. Their suppression was necessary and essential before 100,000 more immigrants would be allowed to enter the country. The bad news was that it left everything in Palestine to be carried out and financed entirely by Britain, the mandatory power. It was not right, Attlee complained, that Britain should carry the whole burden of disposing of the displaced persons. "It was high time," he said, "that the Americans should share some of the cost, both in money and in armed forces."[11]

The next day Attlee appeared before the House of Commons and elaborated on his position. The report, he said, contained long-term commitments the British were not prepared to make. He denied that 100,000 DPs could be sent to Palestine quickly and that Palestine would be able to absorb them. Second, he announced that unless the Jewish "illegal armies maintained in Palestine" were disbanded and their arms confiscated, the British government would not allow any large body of immigrants into Palestine.

The reaction of the British was ironic, given that they had been responsible for the training and organizing of the Haganah in 1941, when a German invasion of Syria and Egypt had seemed imminent. When 27,000 Jews who had volunteered to fight Hitler with the British army returned to Palestine, they remained with the military and were integrated into the Haganah. The British military authorities had then created what they called the Jewish Home Guard, which the Haganah soldiers joined. The unit included sections trained in sabotage and guerrilla warfare, to be used in case of German occupation.[12] As Richard Crossman commented when he heard Attlee's speech, "Everyone in Palestine knew that the disarming of the Haganah would involve full-scale military operations against the Jews" by the British and would "precipitate the crisis which it had been the purpose of the report to prevent."[13]

Outraged, Bartley Crum told the press that the committee had specifically demanded the admission of the 100,000 refugees without any conditions. "It would be indecent and inhuman," Crum insisted, "to try to trade their lives upon condition that the Jews of Palestine surrendered their arms." Frank Buxton recalled that Judge Hutcheson had quoted to them from the U.S. Constitution's clause that gave the people the right to bear arms. If Clement Attlee actually did not know how impossible it would be to disarm the Haganah, he added, he was "an incredibly stupid man."[14] The Haganah, Buxton said, was more like the American revolutionary army, "a rabble in arms in the fine sense." It had to be distinguished from a regular armed force equipped with aircraft, tanks, trucks, and lines of communication.[15]

Most important was the president's reaction. The British were quite mistaken, Truman said, if they thought that the Americans would assist them in disarming the Jewish forces in Palestine.[16]

American Zionists finally decided that it would be a much better tactic to support Truman than to attack him for the report's recommendations. Lauding the president's "statesmanlike and humane spirit," the AZEC leaders sent him a letter, signed by Rabbis Silver and

Wise, Eliahu Epstein (Elath), and other top Zionist leaders. The demand for immediate certificates for the 100,000 Jewish DPs gave them "complete satisfaction," they told Truman, and they pledged to cooperate with the administration in working for its implementation. In both Jerusalem and the United States they put themselves at Truman's disposal to help execute the immigration in any way they could.[17] Silver and Wise then called on AZEC's local committees to mobilize by making appointments with their senators and congressmen and sending the president letters and telegrams urging him to move forward with immigration plans immediately.[18]

As expected, the Arabs denounced the committee's report and the State Department bore the brunt of their anger. Loy Henderson met the foreign ministers of five Arab countries: Egypt, Iraq, Lebanon, Saudi Arabia, and Syria. Speaking for the group, Egypt's minister, Mahmoud Hassan, reiterated their understanding that the United States could not change policy without first consulting the Arab nations. The admission of 100,000 Jewish immigrants would in their eyes be such a basic change. It was their hope that given the hostile Arab reaction to the report, the United States would announce that it was not bound by the committee's recommendations.[19]

In Jerusalem, the Arab Higher Committee called for a strike. Jamal Husseini pledged that Arabs would fight implementation of the report. He announced the possibility that their exiled leader, the Grand Mufti, would come to Syria and take charge, an event that, he threatened, might produce an Arab uprising.[20] The Mufti, called the "Archvillain of the Mideast" by *Newsweek* magazine, had been living under house arrest in Paris. However, because of world pressure, it was expected that he would be extradited and handed over to the British and then prosecuted for war crimes. Then, on June 19, after dyeing his hair and shaving his beard, he had managed to smuggle himself out of his residence and gained asylum in Cairo, Egypt, under the protection of King Farouk.

Mohammad Haj Amin al-Husseini had been appointed Mufti in 1922 by Britain's first high commissioner of Palestine, Sir Herbert Samuel. Ironically, Samuel was Jewish and had given the authority to one of the most anti-Semitic Arab leaders. By giving him this honor, the British had hoped to channel el-Husseini's opposition to the Balfour Declaration and Jewish settlement (which included inciting deadly riots) into more moderate behavior. Husseini quietly built his power base in Palestine, becoming president of the Supreme Muslim Council and controlling much of Arab Palestine's civil and public life through patronage.

Husseini had been in contact with the Germans since 1936. Angered by the influx of Jewish refugees fleeing Hitler, the Mufti and his followers carried out their attacks not only against the Jews but against the British and Arabs who didn't agree with them. Forced to flee Palestine, he eventually ended up in Berlin, where he argued for the destruction of the Jewish national home by encouraging Hitler and his staff to extend its European policy of extermination to Jews living in the Middle East. The mufti, *The Nation* magazine editorialized, was "as much a war criminal as any Nazi still on trial at Nuremberg." In exile in Germany during the war, the editors pointed out, the mufti had supervised propaganda, espionage, and the organization of Moslem military units in the Middle East to oppose the Allied forces.[21] Speaking on Berlin radio in 1944, the mufti said, "Arabs, rise as one and . . . Kill the Jews wherever you find them. This pleases God, history and religion." In his memoirs, he further explained that "Our fundamental condition for cooperating with Germany was a free hand to eradicate every Jew from Palestine and the Arab world."[22] Until the war's end, Husseini worked to recruit Muslim volunteers, mainly from Bosnia, for the Nazi armed forces.

In Europe, J. H. Hilldring, the State Department's assistant secretary who was responsible for the DP camps, eyed Britain's response to the Anglo-American Committee Report from yet another perspective. Unlike his colleagues in the department, Hilldring was more sympathetic to the Zionists. He was anxious to close down the camps, which were very expensive to maintain, and to see the refugees on their way to a better life. Hilldring told Acting Secretary Dean Acheson that the British were obviously stalling on the Anglo-American Committee's recommendation for authorization of 100,000 immigration visas to

Palestine. To deal with Attlee's obstinacy, he suggested that the United States exert pressure on the British government. The president, he thought, should make an urgent statement demanding implementation, and the United States should formally take the responsibility for moving the refugees to Palestine from Europe. The expense would be great but a lot less than maintaining the Jewish DPs in the German and Austrian camps for yet another year.[23]

Could Attlee's and Truman's positions be reconciled? Their divergent responses to the committee's report now hung in the air. Loy Henderson pondered what the government's next move should be. On May 3, he drafted a proposal and sent it to Acheson, writing in the margin, "We are of course playing with dynamite." Acheson was aware of how delicate the situation was. With Byrnes in Paris, the assistant secretary found himself increasingly embroiled in the Palestine issue. Acheson adopted Henderson's proposal and prepared a draft for the president to be used as an answer to Clement Attlee.[24]

Truman worked on the letter, consulted with his advisers, and, on the eighth, wrote Attlee confirming that he thought the committee report was a sound basis for proceeding. He suggested that the British and Americans take two weeks to solicit Jewish and Arab responses to the report and its recommendations, after which the United States would consult with the British government and decide whether the report as a whole was suitable as a basis for Palestine policy. He stressed the urgency of acting quickly, so that they could proceed with arranging the movement of the 100,000 Jewish DPs.[25]

The British thought they would have a hard time coming up with a policy statement in that short a time frame. They were in the midst of difficult negotiations with the Egyptians on a new treaty that included the withdrawal of their forces from Egypt and asked Truman if he could delay action on the 100,000 DPs until May 20. By that time they would know whether they needed to formally request U.S. military assistance. Truman, not wanting to alienate the United States' wartime ally, concurred.[26]

Bevin then presented Truman with a memo including ten points he wanted him to consider. To fulfill the committee's recommendation,

Bevin argued, would require the kind of funds and military resources that Britain could simply not handle. He wanted to know what the Americans were willing to contribute before the British government could make any decisions and asked for suppression of the Haganah and Jewish Resistance and for a guarantee of £60 million to £70 million sent by the United States in the next few years to help in administering Palestine.[27]

In the meantime, Prime Minister Attlee agreed with Truman's suggestion to consult both Arab and Jewish interests before deciding how to act on the report. Attlee now informed Truman that they needed expert officials to study what military and financial obligations both powers would have to take to implement the report. He requested a month for Britain to respond formally to the committee's recommendations.[28] Truman replied that he would give Attlee the time requested and told him that the United States would itself organize an appropriate group of government officials.

Toward the end of May, Attlee finally sent Truman an updated version of his original ten points of discussion. Truman was stunned to find that Attlee had now expanded Bevin's ten points to more than forty-three new subjects that the experts were expected to discuss.[29] Truman had been warned that the British might use this opportunity for delay and obfuscation, and he was experiencing it.[30] The president was growing impatient. On June 5, he wrote to Attlee that although it would take considerable time to find satisfactory answers to all of the questions Attlee raised, he wanted to "begin immediately consideration of the question of the 100,000 Jews whose situation continues to cause great concern."

Realizing that he had to respond to the British demands for U.S. help in dealing with the situation, Truman told Attlee what he had been waiting to hear. "I can assure you now," he told the prime minister, "that we shall take responsibility for transporting these persons as far as Palestine and shall lend necessary assistance in the matter of their temporary housing. We shall be glad to consider also providing certain longer term assistance for them." Furthermore, because of the urgency of the problem, the president wanted to initiate discussions between the United States and Britain "on the physical problems directly connected with their transfer as soon as possible."[31]

Unfortunately, the president's offer did not have the result he desired. On June 10, Attlee turned him down on any quick action for the refugees. Her Majesty's government, he wrote to Truman, would "not feel able to determine their policy on any one of the Committee's recommendations until they have examined the results of the official consultations on the Report at a whole." He added that they would also have to consider the political and military consequences of giving visas to the 100,000 before they could act.[32]

The following day, Truman announced that he was creating a new cabinet committee that would have the authority to negotiate with the British over Palestine. It was to be staffed by individuals chosen by the president, who would be surrogates for members of the president's cabinet. The group would be headed, the president informed the public, by Henry F. Grady, who had been an assistant secretary of state. Its job was to work out an arrangement that would implement the recommendations of the Anglo-American Committee. It was not meant to carry out any investigations of its own.[33]

If the British were becoming more unpopular in America, Ernest Bevin made things much worse when he addressed the annual Labour Party conference in Bournemouth, England, on June 12. The foreign secretary started off well, making the point that foreign policy was the concern no longer just of statesmen but of all people living in a democracy. Turning to Palestine, Bevin explained why he thought so much pressure was coming out of New York City for admission of the 100,000: "There has been agitation in the United States, and particularly in New York, for 100,000 Jews to be put into Palestine. I hope I will not be misunderstood in America if I say that this was proposed with the purest of motives. *They did not want too many Jews in New York*" (our emphasis).[34]

If Bevin had not uttered these words, his speech might have gone over as a simple reiteration of British policy and a reasoned argument as to why he differed with Truman and the Anglo-American Committee Report. The press, of course, focused on those very words, which then provided red meat for a rally that took place only a few hours later at New York City's Madison Square Garden. Speaker after speaker denounced Bevin before the crowd of 1,200 for being not only an anti-Semite but "a traitor to the cause of liberalism and labor." Bartley Crum led the assault. He tore up his speech and demanded that the Mandate be taken away from Britain and given to the United Nations. He called for the immediate withdrawal of British troops from Palestine and called on the Senate to review the pending U.S. loan to Britain. Rabbi Silver called Bevin's statement "a coarse bit of anti-Semitic vulgarity reminiscent of the Nazis at their worst" and told the crowd, "We Jews have had enough. We want a national home for our people." Rabbi Wise publicly told Truman, "it is you who are being insulted" by Britain's refusal to open Palestine's gates.

More troubling for the British, their new ambassador to the United States, Lord Inverchapel, reported, was the participation of Senator Edwin C. Johnson, a Colorado Democrat. His presence suggested that public opinion throughout the United States might be against the British proposals. Johnson too had given a militant speech, Inverchapel wrote, telling the crowd that the British view that more troops had to be sent to Palestine was nothing but an attempt by Britain to continue its "nefarious and shameful policies." Johnson had not been identified with the Zionist cause, and, noted Inverchapel, his own constituency "is not believed to contain any appreciable Jewish population." Yet he had "surprisingly strong feelings against the British position in Palestine."[35] Apparently Inverchapel was unaware that Johnson was a member of the pro-Zionist American Christian Palestine Committee.

The attack on the British loan, which had been raised at the rally, was a loaded issue. The Truman administration, while sympathetic to the Jewish groups' demands, believed it was crucial for postwar relations with Britain and the rebuilding of the West to proceed with the loan. Niles rushed off a memo to Truman's press secretary, Charlie Ross. If the press asked the president what he thought of Bevin's remarks, Niles advised, the president should respond that he simply stood by his previous insistence that the 100,000 refugees be admitted to Palestine. As for the demands that the loan be reevaluated, Niles suggested that Truman say that the fact that Bevin was not moral did mean the United States had to be immoral. What Truman had to convey was that he would not let anything move him from seeking "to serve the humanitarian needs of the world, whether it is to help Jews or any other needy people."[36]

The White House also knew that moderate Zionists, such as Stephen Wise, stood with the president and agreed that Britain should get the loan. Wise had written Niles that he thought the White House should "counteract the Silver mischief" and explained that he did not want to harm the interests of either the Americans or the British, just to "spite a man however nasty his speech and lamentable his conduct against Zionism."[37]

Inverchapel told Bevin that his unfavorable reference about the United States not wanting any more Jews in New York was causing an uproar. His speech was being used to tell the American public that the British government had already made up its mind not to admit the 100,000 refugees and did not ever intend to carry out the recommendations of the Anglo-American Committee. Moreover, since Truman was so closely identified with the committee and the cause of the 100,000 Jews, it was "being argued in anti-Administration circles that you have made the President look foolish." On the positive side, Bevin was not to worry about the loan. Inverchapel had learned that morning that the loan had already successfully gone through a House committee, which had sent it to the floor of the House with a vote of 20 to 5.[38]

The American Zionist leaders, making good on their offer of help to make the transfer of the 100,000 Jews a reality, wrote a lengthy letter to the president under the rubric of the Jewish Agency executive. The letter dealt solely with the technical and financial problems of arranging for the immigration and settlement of 100,000 Jews in Palestine. The Agency had for many years worked on the machinery for selecting, receiving, and settling immigrants, and an additional 100,000 arriving at once would not result in any great problems. Fifty percent of them would be suited to immediate gainful employment, they emphasized, in agriculture, manufacturing, and construction. Temporary public assistance would be necessary and would cost in the vicinity of

$40 million. Transportation from Europe, which the United States had agreed to sponsor, would cost around $10 million. Training, medical care, and the like would cost around $20 million, and clothing, furniture, and household utensils would add on an additional $12 million to $15 million. Housing materials would come to $40 million. Twenty-five thousand refugees would be orphaned children, and the total cost for them for a five-year period would come to $60 million. The Agency noted that the cost should rightfully be assumed by Germany as reparations, but it would proceed with the DPs' settlement even if such funds were never allocated. Its only hope was that Truman would give its requests "sympathetic consideration."[39] Truman received the letter and quickly passed it on to David Niles, with whom he asked to discuss the issues raised.[40] He appreciated the Agency's factual and concrete suggestions.

Truman was trying to find some way around the obstacles the British were throwing up and asked Byrnes to give Attlee the message that although he understood that the British wanted to consider the entire report before reaching a decision on the 100,000 DPs, couldn't they move ahead by ironing out the technical details of how the transfer would proceed? Then when the time came, everything would be in place. To accomplish that goal, Truman told Attlee, he had appointed Averell Harriman, now the U.S. ambassador in London, to initiate discussions with British representatives to deal with the technical issues of the transfer.[41]

These meetings took place as requested by Truman, but Attlee was hardly satisfied. He now requested that a representative of the U.S. Joint Chiefs of Staff come over to London to hear the views of the British Chiefs of Staff "on overall Middle East strategic questions as they related to Palestine." Byrnes moved quickly to put a damper on any expectation of American military help. He informed Attlee that "any military discussions between the British and US on the specific subject of Palestine are most undesirable at this time and that nothing should be done now which might be construed as indicating a US interest in the possibility of US military involvement in Palestine."[42]

Attlee had, from his perspective, good reason to desire that the

United States send its troops to aid British work in Palestine. Terrorist activity by the Zionist cadre had been increasing in Palestine, and the Irgun had managed to kidnap six British officers. "Drastic action," Attlee told Truman, "can no longer be postponed." The office of the British high commissioner for Palestine was now resolved to break up illegal organizations, which, to its eyes, included the Haganah. It was planning to raid the offices of the Jewish Agency shortly and search for incriminating documents that would identify the actual terrorists. Anyone implicated, including Haganah forces, would be arrested and headquarters of illegal groups occupied. "Open defiance" was no longer to be allowed, and establishing law and order would come first. It was the "recent outrages by the Jews" that forced it to act.[43]

The tough British response to Jewish terrorism led Truman to fear that the United States might end up in the position of being forced to lend military assistance to the British in Palestine. He had requested that Acheson confer with the Joint Chiefs of Staff to prepare a report on this question, which was delivered to the president on June 21. The Joint Chiefs were clear and concise in their recommendation: "We urge that no U.S. armed forces be involved in carrying out the Committee's recommendations. . . . the guiding principle be that no action should be taken which will cause repercussions in Palestine which are beyond the capabilities of British troops to control." U.S. troops were already overextended elsewhere. That, however, was not their main concern. Though in theory U.S. troops might help to pacify Palestine, the Joint Chiefs pointed out, "the political shock attending the reappearance of U.S. armed forces in the Middle East would unnecessarily risk such serious disturbances throughout the area as to dwarf any local Palestine difficulties."

Should U.S. troops enter the area, their report continued, both American and British interests would find their role curtailed, which might result in the Soviet Union replacing them in influence and power throughout the Middle East. This would be the equivalent of a Soviet military conquest and would affect the control of Middle Eastern oil and America's standard of living as well as its military strength, which was based on oil. The Middle East had the largest undeveloped reserve of oil in the world, they informed the president, and they feared that the world might come to the limits of its oil resources within one generation. If the United States took part in a trusteeship of Palestine, as the British suggested, it would only lead to military involvement. For all of the above reasons, the key to U.S. policy had to be to keep the Middle East oriented toward the Western powers.[44]

The continual British stalling and the consequent inability of the Truman administration to make progress on the president's promises inflamed both the Jewish community and a wider public that was sympathetic to the plight of the refugees. It was hard for them to believe that the United States could not exert enough pressure on the British to achieve what they thought was a relatively modest goal. Representative Emanuel Celler wanted to take the entire New York congressional delegation, including Democrats and Republicans, to the White House to discuss the refugee crisis. Truman, who was growing weary of this seemingly insoluble problem, was doing everything possible to avoid having another such meeting. Celler had spoken to David Niles but had not received an answer from him or the president. New York's congressional delegation, Niles told Truman's secretary, Matt Connelly, "were becoming impatient" and were about to tell the press that the president would not see them.[45] Niles spoke to Truman about the request, but Truman still wouldn't budge.

Celler again wrote to Connelly, to let him know that he was quite surprised to hear that Connelly could not arrange for a New York congressional delegation to meet the president. He warned that if it got out, it would give political ammunition to upstate New York Republicans before a crucial election. "It is bad politics for the President not to meet with them–even if it is on the Palestine question." If he did not hear back, the congressman threatened, he would take it that "the jig is up" and inform the other members of congress.[46]

Truman finally gave in. But when the delegation entered his office, they found an annoyed Truman shuffling papers on his desk as Celler began to read their joint statement. He had scarcely read four sentences, Celler recalled, when Truman stopped him. "His voice and face were cold as he said, in effect, that he was tired of delegations visiting him for the benefit of the Poles, of the Italians, of the Greeks. I remember his saying, 'Doesn't anyone want something for the Americans?' "[47]

The Zionists tried to find other ways of reaching Truman. The president, however, continued to avoid scheduling meetings on the subject. He felt he was doing all he could and was tired and angry at the constant pressure. He took to referring to Silver and Wise as "extreme Zionists" and Celler and the New York politicians as the "pressure boys."

But now a request was made by his friend and former business partner, Eddie Jacobson. Jacobson, the sixth child of Eastern European immigrants, was born in New York City in 1891. When he was two years old, the family moved to Leavenworth, Kansas, and then, around 1905, to Kansas City, Missouri. There, Jacobson ended his schooling after the eighth grade and went to work to help support his family. Jacobson and Truman first met when Jacobson, who was a fourteen-year-old stock boy in a Kansas City clothing store, would take the store's deposits to the Union National Bank, where the twenty-one-year-old Truman worked.[48]

The two lost touch after Truman returned to work on his family's farm in Grandview, Missouri. They met up again in 1917, when Jacobson enlisted as a private in the 2nd Field Artillery of the Missouri National Guard two months after the outbreak of World War I. Truman was first lieutenant of the battery to which Jacobson was assigned. Their regiment was then mustered into the U.S. Army as the 129th Field Artillery. While they were waiting for their overseas travel orders at Camp Doniphan, Oklahoma, Truman asked Jacobson to help him set up a canteen. They collected two dollars from each of the 1,100 men in the camp, which they paid back within six months, generating a $15,000 dividend. The canteen's success was duly noted by Truman's superiors and helped his career in the military.[49] Admiring what he thought was special Jewish business acumen, Truman wrote home to his girlfriend, Bess Wallace, "I have a Jew in charge of the canteen by the name of Jacobson and he is a crackerjack."[50] Teaming up with Jacobson gave Truman a taste of what it meant to be Jewish. Some of the other officers began teasing him, calling him "Trumanheimer" and a "lucky Jew." Truman answered, "I guess I should be very proud of my Jewish ability."[51]

The two men were separated in France, when Truman was assigned to Battery D and promoted to captain. Returning at war's end, Truman and Jacobson both faced an uncertain future. The one commitment they shared was to their sweethearts, Bess Wallace and Bluma Rosenbaum, whom they would marry in 1919. Truman claimed that it was love at first sight when the six-year-old Truman walked into Sunday school and saw the "little blue-eyed golden haired" five-year-old. Bess Wallace was from the top tier of Independence, Missouri, society, while Truman came from a family of farmers and horse traders. As far as the established Wallace family was concerned, the Truman family was below their station in the social scale. Yet by the time Harry and Bess were in their mid-twenties, the tenacious Truman finally won Bess's hand.

Truman was not just marrying Bess. When she was eighteen, Bess's adored and handsome father, David Wallace, having problems with alcohol and mounting debt, shot himself. His wife, Madge Gates Wallace, fell apart, never fully recovering. Bess took on the role of parent to her three brothers and a "semi-parent" to her mother. Whoever married Bess Wallace, wrote Truman's daughter, Margaret, would have had to be prepared to spend a "great deal of time with Madge Gates Wallace." The average man, she thought, would never have had the courage to take this on because her grandmother regarded all of Bess's suitors as thieves "trying to steal her only daughter, the consolation of her tragic life." She especially thought Harry Truman was unworthy of her daughter.[52]

After their wedding, Bess and Harry moved into Madge Wallace's house. Truman tried hard to get along with his difficult mother-in-law. He accepted her rules, since she was both owner and matriarch of the home. Truman's relatives, one of his biographers points out, rarely set foot in the Wallace household.[53] Neither did Truman's Army buddies. Mrs. Wallace's rules included that Jews would not be welcome in her home. Truman obeyed that rule, but outside the house, he was free to do as he pleased. Truman spent many evenings having dinner and playing cards at Jacobson's house and went on fishing and hunting trips with Jacobson and other Army buddies, where he was the cook.[54]

During the war, Truman's canteen with Jacobson had been such a "profitable experience on limited capital," they decided to team up once again in a business venture. Jacobson suggested a haberdashery.[55] By that time, the twenty-eight-year-old Jacobson had had twelve years of experience in the business. The two pooled their savings and opened Truman & Jacobson's Gents' Furnishings in downtown Kansas City. Truman did the bookkeeping and Jacobson the buying. According to Eddie's wife, Bluma, the partners never signed any agreements but relied on their trust in one another, a trust that lasted a lifetime.[56] The business flourished at first, enthusiastically supported by their war buddies, for whom it became a gathering place. But despite their long hours and hard work, the postwar depression of 1921 forced them out of business in a year. Overnight their stock became worth almost nothing. Truman went on to become a judge of the Jackson County Court and Jacobson a traveling salesman.[57]

Eddie Jacobson was not a Zionist, but, like many Jews, the Holocaust made him acutely concerned for "my suffering people across the seas." He had rejected his parents' Orthodox Judaism, and he and his wife, Bluma, joined a Reform congregation, B'nai Jehudah, where he faithfully attended services every Friday evening. During the war, his congregation, led by Rabbi Samuel Mayerberg, actively supported the rescuing of European Jews from Hitler and getting them admitted to Palestine. Like many Reform rabbis in those years, however, Mayerberg opposed the creation of a Jewish state.[58]

Shortly after FDR died and Truman became president, Jacobson opened another store in Kansas City, called Westport Menswear. Now he was suddenly famous. Soon people began calling on the obscure Jacobson. They hoped that he would be willing to use his connection with the president to help them gain office or receive support for various projects. Jacobson declined. He was very protective of his friend, didn't want to be used, and let it be known that he would not ask the president for any personal favors for himself or anyone else.

Jacobson was also approached by Zionists seeking access to Truman. Among the first was Dr. Israel Goldstein, the president of the Zionist Organization of America, who was in Kansas City in May to give a report to the Jewish community on the recent U.N. proceedings in San Francisco. Jacobson turned him down as well. In truth, Jacobson was not particularly well versed in the situation of the refugees or Zionism. But this would change. Jacobson was invited to a meeting at the home of Mr. and Mrs. Ernest Peiser, where he was introduced to the Reform Rabbi Arthur Lelyveld. The thirty-three-year-old rabbi was the director of the committee on Unity for Palestine, an arm of the Zionist Organization of America. His job, according to his son, was to "stump from one major Jewish community to the next converting anti-Zionists and winning over those who had yet to commit themselves."[59] The rabbi had several intense meetings with Jacobson, to whom he presented the Zionist case. Jacobson was so impressed with his arguments that he agreed to take the rabbi with him to see Truman.[60]

Jacobson's education was also aided by his friend and attorney, A. J. Granoff. Granoff was heavily involved in Jewish affairs and had joined the Jewish fraternal organization B'nai B'rith in 1924, eventually becoming president of the Kansas City Lodge and the president of District Grand Lodge No. 2.[61] He met Jacobson in the mid-1930s, when their children attended the same Sunday school at B'nai Jehudah. Their relationship developed further when Jacobson opened his store and Granoff did his legal work. Granoff never sent him a bill, but Jacobson insisted on paying him with merchandise from his store: suits, shirts, and socks.[62] Later, B'nai B'rith's international president, Frank Goldman, and secretary, Maurice Bisgyer, arranged to meet Jacobson through Granoff.[63] This connection would later prove to be invaluable to the Zionist cause.

A year passed before Jacobson made good on his offer to Lelyveld. On June 26, 1946, Jacobson and Lelyveld met with Truman at the White House. Jacobson also brought Charles Kaplan, the vice president of the Shirtcraft Corporation of New York. As Jacobson, Lelyveld, and Kaplan left the White House, Jacobson, speaking for the group, told the press that they had "wanted to clear up several things" with Truman regarding Palestine. Looking at his two colleagues, he joked, "Kaplan sells shirts, I sell furnishings, and the Rabbi sells notions."[64] Despite Jacobson's quip, the meeting was important. Truman spoke to them at length about his views on the Palestine issue. Lelyveld reported to Rabbi Silver that Truman had told them he thought he knew all there was to know about Palestine and wanted to deal only with the operational level, avoiding "long-term objectives as being beset with too many difficulties and complexities."

Nevertheless, Lelyveld came out with "encouraging impressions." Truman seemed committed to action on the 100,000 DPs and had instructed Grady to move on it. Again, Truman said that after the first wave arrived, they would think about the next 100,000. On the discouraging side, the rabbi was concerned that Truman was too impressed by the threats coming from the Arab League. He seemed to be repeating what he had been told by State, that Arab guerrilla warfare could interfere with access to the oil supply lines. Moreover, Truman was upset that both Silver and Wise focused solely on Zionism and seemed not to be concerned with the larger issue of world peace and Soviet expansionism. Truman, they learned, had seen Dr. Judah Magnes and considered him to be "the finest Hebrew" he had met. Magnes, of course, advocated a binational state.

Finally, Truman told them, he did not regard Bevin's outburst at Bournemouth as an insult to him or to America. Bevin, Truman said, had understandably "blown up" because he knew that he himself was "often tempted 'to blow up' because of the pressure and the agitation from New York."[65] Truman had had his own outbursts due to his frustration with the seemingly unsolvable problem of the refugees and Palestine and the pressure on him to find a satisfactory solution. As much as he blamed all the parties involved, often his chief target was the "extreme Zionists." After one such outburst, his aide George Elsey said he knew "it certainly was not an accurate expression of his real feelings, but it was in character. Whenever he felt unduly pressed, he would let off steam verbally or on paper and then return to normal."[66]

After Attlee's May 1 speech and Bevin's statement, Jews in Palestine were devastated. Their hopes had been raised when the Anglo-American Committee had recommended the immediate transfer of the 100,000 to Palestine. Bevin had said he would implement a unanimous report, but now it was clear that he hadn't meant it. The British were making acceptance of the committee's recommendations contingent on a condition that the Jewish Agency could never accept, the dissolution of the Haganah. On June 19, the Haganah reacted. It blew up eight bridges on the Palestine frontier, paralyzing communications with neighboring territories. At the same time, the Irgun kidnapped five British officers and held them hostage.[67] The purpose of the action was to show the Arabs that if they went to war, they would have to fight a disciplined, effective, modern armed force, capable of taking on even Britain's battle-tested troops. "The Jews," a reporter commented, "are no longer the clay pigeons they had been in '36."[68]

"It is an undisguised fact," an anonymous reporter wrote, "that Britain is on the verge of an Anglo-Jewish war. Palestine is a British armed camp. Arab and British sentiment was completely against the Palestinian Jews, while the Jewish sentiment for moderation was evaporating, and the moderate Jews who used to back up Weizmann are diminishing" and moving over to the militant leadership of Ben-Gurion. Though the majority condemned the Irgun and Stern gang, he thought it significant that the Haganah, not the terrorist wing of the movement, had blown up the Jordan bridges.[69]

The Jewish Agency economist David Horowitz returned from England to Palestine "filled with a sense of disappointment, pessimism, dread, and an irksome foreboding of the storm to erupt." He brought back with him substantiated reports of impending British attacks that were scheduled against the Jewish Agency, the Haganah, and in effect the entire Yishuv. He told the Agency leaders that British plans included mass arrests, forced disarming of the Haganah, and detention of Agency executive committee members. The operations could even, he thought, put the entire Yishuv "in danger of collapse." Golda Meyerson (later Meir) told Horowitz they already knew about it. Haganah intelligence had gotten hold of secret British plans, which indicated that British troops planned to move with "one massive and lasting blow."[70] The diplomatic records confirm the reasons for their fear: Attlee had cabled Truman that the British were going to take drastic action in Palestine, necessary because of increasing Jewish terrorism.[71]

By now Palestine had become virtually an armed camp, a military zone of occupation run by the British armed forces. Writing in *The Nation* magazine, the liberal, pro-Zionist editor of the weekly, Freda Kirchwey, reported that Palestine's population was living under "massive rolls of barbed wire." The British Mandate authorities had instituted massive censorship. Political prisoners were held in camps protected by more barbed wire. In effect, Kirchwey wrote, "Palestine is an occupied country from end to end. The Jews and Arabs alike live under military rule while civilian officials take shelter behind sandbags and armed guards. . . . Press censorship is complete. . . . Arrests are frequently made under similar emergency decrees." Military and police forces were so concentrated that she regularly witnessed "convoys of British tanks and trucks moving along the roads holding up civilian traffic." Public barracks were really forts, "concrete structures formidable in size and solidity." In addition, Arab troops from the Trans-Jordan Frontier Force were being used to police towns and highways, which infuriated Palestine's Jews and was "calculated to provoke Jewish resentment."[72]

A few days after Kirchwey wrote her article, all hell broke loose. On the morning of the twenty-ninth (afterwards referred to by the Jews of Palestine as the Black Sabbath), Horowitz awoke to the sound of explosions and gunfire on Tel Aviv's streets. He and all the other residents of the city turned on their radios and heard an announcement repeated in Hebrew, Arabic, and English. The British high commissioner, Sir Alan Cunningham, announced major military operations against the Yishuv, the Jewish Agency, and the Haganah. Buildings were occupied, including the Jewish Agency headquarters. Jewish settlements were searched for arms, and some, such as the largest one,

Yagur, were destroyed. Tel Aviv became a military center occupied by British troops, who marched up and down the streets, sometimes firing rifles into the air to keep the population indoors. Phoning friends, Horowitz learned that two thousand people, including the major Jewish leaders in Palestine, had been arrested. They had made one exception: Chaim Weizmann had been left unharmed in his Rehovoth home. David Ben-Gurion, who was in Paris that day, also avoided arrest and detention.[73]

To make sure Weizmann understood that the British were serious, Cunningham phoned him at Rehovoth. The underground had to cease all military actions against the British, he told him, or "we will destroy Haifa." Referring to the mass rallies in America, he chided Weizmann, "Who do you think will stop them–Madison Square Gardens, mass meetings. . . . Thousands of people are killed daily in China, Calcutta, etc., and does the world get very much excited about it?"

Attlee defended his actions by arguing that the Anglo-American Committee had recommended that the Jewish Agency help curb the terrorist organizations in Palestine and that it had failed to do so. Even so, the British had carried out their raids and arrests with a minimum of force, killing only three Jews and injuring only thirteen. Moreover, Attlee said, the raids had already uncovered arms, ammunition, and explosives belonging to the Jewish Resistance.[74]

Moderates like Weizmann and Goldmann knew that there would be reprisals for terrorist activities and that it had been reckless to give the British an excuse for such actions. Two days before the arrests, Goldmann had phoned Ben-Gurion, urging him to have the Jewish Agency take its own action against Zionist extremists and terrorists, particularly the Irgun and Stern Gang.[75]

The British raid on the Yishuv and the arrests of the Jewish Agency leaders produced outrage in the ranks of America's Jews. Rabbis Silver and Wise called the British force "nothing less than an act of war against the Jewish people." Its very purpose was not to stop terrorism, they charged, but "to liquidate the Jewish national home." Significantly, they did not believe that President Truman had been warned in advance about the raids, a position that only exemplified their naiveté about how U.S.-British relations worked. They found it incredible that the Truman administration "would be accessory to this vicious and tyrannical act."[76]

At the White House, the president received Rabbis Wise and Silver, Goldmann, and the Zionist leader Louis Lipsky. They asked him to intervene and told him they feared the British were out to "destroy generations of labor and achievement of the Jewish pioneers." The group was surprised that the British had carried the arrests out while in the process of negotiating with the United States over the issue of the 100,000 DPs. They told Truman about the arrest of Jewish leaders, the destruction of Yagour, and the herding of Jewish males into detention camps. According to Goldman, Truman prevented the situation from getting any worse by phoning Attlee. He "talked very bluntly" to the prime minister, who then, without consulting the British colonial secretary or the cabinet, gave orders for the army to cease its military action in Palestine.[77]

After Silver and the rest of the group had left, Abe Feinberg came in to see him. Apparently Truman felt that Rabbi Silver's behavior towards him had been especially disrespectful. Feinberg found the president "red in the face." Was anything wrong? Feinberg asked him. Truman answered, "Yes, damn it, the presidency is something to be respected, and that clown had the nerve to shake his finger in front of me. . . . I told him he'd never be welcome here again."[78]

Nevertheless, Truman issued the following official White House statement:

> The President expressed his regrets at these developments in Palestine. He informed the representatives of the Jewish Agency that the Government of the United States had not been consulted on these measures prior to their adoption by the British Government. He expressed his hope that the leaders of the Jewish community in Palestine would soon be released and that the situation would soon return to normal.[79]

Truman was dissembling more than lying. He had told the Zionist representatives that he had not been consulted and they and the public took his statement to mean that the British had acted without his prior knowledge. But Truman had not said that he had not known in advance that the British were about to do something. In fact, Attlee had told him on June 28 that they intended to take some action. Truman had sent a note back the next day "regretting that drastic action is considered necessary by the mandatory government," and expressing his "hope that law and order will be maintained."[80]

It was against this backdrop that the president's cabinet committee, headed by Henry F. Grady, would make its way to London. This new committee was charged with the challenging task of reconciling American and British differences and negotiating an agreement on implementing the Anglo-American Committee's recommendations. Grady had asked Professor Paul Hanna, a Middle East expert sympathetic to Zionism, to be part of the mission. Emanuel Neumann was worried and expressed his concerns to Hanna. "Once more," Neumann wrote to Hanna, "there are new men with almost no background on the question, and from what I hear, not too sympathetic. I understand they want to consider the recommendations 'in the light of American policy in the Near East.' What that policy is, God only knows." The State Department did not see things differently from the British Colonial Office, he observed, and if no action occurred, there would be mass suicides in the DP camps while statesmen "kept on deliberating, investigating, studying and deliberating over all again."[81]

SEVEN

CONFLICT BETWEEN ALLIES: THE MORRISON-GRADY PLAN

On July 12, the president's new cabinet committee's representatives flew to London in the president's official plane. Henry F. Grady, a former assistant secretary of state, served as chair. Joining him were Goldthwaite Dorr, a New York City lawyer temporarily working in the War Department, and Herbert Gaston, a former assistant secretary of the Treasury. None of them had any experience in dealing with the issue of Palestine. Accompanying them were eight expert advisers, who, they hoped, would fill in the gaps in their knowledge. Their mission was to decide how the recommendations of the Anglo-American Committee could be implemented.

The New York Times predicted that although Truman had instructed his committee to seek ways to transfer the 100,000 DPs to Palestine, the British would insist that all of the Anglo-American Committee's recommendations be accepted, not just one. Grady, the journalist Herbert Matthews wrote, "is going to have a tough time of it."[1]

The mood in London was discouraging. The influential news magazine *The Economist* greeted the Americans with sarcasm. "Jews and Arabs and Americans call the tune," the magazine's editors commented, "while the British pay the piper."[2]

The successful resolution of the refugee crisis was a top priority for Truman. This time, he believed, he would finally get the results that had so far eluded him. He assured the Jewish Agency that he had told Grady of the urgency of moving the refugees out of the camps.[3] And at the start of the negotiations, Grady tried to press for the immediate admission of the 100,000 DPs.[4] But the British had other things in mind. They presented the committee with a plan for provincial autonomy, which looked suspiciously like one that had been submitted to the Anglo-American Committee by Sir Douglas Harris of the British Colonial Office and that the Anglo-American Committee had rejected. The plan divided Palestine into two partially self-governing Arab and Jewish provinces with a British-controlled central government. Jerusalem and the Negev would also be under the direct jurisdiction of the British mandatory power.

Ambassador Harriman recognized it for what it was. Nevertheless, he wrote to Byrnes, it "seems to offer the only means now apparent of moving the 100,000 into Palestine in the near future." A positive feature, Harriman pointed out, was that the British hadn't asked for U.S. military aid or participation in a trusteeship.[5] Byrnes was wary. He shot questions back to Harriman from Paris: Would the Arabs have to approve of Jewish immigration? Who would control immigration and land sales? Did the plan envision eventual partition? What most concerned Byrnes was that once again "the transfer [of] these Jews will be almost infinitely delayed." He had hoped the United States and Britain could reach an agreement to start the transfer in the immediate future.[6]

It took only one week for Grady to embrace the British plan. On July 24, Grady informed Harriman about the results of his group's negotiations with the British. Bubbling over with enthusiasm, Grady wrote, "I believe it merits most expeditious consideration and acceptance." Best of all, the British government was anxious to implement it immediately. The agreement, he argued, was the only realistic solution, particularly if Jewish immigration were to be allowed. The agreed-on plan would leave room for a federation if both Arabs and Jews went along. In the interim, both would be segregated into separate areas.[7] Further details of the plan quickly emerged. It provided for a Jewish and an Arab province in a federal Palestine, with both groups having only limited local authority to make decisions. The strong federal government would be controlled by the British Mandate and would have complete authority in Jerusalem, Bethlehem, and the Negev. It would also have jurisdiction over defense, foreign affairs, taxation, and immigration. Most important, admission of the 100,000 European Jews was made dependent upon acceptance of the plan.[8]

Grady argued that the Jews should be pleased, since, in his opinion, it gave them the "best land in Palestine, practically all citrus and industry, most of the coast line and Haifa port." With the exception of the Negev and Jerusalem, he thought, all that was needed to build a national home had been granted. Grady felt that if it was unfair to anyone, it was the Arabs.[9] The plan, which came to be known as the Morrison-Grady Plan after the British head, Herbert Morrison, would, however, not be put into effect until after the British held conferences about it with the Arabs and the Jews. The conference with the Arab League was to take place in mid-September. Grady told the press that he was pleased with the results and found the British to be most cooperative.[10] As he cabled to Loy Henderson, he saw "no practical alternatives to our recommendations." The British would simply not negotiate on any other terms."[11]

On July 25, *The New York Times* reported that in reality the Jews would get only 1,500 square miles under a tight federal rule. The Jewish area, the article pointed out, was less than the British Peel Report had proposed in 1936, when that commission had recommended 2,600 square miles be given the Jews and that was less than the 45,000 square miles allotted to the Jews when they had originally been promised a national home. Moreover, the power in the hands of the British government was greater than that proposed in India, and the British would still have control over Jewish immigration.[12]

Truman had not yet read the *New York Times* story when his secretary of commerce, Henry A. Wallace, called to discuss it. He told the president that the Jews thought the British were trying to set up a Jewish ghetto in Palestine. Truman denied it, saying "It doesn't do anything of the kind–the Jews get the best part of Palestine as their province, . . . 30 miles by 50 miles." Moreover, he believed it set up autonomous Jewish and Arab provinces that both would approve and that the *Times* story was inaccurate. Wallace then read portions of it out loud to Truman, who replied that the only thing Britain would control was foreign relations and taxes. "There is poison being brewed in the New York *Times* story," Truman added. He saw Morrison-Grady as the only solution for Palestine. Grady had been in touch with him and had followed his instructions exactly.[13]

Truman's optimism soon gave way to disappointment. New York Senators Robert Wagner and James Mead came to see him, along with James McDonald. McDonald brought a long memorandum, to which he referred during their meeting. The Morrison-Grady Plan, McDonald argued, was actually a repudiation of the president's own views as well as being completely at odds with the Anglo-American Committee's report. He told Truman that the Morrison-Grady Plan "would establish in Palestine a Jewish ghetto wholly inacceptable [*sic*] to Jews throughout the world and to the conscience of mankind." It made admission for the 100,000 DPs contingent upon agreement on long-term political issues and upon Arab acceptance of the new plan. The Jews would never agree to it since it was a "whittling down of the territory of the Jewish National Home" and a surrender of everything that Balfour had promised them.[14]

Despite the president's good intentions, McDonald bluntly told him, he was "losing everything." McDonald thought that Truman did not understand the essence of the Morrison-Grady Plan. The Jews were so opposed to it, McDonald said, that they would rather not have 100,000 Jewish refugees go to Palestine than acquiesce to Morrison-Grady. "If we get the 100,000 at the price of this," he told Truman, you "will go down in history as anathema." At that point, Truman exploded. He insisted that he was not underwriting anything as a price for getting the 100,000 DPs into Palestine. McDonald did not cave in. If you accept this plan, he told Truman, you "will be responsible for scrapping the Jewish interests in Palestine." Bristling at McDonald's accusation, Truman responded, "Hell you can't satisfy these people. . . . the Jews aren't going to write the history of the United States or my history."[15]

FDR understood the problem, retorted McDonald. "I am not Roosevelt," Truman replied. "I am not from New York. I am from the Middle West." But you can still win Jewish support, McDonald responded. The problem, McDonald continued, was that the president had been badly served and had sent bad men to deal with the British. Grady in particular was completely green on the issue, and had he sent the American members of the Anglo-American Committee instead, they would have refused to go along with any such plan. "You must refuse to be a party to it," McDonald told him. Wagner and Mead blanched, noting that they had never seen anyone who was not a politician speak so frankly to Truman.

At that point, McDonald cited the Anglo-American Committee's recommendations. "I have an obligation to come and tell you what I feel about" Morrison-Grady, he told the president, reminding him that almost 98 percent of American Jews agreed with him. Truman only kept complaining about "how ungrateful everybody was." He kept looking at the clock, and McDonald thought it was time for the group to depart. McDonald concluded that Truman was convinced that Morrison-Grady offered a real solution and did not want it questioned. All he concentrated on was that if the plan were accepted by Britain and the United States, he would achieve immigration to Palestine for the 100,000 DPs. McDonald thought that Truman might have been "shaken in his faith" a bit because he had stood his ground and McDonald had been a member of the Anglo-American Committee whose report Truman had publicly endorsed. McDonald raised the issue of Truman's attitude toward Zionism in his notes, observing that the president "referred only to the Jews generally and not to the Zionists." The problem was, wrote McDonald, that Truman did not always distinguish between them, since the "battle of voices is too much," and even AZEC did not "always speak as one voice." As for himself, he noted that "if a man lives with the Jews and knows their history, he can understand. The Jews are always blamed for the weaknesses of others." On one level, McDonald thought, Truman had understood his arguments, but the problem was that the president was "not a scholar [and] is not interested in following the thing through." The

Jews would have to make a case for what was acceptable to them. Truman had cooled off by the time he wrote McDonald a note. "I hope I wasn't too hard on you," he wrote; "it has been a most difficult problem and I have about come to the conclusion that there is no solution, but we will keep trying."[16]

Truman quickly learned just how unpopular the Morrison-Grady Plan was. Senators Wagner and Taft both blasted the plan on the Senate floor. It meant "deep despair for the million and one-half surviving Jews in Europe," Taft charged. Wagner termed it a "deceitful device." Nine House members, led by Representative Emanuel Celler, went to the White House to protest the plan. Celler and his House colleagues told Truman that acceptance of the plan "would be approving a ghetto in Palestine" and that Britain knew that neither the Arabs nor the Jews would ever accept it. Truman was not happy. Soon he told them that he did not have time to listen, that he knew everything about the subject, and that he was working on a broader plan to gain admission for DPs to go to South America and to British possessions. Finally, he insulted the congressmen by telling them that he understood the visit was political. He knew, Truman said, that they were all up for reelection in the fall.[17]

Truman was correct that political issues were involved. The chairman of New York's Democratic Committee, Paul Fitzpatrick, wired him that if the Morrison-Grady Plan became policy, "it would be useless for the Democrats to nominate a state ticket for the election this fall." He said this "without reservation," Fitzpatrick emphasized, and he could easily substantiate the claim.[18] Truman heard as much from Ed Flynn, the old-style Democratic political boss from the Bronx. Flynn forwarded Truman a lengthy letter he had received from Judge Bernard A. Rosenblatt, a member of AZEC. In it, Rosenblatt argued that the British proposals were an attack on the "honor of our country and its President." As far as he was concerned, the British were playing "the old political game of delay and procrastination, so as to postpone the admission of the 100,000 Jewish refugees until they are no longer on this earth." For the president, he argued, it should be a matter of both "justice and political expediency." Only heroic measures in defiance of the Morrison-Grady Plan would save the New York State Democrats in the coming fall election. Flynn added that what Rosenblatt said about New York's reaction to the new plan was absolutely true.[19] Truman brushed it off and commented that Rosenblatt's analysis "follows the usual line." The problem, Truman explained, was that "the British control Palestine and there is no way of getting One Hundred Thousand Jews in there unless they want them in."[20]

David Niles also urged Truman not to accept Morrison-Grady. Agreeing with McDonald, he told Truman that if he accepted it, he would be accused of giving up everything just for the 100,000 DPs. If he rejected it, his opponents would say he had achieved nothing.[21] Instead, Niles suggested the president recall the American members from London and work with them to come up with another proposal. Niles also suggested that Truman stay neutral until he could convene a meeting of both the Grady team and the American members of the Anglo-American Committee. Then they could try to reconcile their positions before making a statement.[22]

In the meantime, on July 29, Byrnes drew up a statement for the president to issue, accepting the Morrison-Grady Plan as the basis for future negotiations. He advised Truman to issue this statement immediately after Attlee gave a similar one to the British government on Wednesday afternoon.[23] Truman now had to make a decision. The next day, the entire cabinet meeting was devoted to the Palestine issue. Truman came in with a sheaf of telegrams from various Jewish groups and citizens arguing against it. Dean Acheson and Secretary of the Navy James Forrestal wanted Truman to accept the British plan, and Henry Wallace and Secretary of the Treasury John Snyder were opposed. "The whole matter was loaded with political dynamite," Wallace told the cabinet; ". . . the Jews expected more than 1500 square miles; . . . they hoped to be in on a part of the Jordan River Development." After substantial discussion, Truman told them he would wire Byrnes not to go along with Attlee.

Truman then surprised his cabinet, according to Wallace's diary notes, by taking out his anger at the Jews for the sinking of the plan. "Jesus Christ couldn't please them when he was here on earth," Truman explained, "so how could anyone expect that I would have any luck?" Truman then said "he had no use for them and didn't care what happened to them." Wallace responded by arguing that American Jews all had relatives in Europe and knew that most of them had been killed by Hitler's extermination machine. "No other people have suffered that way," he added. Forrestal disagreed, the "Poles had suffered more than the Jews" and again noted the need for Saudi oil should another war break out. Truman had heard enough and answered that he didn't want to handle this from the standpoint of oil "but from the standpoint of what is right." Looking at the map and the territory that would go to the Jews in Palestine if Morrison-Grady were accepted, Wallace could see why Taft had called it a "splintered area." Truman, however, thought the Morrison-Grady Plan was "really fair" and was disappointed that he had to abandon it.[24]

A slightly different version of what happened at that cabinet meeting came from Dean Acheson. Years later, soon after the founding of Israel, a BBC journalist asked Acheson when he thought the last possibility had existed of a common British-American position on Palestine. "The fatal step," Acheson replied, "had been taken . . . at the time of the Morrison-Grady proposals." A cabinet subcommittee consisting of himself, Secretary of War Robert Patterson, and Secretary of the Treasury John Snyder came to the conclusion that the Grady-Morrison Plan was sensible and was a "viable solution of the problem." Acheson acknowledged that American Zionists would have protested but argued that two years remained before the 1948 elections and hence "the outcry could be ignored."

According to Acheson, Truman had approved the subcommittee's recommendation. Just at that moment, Truman received a telegram from James Byrnes, who, in a last-minute change of heart, informed him that to accept the plan would lead to "serious domestic repercussions" and asked Truman to ignore any messages he had received bearing Byrnes's signature that argued a contrary position. He did not explain what had changed his mind or what those "repercussions" might be, but one could surmise they were political.

After the subcommittee's meeting broke up, Truman immediately went to a cabinet meeting. Acheson thought that after he spoke, Truman was ready to follow his arguments and accept the Morrison-Grady Plan. But when the president read Byrnes's telegram out loud to the cabinet members, they all got "cold feet." A voice vote was taken, leaving Acheson alone as the sole advocate. Truman then told Acheson it was up to the State Department to think up something else. Had Byrnes's telegram arrived a few hours later, Acheson later said, the cabinet would have approved Morrison-Grady, and the "history of the Palestine problem during the last two years would have been far different and much happier."[25]

Now it was Acheson's unhappy job to tell the British of Truman's decision. He went to see Ambassador Inverchapel and told him that Truman had concluded that he could not make the statement supporting the recommendations because he did not have support for it at home. Elaborating, Acheson explained that "in view of the extreme intensity of feeling in centers of Jewish population in this country neither political party would support" Morrison-Grady and any statement by the president would be purely personal and misleading. This development, Inverchapel said, would cause an embarrassing situation for Attlee, who had expected an agreement.[26] Reporting back to London, Inverchapel called the U.S. decision a "deplorable display of weakness," which he attributed solely to "reasons of domestic politics." He had spoken to Loy Henderson that evening, who told him, "But for the attitude of the Zionists, [Henderson] declared, there was nothing in the joint committee recommendations which would not have been acceptable to the United States Government."[27]

American observers agreed that politics had played a part. James Reston, Washington's insider columnist at *The New York Times,* reported that Washington political leaders were advising Truman not to accept federation as a basis for negotiation, because they were thinking of the congressional elections of 1946 as well as the 1948 presidential campaign. Reston acknowledged that there were serious humanitarian, strategic, and international questions that had to be considered, but he thought they might be less decisive than domestic political issues. Publicly, only the broad issues were being raised, he wrote; privately, within the White House and Congress, the administration talked about the effect of its decision on the electorate in the key states of New York, Illinois, Pennsylvania, and Ohio. And within those states, the concern was about "the votes of the Jews."[28]

At the end of the month, with a stalemate and no path forward having been decided, Truman recalled the cabinet committee from London. He asked that its members discuss the issues with him in Washington in greater detail, in order to reach decisions that would contribute to a solution.[29] Inverchapel wired London that Truman's decision was seen by Americans as a victory for opponents of British proposals. It was a reprieve from Byrnes's earlier approval of the British plan, which Truman had now repudiated, and the Zionists were taking the political credit for Truman's move.[30]

In any case, the Arabs vehemently rejected the Morrison-Grady Plan. Speaking for the Arab League, Abdul Rahman Azzam of Egypt declared its "unalterable opposition" to what it called a federal solution and partition of Palestine. If accepted, Azzam claimed, it would be nothing less than the foreshadowing of a Jewish state. And as far as negotiating in London, they would not do so if Jews were present at any conference.[31]

Truman felt defeated over the failure of the Anglo-American Committee and Morrison-Grady. His best efforts to find a solution had gone nowhere. The Jewish Agency leader, Nahum Goldmann, was in Paris attending a meeting of its executive committee when he received an urgent phone call from David Niles. Truman, Niles told him, was fed up with both the British and the American Zionists and was "threatening to wash his hands of the whole matter." The only thing that would stop him was if the Jewish Agency came up with an alternative, realistic plan to substitute for Morrison-Grady.

Goldmann and other members of the Jewish Agency executive had been considering distancing themselves from the Zionists' Biltmore Declaration of 1942, which called for a Jewish state in the entire area of Palestine, and adopting the more realistic goal of partition. Unless they accepted partition, they reasoned, Jews were destined to be a minority in an Arab state. Wouldn't it be better to have a smaller state, where they would be in control of their destiny? If Truman actually washed his hands of the Palestine issue, it would be, in Goldmann's eyes, "a worse catastrophe than an open state of war with England."[32]

After a few days of discussion, the Jewish Agency executive adopted a resolution stating that it was "prepared to discuss a proposal for the establishment of a viable Jewish state in an *adequate* area of Palestine." That area would have to have full autonomy, including control over immigration.[33] The executive charged Goldmann with presenting their proposal. The first step was for him to get support for it from the various representatives of the American Jewish community, both Zionist and non-Zionist, and from the government. First he met with Rabbi Silver and laid out the executive's plan. Silver, who resented Goldmann's interference in American Zionist affairs and was a supporter of the Biltmore Program, nonetheless agreed to go along with the majority resolution and not to interfere with his negotiations.

It took Goldmann three meetings with Dean Acheson before he convinced him that partition was feasible and perhaps the only way out. Acheson thought he could support such a plan and advised him on how to proceed. He must convince David Niles, Secretary of the Treasury Snyder, and Secretary of War Patterson to support the partition plan. Goldmann knew that David Niles was very important to Truman and had been one of the moderate Zionists' "best and most loyal friends in Washington." However, he also believed that Niles "is not very much in favor of a Jewish State. . . . He helps us very much, but he never was ideologically a Zionist." It was difficult to sit down and talk to Niles during the day, so Goldmann went to see him at the hotel where he kept a room. According to Goldmann, after a two-hour talk, Niles left convinced that the Agency's new resolution was the only way out of the impasse.[34] Goldmann also convinced Snyder and Patterson as well as Judge Joseph Proskauer, the anti-Zionist head of the American Jewish Committee, to support the partition plan.[35]

Now that Goldmann had gotten the necessary parties to agree, it was decided that Niles and Acheson would present the partition plan to Truman with their endorsement. On August 9, 1946, Goldmann once again went to see Niles in his hotel room. Niles had "tears in his eyes," when he told him "that the President had accepted the plan without reservation and had instructed Dean Acheson to inform the British government."[36] Even though the British rejected it, this proposal marked *a major turning point in the evolution of Truman's view of the Palestine crisis. For the first time, Truman adopted the idea of partition, a view now shared by a consensus of the important Jewish leaders in Palestine and America.*

On August 12, Truman informed Clement Attlee that he had "reluctantly" concluded that he could not support Morrison-Grady in its current form as an acceptable Anglo-American plan. There was so much domestic opposition to the plan, he told Attlee, that it would be impossible to rally sufficient public opinion behind it "to enable this Government to give it effective support." Because of the crisis facing the homeless European Jews, however, Truman assured the prime minister, he wanted to continue a search for a solution. Ambassador Harriman had given Ernest Bevin the details of Goldmann's partition proposal, and Truman hoped the conferences that the British were planning to hold with the Arabs and the Jews could be broadened to include these suggestions. The president also hoped that the British could decide on some course of action that would enable Truman to get the necessary support in America and in Congress so that "we can give effective financial help and moral support."[37]

Meanwhile, in London, Goldmann tried to work his magic on Bevin. Meeting with him on August 14, the Zionist leader once again laid out the Jewish Agency's partition plan along with his arguments in support of it. Bevin brought up the issue of terrorism in Palestine carried out by Zionist factions, the worst of them being the July 22 bombing of the King David Hotel by the Irgun, which killed forty-one people and injured forty-three. Given such provocations, Bevin thought that "Her Majesty's Government had acted with great clarity and coolness." Bevin went on to warn Goldmann that the Jews were "creating a situation in which they were likely to lose the one great friend they always had in the world." When Goldmann responded that the British had not been communicating with the Jewish Agency, Bevin reprimanded him. It was very difficult "to cooperate with people who were riding two horses; i.e., having talks with us on the one hand

and arming their people on the other." Goldmann asked whether they could at least let in 10,000 extra immigrants. Bevin refused, explaining that "we were committed to the White Paper" and that the British had already inflamed the Arabs by agreeing to let more refugees in. "Not a word of appreciation," Bevin complained, "had ever been received from the Jews for this." Goldmann ended by pleading with Bevin that Britain not "drive the Jews again into desperation."[38]

According to Goldmann, after many such meetings, producing hope, then despair and back again, in the end Bevin told him that although he personally wasn't opposed to partition, he could never accept it without the assent of the Arabs. The Jewish Agency sent the chief of its Arab Department to Cairo to talk with "responsible" Arab leaders, who, they reported, favored secret preliminary meetings with all sides to consider Morrison-Grady and other proposals, including partition. These meetings would have to be closed, Goldmann told Acheson, since "fanatical extremists [in] Arab countries would render it difficult for them to begin to openly make compromises."[39]

Attlee informed Truman that the British would be going ahead with the planned London conference, which would include both Arab and Jewish representatives. The basis of the talks, however, would have to be the Morrison-Grady Report, which Truman and the Jewish Agency had already rejected. The plan would be used, Attlee promised, only as an initial basis for discussion, and the British would not hold an "immovable" position in advance of the conference.[40]

The London conference on Palestine started without Zionist participation in September. The Jewish Agency said it had to decline the invitation but would not make its decision public, so as not to embarrass the British government. Its chief reason was the insistence of the British to use the Morrison-Grady Plan as the basis of discussion. Attlee opened the meeting with a speech to the Arab delegates. Speaking candidly, he told them that they would have to agree to a separation of the two communities. Moreover, they could not dismiss any further Jewish immigration into Palestine, which was essential to the situation of the DPs in Europe and the general plight of the Jews. Any solution, he told them, would have to consider the political rights of the 600,000

Palestinian Jews, agreement to Jewish immigration into Palestine, and the establishment of institutions that would enable both Arabs and Jews to govern themselves.[41]

The Arabs would have none of it. They handed Bevin a list of counterproposals. They demanded that the Mandate be ended; that a unitary Arab state in Palestine governed by an elected constituent assembly be established in which the Jews would be limited to one third of the total membership; that the Jews would have all rights consistent with those usually granted to minorities; that Jewish immigration end; that the new Arab governments institute a treaty of alliance with Britain; and that guarantees be given for the sanctity of all holy places.[42] A week after making the initial demands, Azzam Pasha, speaking on behalf of the Arab League at a London press conference, made these demands public. Its plan made clear that an independent Palestine, as they saw it, would prevent any future Jewish immigration into Palestine as well as the sale of land to Jews, except by consent of the majority of the Arab population. As expected, Zionist spokesmen condemned it as a veiled plan to continue the White Paper as well as a refusal to acknowledge the existence of two actual nations existing in Palestine. The Haganah, still underground, issued its own manifesto declaring that "we will not allow Arab effendis to decide the number of Jewish immigrants for Palestine; we will continue to fight to bring our brothers here."[43]

Though it dragged on, for all practical purposes the British attempt to convene a conference on Palestine had come to a dead end. The Zionists refused to participate, the Arabs made impossible demands, and the U.S. government's various attempts to develop a working plan in conjunction with Britain had failed. Harry Truman would now have to decide on an appropriate response and continue to wrestle with the problem.

EIGHT

TRUMAN'S OCTOBER SURPRISE

In the fall of 1946, the refugee crisis continued to haunt Truman, from both a humanitarian and political standpoint. As the head of the Democratic Party, he could not ignore the political fallout from the lack of progress. In a letter to Bess, he complained that all his top men were focused on politics. Henry Wallace had his eyes only on 1948 and the presidential election. Democratic Senator Jim Mead, who was running for governor in New York, had come to him about the New York elections and "shot off his mouth." Worst of all, he confided to his wife, "the Jews and the crackpots seem to be ready to go for Dewey," the Republican gubernatorial candidate in New York. If they did, the Democratic Party would be left grasping at straws. "There's no solution for the Jewish problem," he sadly confided to Bess.[1]

Nevertheless, Truman would have to keep trying. Ever since the Jewish Agency's partition proposal had received a favorable reception in the United States, Truman had been prevailed upon to give it his support. At Goldmann's instructions, Eliahu Epstein had been working at trying to get a statement from Truman.[2] Robert Hannegan joined the effort. Hannegan was a St. Louis politician and Truman's friend, whom Truman had proposed be appointed chairman of the Democratic National Committee early in 1944. Hannegan's subsequent support had helped Truman gain the nod to be FDR's vice presidential choice at the 1944 Democratic convention.[3] Hannegan now sent Truman a note with an attached letter from Bartley Crum. Crum urged Truman to request that the British immediately issue visas for the 100,000 DPs on humanitarian grounds, so as not to prejudice the ultimate political status of Palestine. The president, Crum advised, should also make it clear that U.S. policy was to back the partition proposals of the Jewish Agency. All the Jewish groups–with the exception of the anti-Zionist American Council on Judaism–were making "a united Jewish front" on behalf of partition.[4]

On October 1, Truman met with Max Lowenthal. Truman had first met Lowenthal as a senator, when he had joined Senator Burton Wheeler's subcommittee, then investigating the railroad industry's finances. Lowenthal had served as counsel to the subcommittee. A protégé of Justice Louis Brandeis, Lowenthal had taken Truman to a salon in the form of teas at Brandeis's home in the 1930s.[5] Later, Lowenthal had also been helpful in securing Truman's nomination as vice president, especially in gaining the support of labor. Born in Minneapolis in 1888, Lowenthal went on after college to study law at Harvard Law School, where he became editor of the *Law Review.* After graduating, he became law clerk for U.S. Circuit Judge Julian W. Mack and also worked with Felix Frankfurter on labor cases and on the War Labor Board during World War I. Frankfurter found his young protégé to be a "very sensitive fellow, particularly responsive to cruelty and hardship, and a very fine disciplined brain."[6]

Lowenthal was, Freda Kirchwey wrote in her *Nation* obituary in 1971, "a notoriously modest man. His benefactions, to individuals and institutions, were always anonymous. He did not want it known that most of his many services for government agencies had been performed without compensation . . . [and] he liked to keep up the amiable pretense that he had done nothing of serious public consequence." Over the years he worked as counsel to various government commissions and as a consultant for individuals in all the branches of the government. After World War II, he was legal adviser to General Lucius D. Clay, the high commissioner of Germany. Later, in 1950, Lowenthal wrote the first critical book on J. Edgar Hoover and the FBI, *The Federal Bureau of Investigation,* an act that infuriated Hoover and led to

Lowenthal's being called to testify before the House Un-American Activities Committee.[7]

"Is the situation as bad there as I have been told?" Truman now asked him, referring to the plight of the European DPs. It would not be so bad, Lowenthal answered, "provided that they could have an exit soon and . . . be given hope." The issue, Truman told him, was being taken up by the Jewish organizations in America, and he was working on the problem right at this moment. Then Lowenthal turned to domestic politics. The important thing politically, he emphasized to Truman, "was to have authorization before the election for the admission of a large number of the displaced persons in Palestine." If they managed to get this done, Lowenthal was certain, "it would bury the Republican party in this election." The situation in New York was very bad, he told the president, "and we need real help."[8]

Around the same time, Abe Feinberg went to see Truman at the White House. The wealthy businessman and Democratic contributor had met then–Vice President Truman in 1945, when Robert Hannegan had introduced them at a cocktail party in New York City. When Feinberg asked Truman how he would like to be addressed, Truman joked, "Call me Senator, I liked that job best." They got to talk further over dinner, and Feinberg found that Truman "was a very warm man, if he liked you." When Truman became president, Feinberg was welcome at the White House, where he got to know David Niles and Truman's appointments secretary, Matt Connelly.[9]

Feinberg had met Ben-Gurion in 1945 during one of Ben-Gurion's visits to the United States. He had listened carefully to the scenario that Ben-Gurion laid out about the probability of Arab attacks on the Yishuv and responded to his call to create an American branch of the Haganah that would raise money and supplies. Feinberg had been horrified by the Holocaust, strongly believed that the Jews should be able to defend themselves, and wanted to see the conditions in the refugee camps for himself. With the Haganah's help, he went to visit the Jewish DPs in various European camps, where he saw them being held in unbearable conditions. Almost everyone he spoke to wanted only to go to Palestine. Feinberg then took direct action: he became involved in purchasing and outfitting ships that could take the refugees from France and Italy to Palestine. He was actually in Palestine when the British made the decision to round up the Jewish leadership. They also arrested him, on the accurate grounds that he was smuggling out British military information to give to the Haganah. According to him, it was only because of his personal relationship with President Truman that the British military authorities released him. When he returned to the United States, he gave David Niles and Truman a firsthand account of the repressive measures the British were using.

Now Feinberg gave Truman some advice: if he wanted to make his position known to the Jewish people, a good time to do it would be "just before the holiest day in the Jewish year, Yom Kippur, the day of atonement. Even unobservant Jews, such as myself," he told the president, "tend to go to the synagogues on the night before, which is a most somber night." The services included a liturgy, and he told Truman that "the rabbis use their most dramatic efforts in the sermons. . . . if you will make the announcement before that night, every single Rabbi in every single synagogue will broadcast what you say." Truman, he added, should forget the newspapers and other media; he would be getting word directly to the Jewish people.[10]

On October 3, the British announced that they were going to postpone the conferences with the Arabs and the Jews (with whom they were still negotiating) until December 16. This meant yet more delays on the refugee issue. That evening Acheson sent for British Ambassador Lord Inverchapel and gave him a copy of a statement that Truman planned to make public the next day. Acheson told him that "elements within the Democratic Party" had "blown up" when news of the British delay was announced and that Truman anticipated political attacks from Republican candidates in the upcoming midterm New York elections.[11] Truman then sent Attlee a direct communication. In view of the deep sympathy Americans had for the Jewish victims of Nazism, he wrote the prime minister, he found it necessary to make a further statement on the problem immediately.[12] Truman included in his telegram the text of the statement he would make public on Yom Kippur.

Truman began his statement by saying that he had learned with deep regret that the Palestine meetings in London had been postponed. In light of this, he felt, it was appropriate to examine the record of his administration's efforts and to state his views on the situation. Truman reminded Americans of his many attempts to gain admission to Palestine for the 100,000 Jewish DPs, recommended to him by Earl Harrison's investigation of the DPs and the unanimous report of the Anglo-American Committee of Inquiry. He had then formed a cabinet committee on Palestine, which had traveled to London to see how the Anglo-American Committee's recommendations could be best implemented. The result had been the Morrison-Grady scheme for provincial autonomy, which would lead to either a binational state or partition. This plan, he reminded the country, was strongly opposed by a majority of Americans and faced major opposition in Congress. He therefore could not give his support to it.

The British then announced that they would hold a conference on Palestine in London. With the participation of both Arabs and the Jews, they would see if they could come to some kind of agreement using the Morrison-Grady scheme as a basis of negotiations. Truman continued:

> Meanwhile, the Jewish Agency proposed a solution of the Palestine problem by means of the *creation of a viable Jewish state in control of [a part of] Palestine* instead of the whole of Palestine. It proposed furthermore the immediate issuance of certificates for 100,000 Jewish immigrants. This proposal received widespread attention in the United States. . . . From the discussion which ensued *it is my belief that a solution along these lines would command the support of public opinion in the United States*. I cannot believe that the gap between the proposals which have been put forward is too great to be bridged by men of reason and goodwill. To such a solution our Government could give its support.[13] (Our emphases.)

When he learned what Truman was going to say in his statement, Attlee asked him to postpone it until he had time to meet with Foreign

Secretary Bevin.[14] Truman immediately let Attlee know that he would not comply with his request and that he felt it "imperative" to make it public as scheduled.[15] The usually mild-mannered prime minister, *Time* reported, was in a "towering rage."[16] Reiterating that he was in the midst of meeting with Jews and Arabs in an attempt to defuse the dire circumstances in Palestine, Attlee told Truman he was stunned that Truman did not give his foreign secretary "even a few hours grace," even though Britain had "the actual responsibility for the government of Palestine." And, he added, he was "astonished that you did not wait to acquaint yourself with the reasons for the suspension of the conference with the Arabs." Truman, he claimed, obviously had incorrect information about the events. Conversations with leading Zionists were taking place informally with a view to their participation that showed good prospects of success. With sarcasm, Attlee concluded that he would wait with interest to learn what the imperative reasons were that compelled Truman to issue his statement when he did.[17]

The next day, Inverchapel had an answer for his prime minister. In his opinion, Truman would not hold up his statement because Dewey was preparing to issue a statement of his own on October 6 "designed to catch the whole Jewish vote in the five eastern states." That is why, he explained, Truman "dare not keep quiet."[18] This became the stock answer as to why Harry Truman issued his Yom Kippur statement, a statement of great importance because it was the first time an American president had mentioned partition as a solution to the Palestine situation. *He did it to gain the Jewish vote.*[19]

Truman further laid out his reasons, and his frustration with British policy, in his response to Attlee's letter. It had been more than one year, Truman emphasized, since he had brought the Harrison report to the British government's attention. Yet even after the recommendation of the Anglo-American Committee nothing had been accomplished. He had tried to be restrained, not doing anything to interfere with British attempts to carry on negotiations on the issue. Now he had no alternative but to "express regret at this outcome." The feeling of despair and hopelessness faced by the Jewish DPs was intensified with the approach of Yom Kippur, and he told Attlee he was certain that "you will agree that it would be most unfair to these unfortunate persons to let them enter upon still another winter" without their being allowed to proceed to Palestine, "where so many of them wish ardently to go." He realized that Britain alone had the mandatory responsibility for Palestine, but he reminded Attlee that the purpose of "the Mandate was to foster the development of the Jewish National Home," a step in which the United States maintained "a deep and abiding interest." That had no meaning without new Jewish immigration and settlement into Palestine.[20]

Dean Acheson denied that the statement was politically motivated. Truman was very serious about resolving the refugee issue and had been working at it for a whole year. Acheson supported the president's desire to give the speech and had helped him draw it up. He thought it was important that Truman clarify his views and recount his actions for the American public and the international community. Because Yom Kippur was such a major day in the Jewish religion, Acheson argued that Truman "chose it as a fitting occasion to announce that he would continue his efforts for the immigration of the one hundred thousand into Palestine," as well as support "some plan for Palestine based upon partition." Since that time, Acheson wrote, "the statement was attacked . . . as a blatant play for the Jewish vote in Illinois, Ohio, Pennsylvania, and New York in the congressional elections only a month away and an attempt to anticipate an expected similar play by Governor Dewey." He did not believe that it had had any such purpose. Acheson's bottom line: Truman "never took or refused to take a step in our foreign relations to benefit his or his party's fortunes. This he would have regarded as false to the great office that he venerated and held in sacred trust."[21]

Obviously, political considerations were important, and those around him constantly reminded Truman of the political impact of his actions. The timing of an important presidential statement could be crucial. But in the broad sense, Acheson's point holds. It was cathartic for Truman to lay out all he had tried to do for the DPs and to try to facilitate some kind of settlement in Palestine, even if his efforts had not yet met with success. Harry Truman did not say anything he didn't believe. He was still trying to get the 100,000 DPs out of the camps. To his regret, the Anglo-American Committee and Morrison-Grady recommendations had been shot down, but now there was a consensus around partition, which he acknowledged. Truman's words had been based on his assessment of what was needed to break the deadlock on Palestine, not because of the sole desire to get the Jewish vote. But if the timing of the statement helped the Democrats at the polls, all the better.

That the refugee issue was very much on Truman's mind is revealed in a letter he wrote to Senator Walter F. George of Georgia the day after he delivered his statement. George wrote to him about another matter but added his view on Palestine: "It would be decidedly unwise for our government to take any position with reference to Palestine which will call for a large appropriation of money or especially the use of American troops in Palestine." Truman answered, "I sincerely wish that every member of the Congress could visit the displaced persons camps in Germany and Austria and see just what is happening to Five Hundred Thousand human beings through no fault of their own." He continued, "We must make every effort to get these people properly located." Only 20 percent of them were Jewish, he noted, and the rest came from throughout Eastern Europe. "There ought to be some place for these people to go," Truman repeated. Addressing the issue of the Jewish refugees, he added, "There isn't a reason in the world why one hundred thousand Jews couldn't go into Palestine" and that the United States couldn't allow others to come to America. As he wrote, he noted the government was trying to arrange for 100,000 to go to South America. He told the senator, *"I am not interested in the politics of the situation, or what effect it will have on votes in the United States. I am interested in relieving a half million people of the most distressful situation that has happened in the world since A. Hitler made his invasion of Europe"* (our emphasis). Expressing his frustration at the failure of Congress to loosen America's immigration laws, Truman reprimanded the senator, "Your ancestors and mine, if I remember correctly came to this country to escape just such conditions."[22]

At any rate, Truman's statement did not help the Democrats, who were trounced in the New York elections. Governor Thomas Dewey not only won, he received 650,000 more votes than Senator Mead, who had hoped that a statement by the president would turn the tide in his favor. And Herbert Lehman was bested by Irving M. Ives in his race for the Senate by 250,000 votes.[23] New York City, the Jewish capital of the world and a bastion of FDR's old labor-liberal coalition, had gone with the rest of the nation to create a Republican Congress. The vote clearly indicated a rejection by the nation's voters, including those of New York, to the traditional liberal agenda of the East Coast Democrats.

The Saudi monarch wrote an irate letter to Truman protesting his statement. Ibn Saud reminded Truman that his position had not changed since he had written to FDR claiming the Arabs' "natural rights" to Palestine and declaring the Jews as "only aggressors, seeking to perpetrate a monstrous injustice . . . in the name of humanitarianism," who would later try to get their way by "force and violence." Their demands, he told the president, included not only Palestine but all the Arab lands, including their holy cities. Because of this, Ibn Saud said, he had been astonished to read Truman's Yom Kippur statement, which, if put into policy, would "alter the basic situation in Palestine in contradiction to previous promises." He did not understand how the American people could support "Zionist aggression against a friendly Arab country."[24]

Truman, writing in cordial and appropriate diplomatic niceties ("I feel certain that Your Majesty will readily agree . . ."), did not seek to assuage Ibn Saud as FDR had. The Jews of Europe, the president reminded him, "represent the pitiful remnants of millions who were deliberately selected by the Nazi leaders for annihilation," and many of them "look to Palestine as a haven," where they could assist "in the further development of the Jewish National Home." It had always been American policy to support such an objective and therefore was only natural that the United States would favor the entry of Jewish DPs into Palestine. For that reason it had been trying to get the British to agree to the immediate entry of 100,000 Jewish DPs into Palestine. His statement of October 4, he told Ibn Saud, was in no way inconsistent with any previous American policy statements. "I am at a loss," he wrote, "to understand why Your Majesty seems to feel that this statement was in contradiction to previous promises or statements made by this Government." It did not in any sense represent a contradiction to previous promises made by the United States, nor was it "an action hostile to the Arab people."[25]

Truman was facing burnout on the Palestine issue when, at the end of October, he received a letter written earlier in the month by Edwin W. Pauley, the Democratic National Committee treasurer and wealthy California oil magnate. Pauley enclosed a report by Joseph Dubois on the situation facing the Jews still in Europe that he had helped Dubois edit but that bore Pauley's name as author. The report seconded the observations made a year earlier by Harrison, though in some respects it was even stronger. "The more one hears about the wanton murder of the Jews by the Germans," Pauley wrote, "the more one wonders why this great crime has not shocked the conscience of mankind more than it has." Pauley's report concentrated not on the situation of the DPs, but on the nature of the crimes of the Nazis. One report Dubois had seen, signed by Heinrich Himmler, bore the account of "progress" made in one day in Warsaw: "Jews disposed of–59,340–our losses none." "We cannot bring five million dead bodies to life," Pauley told Truman, ". . . but we can . . . make certain that the over one million European Jews who survived the Nazi terror are given a chance to live." The Allies had saved them from death, but they "have not yet given them a chance to live." Like Harrison and Crum, Pauley asked for the immediate settlement of the 100,000 DPs in Palestine, as well as financial aid and economic assistance from Germany, given as reparations, restitution, or an export payable in currency.[26]

When Truman had read the Harrison Report, he had been both shocked and deeply moved. Now his response changed. Truman was not happy to get Pauley's report, which had taken almost two weeks to reach his desk. "I am sorry you took the trouble to send me the Jewish report," he wrote Pauley, "although it was an interesting one. "I have spent a year and a month trying to get some concrete action on it."

The British were "muddling the situation," Truman added, "but the Jews themselves are making it almost impossible to do anything for them. They seem to have the same attitude toward the 'under dog' when they are on top as they have been treated as 'under dogs' themselves," which he attributed to "human frailty." To Truman, the Jews were acting in much the same way as big business and labor in the United States. The only one suffering, he thought, was "the innocent bystander who tries to help," and one could guess who that was. As for himself, he told Pauley, he would spend the rest of his time as president working for the best interest of the whole country, a phrase that indicated that he thought following the Zionist agenda would be serving a particular interest.[27]

Although the British generally were attributing Truman's statement to electoral politics, Hugh Dalton, the chancellor of the Exchequer, visiting America in October, had his doubts. "Truman isn't just electioneering in this endless repetition on the one hundred thousand," he wrote in his diary. "This is part of the general outlook of Americans of both parties."[28]

As December approached, a meeting of foreign ministers was scheduled to take place at the Waldorf-Astoria hotel in New York City, and Ernest Bevin would be attending. The foreign secretary had been briefed on what to expect on the Palestine issue in the United States. Ambassador Inverchapel had sent him a report, anonymously written, titled "Jewish Affairs in the United States," providing him with an analysis on the one major issue–Palestine–that was increasing tension between the two Western allies. Zionist sentiment, Bevin was told, was concentrated primarily on the two coasts and had general support among liberal intellectuals–professors, teachers, social workers, and the like–all people whose humanitarian concerns motivated them to act on behalf of the remaining European Jews. In the other regions, the Midwest and the South, any concern with Palestine was either absent or tainted by very real anti-Semitism.[29]

Moreover, the report's unnamed author claimed that Truman's motivation on behalf of the Jews was "prompted by domestic politics, rather than innate conviction." Truman would have supported

Morrison-Grady but had been told that all support for Democrats running for reelection would vanish if he did. Truman was "so irritated and exhausted," it went on, that he "went on a cruise, mainly to escape from the storm." So strong was the pro-Zionist sentiment that newspapers would not print Arab League advertisements. And the picture painted of Arabs was that their countries were controlled by "feudal overlords" and that Arabs lived amid "dirt, disease and ignorance." Commentators always seemed to portray Palestinian Jews as progressive and modern, a people whose young men had fought with the Allied powers during the World War. Arabs, in contrast, were being portrayed as backward and pro-Axis, as the wartime role of the Grand Mufti proved.

The writer maintained that Truman's sentiments, whatever its causes, were genuine and could not be ignored by Britain, if the British government wanted good relations with the United States. The British could not feel happy about the election results either. The Republican victory in Congress meant that now the Republican Party would promise Jews the moon, since their defection in New York to the Republican candidate obliged the Republicans to cater to them. That meant the Democrats would have to go even further to gain back their vote. "Both parties," it concluded, "may overestimate the importance of Palestine to the Jewish electorate, but in present circumstances will not feel able to take a chance on it."

Thus prepared, Bevin anticipated meeting with James Byrnes and perhaps the president to discuss the issue. Bevin was not very popular in New York, as he quickly found out. New Yorkers had not forgotten his unfortunate remark that the only reason America wanted to get the Jews into Palestine was that they didn't want any more of them in New York. They also blamed the British government for the continued suffering of the Jewish DPs still in the detention camps–more than a year after their liberation. Bevin was taken aback by the hostility shown him and, by extension, his Labour government on the part of American organized labor. Dockworkers from the Communist-led National Maritime Union would not touch his luggage. When he attended a football game, the crowd booed when his presence was announced on the loudspeakers. Bevin was also upset to find full-page ads in the New York papers from Zionist groups, loudly denouncing him for Britain's immigration policy in Palestine, as well as its repressive military action against the Yishuv.[30] Looking at New York's liberal, pro-Zionist tabloid the *New York Post,* he saw an ad from the United Zionist Revisionists of America, one of the militant factions on the hard-line Zionist right wing that supported the Irgun. "We want Mr. Bevin to know," it stated, "that many Americans, irrespective of race and creed, and above all, the overwhelming majority of American Jews . . . do not believe that he is entitled to the traditional hospitality of this great city of New York. We believe he deserves to be told the plain truth: 'If anyone is not wanted here, thou are the man.' "[31]

On December 8, Bevin finally had the chance to meet Truman at the White House. The president informed the foreign minister about his meetings with Ben-Gurion and Nahum Goldmann but quickly turned to complaining about the pressures put on him by American Zionists. "I can get nowhere with Dr. Silver," Truman said; "he thinks everything I do is wrong." In contrast, Truman agreed with Bevin's observation that Chaim Weizmann was "the most intelligent of all the Jewish leaders." Bevin was most sympathetic to Truman's plight. The Jews, he told the president, "somehow expect one to fulfill all the prophecies of all the prophets. . . . I tell them sometimes that I can no more fulfill all the prophecies of Ezekiel than I can those of that other great Jew, Karl Marx."[32]

The other problem they had, Bevin continued, was with the Arabs. The truth was that "they would not be able to maintain the status quo in Palestine as they wanted." It would be easier for him to help move towards a settlement, Truman responded, now that the midterm elections were over. Truman, the recording secretary wrote, "then went out of his way to explain how difficult it had been with so many Jews in New York." "If the British Government reached an accord over Palestine," Truman said, "the U.S. Government would be very pleased to give any help they could, including finance." Bevin was not optimistic. "It would never be possible to get Jews and Arabs to agree with each other over this," he answered. The two leaders concurred that their countries had possibly made a major error: both had given contradictory pledges to both sides regarding the dispute. Their aim, Bevin replied, had to be to try to narrow the difference so that they could arrive at a solution both Jews and Arabs could live with. Finally, Bevin turned to the issue of the 100,000 DPs and Jewish immigration. The problem was that everyone was concentrating on Palestine, since the European Jews "had nowhere else to go." If only they had another destination, he pleaded with Truman, "this would reduce the tension."

A week after Bevin's visit, Truman and Acheson met with one of the most militant of Ibn Saud's sons, His Royal Highness Prince Faisal. Faisal, Acheson wrote, "striking in white burnoose and golden circlet, which heightened his swarthy complexion, with black, pointed beard and mustache topped by a thin hooked nose and piercing dark eyes, gave a sinister impression, relieved from time to time by a shy smile."[33] Faisal told the president that the Zionist goal was to take away the land from those who had lived on it for centuries. His goal, Truman replied, was simply to help create peace in the region and a fair and just settlement of the Palestine issue. Moreover, Truman answered, "the plight of hundreds of thousands of displaced persons in Europe . . . must appeal to all men of good will." He was trying to get Congress to allow many of them to come to the United States and was hoping others would go to South America and South Africa. But some of them, he told Faisal, "desired to go and . . . could well be received in Palestine." To that, Faisal argued that they were "bad people," which had been proved by the terrorist attacks against the British in Jerusalem. "The pitiful remains" of the Jewish people, the president responded, could not be among those of whom Faisal had just spoken. When he spoke as president, it was "for the oppressed who had suffered so cruelly before and during the war and who were now seeking a home." Faisal argued that the real problem was their desire to create a Jewish state, when Palestine should become independent and controlled by the majority, who happened to be Arab. Until then, "immigration should cease." Truman ended by striking a positive note, ignoring his last remarks and saying he agreed with his sentiment that Jews and Arabs could live together in the region.[34]

Listening to the two talk, Acheson observed that "their minds crossed but did not meet." Faisal was concerned with the Near East and Truman with the plight of Europe's dispossessed Jews. Both men, he thought, could not grasp "the depth of the other's concern; indeed each rather believed the other's was exaggerated." The meeting ended with platitudes by both, Acheson thought, which the president seized upon to proclaim an agreement. Acheson found the meeting disturbing. Faisal impressed him "as a man who could be an implacable enemy and who should be taken very seriously."[35]

While the Arabs presented a defiant united front, the Jews, holding their twenty-second World Zionist Conference in Basel, Switzerland, struggled to formulate their position. The first world conference to be held since 1939 was a grim reminder once again of the decimation of European Jewry. The majority of delegates were from Palestine and America because most of the European delegates who had previously filled the seats were gone. The conference was taking place against the backdrop of turmoil in Palestine as the British used repressive measures against Zionist terrorists, who were employing road bombs, explosives, and kidnapping and even hanging of captured British troops in a dangerous cycle of violence.

The mood at the conference was militant. The Jews of Palestine considered Britain's actions to stop the refugees from landing in Palestine illegal. Its mandate was to facilitate the building of the Jewish national home, not to prevent Jews from reaching it by detaining them and shipping them to camps in Cyprus. Moreover, intelligence suggested that the British army was gearing up for a war against the Yishuv and its leadership. Beloved as Chaim Weizmann was, his continual pleas for moderation and patience were wearing thin. As he himself acknowledged at the meeting, the absence of "faith, or even hope, in the British Government" led Palestine's Jewish leaders to "rely on methods never known or encouraged among Zionists before the war." Some may have called their tactics resistance; others defense; still others a needed offensive. For Weizmann, whatever terms they used, the reality was that of a call to fight with arms against British authority in Palestine.

Weizmann rejected terrorism. For his position, Weizmann said, he had become "the scapegoat for the sins of the British Government." Not only was it tactically wrong, he argued, it was philosophically immoral. It was not the Jewish way. His words to the conference were bitter, passionate, and eloquent:

> I warn you against bogus palliatives, against short cuts, against false prophets. . . . If you think of bringing the redemption nearer by un-Jewish methods, if you have lost faith in hard work and better days, then you are committing idolatry and endangering what we have built. . . . Zion shall be redeemed in righteousness and not by any other means. . . . Masada for all its heroism, was a disaster in our history. It is not our purpose or our right to plunge to destruction in order to bequeath a legend of martyrdom to posterity. Zionism was to mark the end of our glorious deaths and the beginning of a new path, leading to life.[36]

Weizmann had carried the struggle far and would be called upon again, but Rabbi Silver, Ben-Gurion, and their supporters believed he was obstructing progress and had to go. The issue that allowed this split to surface was whether the Zionist conference would endorse formally taking part in the British conference on Palestine's future in London. Weizmann's motion to participate was defeated by a small majority.

Silver told the conference that the Jewish Agency's office in Paris had been derelict in proposing its partition plan without having put the issue to a vote at the World Zionist Conference. Nahum Goldmann came under particular criticism by Silver, since he had previously agreed never to tell any American government official that AZEC would support partition. But much of this opposition to partition and calls for fidelity to the Biltmore Declaration was hyperbole. Ben-Gurion and Silver both spoke out against partition publicly, though privately they were willing to consider it as the basis of a settlement. They adhered to the position that if they gave their support to partition, the Arabs would certainly reject it and the British would whittle it down further. You could always give up something, but you could not get back what you had already relinquished.

The conference voted Rabbi Wise out of the leadership and made Rabbi Silver president of the American branch of the Jewish Agency. Out of respect for Weizmann and his years of service to the movement, however, the conference left the World Zionist Organization's office of president vacant. It was clear enough that Silver and Ben-Gurion were now in command.[37] Although Weizmann and Wise were removed from power, an observer reported to the State Department that Ben-Gurion and Goldmann, with the support of Moshe Shertok, were able to elect an executive board of which a majority actually supported partition as well as attending the forthcoming London conference.[38]

At home, a major change was about to take place in the State Department. Secretary of State James Byrnes had handed in his resignation. The president had not accepted an earlier resignation letter from him, but this time he did.[39] In his private diary, Truman noted, "Byrnes is going to quit on the tenth and I shall make Marshall Secretary of State. . . . Marshall is the ablest man in the whole gallery."[40] When the papers broke the story, Truman reflected that he was "very sorry Mr. Byrnes decided to quit. I'm sure he'll regret it–and I know I do. He is a good negotiator–a very good one." Referring to Byrnes's doctor's recommendation that he retire for reasons of health, the president wrote, "I don't want to be the cause of his death. . . . So much for that." His regret was offset by the news that General George C. Marshall was confirmed with the unanimous consent of the Senate, "a grand start for him." Marshall, Truman thought, was "the greatest man of World War II," who had managed to get along with "Roosevelt, the Congress, Churchill, the Navy and the Joint Chief[s] of Staff." Alluding to all the troubles he had had with State, he wrote, "We'll have a real State Department now."[41]

The atmosphere in London was dark and dreary as the British finally resumed their conference with the Arabs on January 27. England was in the midst of a coal crisis, which meant that buildings had no electricity and people were working by candlelight. It literally was "darkness at noon," observed Emanuel Neumann. Elevators did not work, factories were at a standstill, and millions were out of work.[42] The British-Arab talks were generally gloomy as well. Faris al-Khoury, the head of the Syrian delegation, informed Bevin that the Arabs had already submitted their views to the British government and did not want to have any more discussions until they had received a response to their proposals. At the talks, the Palestinian Arabs were represented by the Arab Higher Committee, led by Jamal Husseini, who was operating under the direct supervision of the Mufti, who was giving him orders and directives from his exile in Egypt. There would be no compromise forthcoming from Jamal Husseini. Azzam Pasha and some other members of the Arab League did not even attend, because they believed that the conference was bound to fail. Indeed, their own private view was that Britain had secretly decided upon partition as a solution, which to them meant "violence and law-breaking" since it proved that Britain had yielded to Zionist demands.[43]

Meanwhile, the Jewish Agency and the British government were carrying on private negotiations to see if they could form a basis for more formal talks. The British were represented by Colonial Secretary Arthur Creech-Jones, Hugh Dalton, the Labourite Aneurin Bevan, Prime Minister Attlee, and Herbert Morrison. The Jewish Agency contingent included David Horowitz, Moshe Shertok, Nahum Goldmann, Aubrey (later Abba) Eban, David Ben-Gurion, and the Haganah chief, Moshe Sneh. The Zionists disagreed among themselves. Sneh favored an uncompromising demand for all of Palestine, as outlined in the Biltmore Declaration. Goldmann supported partition, which he saw as the last hope for any kind of Jewish state. Ben-Gurion and Shertok were ready for armed action against the British in Palestine but not before publicly seeking a peaceful resolution. But even though they would support partition, they did not want to come out for it at the start.[44]

When the Zionists sat down with the British at the Colonial Office for informal talks, Bevin declared that he was against partition, which he saw as not being in the Mandate and unfair to the Arabs. David Horowitz thought that the foreign secretary had really lost any interest in compromising with the Jews when he saw how uncompromising the

Arabs were. The Arabs demanded that the British leave Palestine, preferring to solve the Palestine problem by way of an armed conflict between Arabs and Jews. Bevin's hope, Horowitz thought, was to scare the Jewish Agency into submission, out of fear of having to face the Arabs alone. Finally, Horowitz decided that what Bevin really wanted to do was to rid Britain of the problem and put Palestine into the hands of the United Nations.[45]

On February 7, Bevin made his last suggestion for a solution. He proposed a modified version of the Morrison-Grady Plan as the only realistic alternative. It was a variation on a binational state: division of Palestine into Arab and Jewish provinces, which would have local self-government. Britain would remain as the mandatory for five years. Immigration would be permitted for the first two years at the rate of 4,000 a month. The Arabs would have to be consulted about any further immigration, but the final decision would remain with the high commissioner and the Trusteeship Council of the United Nations. The Arabs and the Jews both rejected the plan.

There was much talk but no compromise and no solution. The Jewish Agency did not officially attend the conference, and its representatives insisted that any plan they could accept would have to leave open the eventual creation of a Jewish state. The Arabs made it clear that they would accept nothing less than an Arab Palestine, with a Jewish minority living under its laws. "We have now decided," Bevin wrote to Marshall, "that a time has come when the peoples living in Palestine must be made to accept responsibility for their own fate. We cannot," Bevin stressed, "go on forever maintaining an alien rule over that country."[46] The costs of doing so were becoming too steep. By the end of January, when Jerusalem stood at the boiling point, the British government announced that it would remove all women, children, and nonessential males working for Great Britain from Palestine, in order to free the army to devote all its time to restoring order.[47]

On February 18, Ernest Bevin announced in the House of Commons that Britain was referring the Palestine question to the United Nations without recommendations. He had once said he would stake his political future on his ability to solve the Palestine issue. By his own terms, he had failed. Bevin did not, however, want to take the blame. Seeking a scapegoat, he found an easy one: Harry S. Truman, his administration, and the U.S. government. U.S. policy, he told the British parliamentarians on February 25, had "set the whole thing back." The United States, he charged, seemed not to realize that Britain was the mandatory power and thus had the responsibility for issues such as immigration. "If they had only waited to ask us what we were doing we could have informed them but instead of that a person named Earl Harrison was sent out to their zone [of Germany]." The Harrison Report had "destroyed the good feeling which the Colonial Secretary and I were endeavoring to produce in the Arab states." Britain had been magnanimous. Even though it held the Mandate for Palestine, its government had invited the United States to join it in establishing the Anglo-American Committee of Inquiry. Then Bevin had been criticized by Americans for not accepting their final report. Yet, he argued, "none of the report was accepted by the United States except one point, namely, admission of 100,000 immigrants." He had pleaded with James Byrnes to dissuade Truman from issuing the demand for their admission to Palestine; his attempt had been futile. Then, just as he was on the verge of success in negotiating with the Arabs and the Jews, Truman had issued his declaration demanding their admission, thus spoiling the talks in progress with Jewish leaders. Bevin told Byrnes, "I believed we were on the road if only they would leave us alone." I begged him that Truman not make his statement public, he told Parliament, but was then told by Byrnes that the president's act was purely political: if he did not issue it, "a competitive statement would be issued by Mr. Dewey." Thus the British problem had been made "the subject of local [American] elections."[48]

Bevin's speech, Truman later wrote, was "a very undiplomatic–almost hostile–statement for the Foreign Secretary of the British Government to make about the President of the United States. He knew this had been my position all along." Therefore, Truman explained, he was "outraged by Mr. Bevin's unwarranted charge." Truman had his press secretary, Charlie Ross, draw up what Truman called a "very moderate, entirely impersonal statement that pointed out that the matter of getting one hundred thousand Jews into Palestine had been the cornerstone of our Palestine policy since my first letter to Attlee in August 1945." In addition, Truman condemned Bevin's February 18 statement, in which he had said of the refugees that "after two thousand years of conflict, another twelve months will not be considered a long delay." For Truman, "the callousness of [Bevin's] statement and its disregard for human misery" were what led Americans to demand immediate action.[49]

Bevin's frustration at Truman was due in part to his having to oversee the dismantling of the British Empire. At the end of March 1947, the Attlee government gave up control of Greece and Turkey and would give India its independence by June 1948. With U.S. help it might have been able to forestall the same outcome in Palestine, but the Americans had not cooperated.

For the British army, Palestine had become a giant nuisance and fiscal drain. "One gets the impression," Clifton Daniel wrote in *The New York Times,* "that the British are besieged in their own fortress and are appealing to the United Nations to rescue them." Daniel continued:

> With thickets of barbed wire and blockhouses of sandbags they have created in Jerusalem a modern version of a medieval walled and moated city. . . . Only eleven Britons in the whole country are allowed to live outside security zones, and two of them are under constant guard. Those within the zones are confined from nightfall until daylight. . . . The Jews find a certain sardonic humor in the British predicament. Jerusalem is brimming with jokes about British ghettos, British displaced persons and Britons behind barbed wire.[50]

All these precautions, including the creation of barbed-wire-enclosed zones in Jerusalem, were a response to continuing terrorist attacks. In addition, all of Palestine's Jews, moderates as well as radicals, supported the continuing illegal immigration of European Jews, who were being smuggled into Palestine. Their argument was simple: Britain, and not the Jews, was violating the terms of the original Mandate, which called for Jewish immigration and the establishment of a Jewish national home. In response, Britain announced a new martial law. Communications and transportation ground to a halt, and Tel Aviv's thriving commercial life came to a standstill. Closing down Tel Aviv, Jewish Palestine's industrial hub, meant that entire areas would be paralyzed. Rather than working to the satisfaction of Britain's military authorities, the increased repression hardened the public temper and worked to the benefit of the Jewish extremists.[51]

Despite Britain's travails in Palestine, the Jewish Agency was not sure what to make of its desire to hand it over to the United Nations when it had gone to such great lengths to keep it. David Horowitz scheduled a lunch in London with Harold Beeley, the Foreign Office's desk officer for Palestine and Bevin's chief adviser on the Middle East. Why did Britain want to turn the Mandate over to the United Nations, Horowitz asked? Beeley answered that the U.N. Charter dictated that policy could not be made until the United Nations achieved a two-thirds majority of the members' votes. The only way a majority vote on Palestine could take place was if the United States and the Soviet bloc nations joined together and supported the same resolution. "That has never happened," Beeley proclaimed, "it cannot happen, and it will never happen!" Horowitz thought he understood Britain's strategy. The United Nations would do nothing; Palestine would be lost in a fight between the East and the West. Then the "White Paper would be upheld and the Mandatory [Britain] would return to discharge its functions, this time with all the force and authority of the United Nations behind it."[52] The United Nations, in other words, would be forced to come to Britain's rescue. The question was, would it?

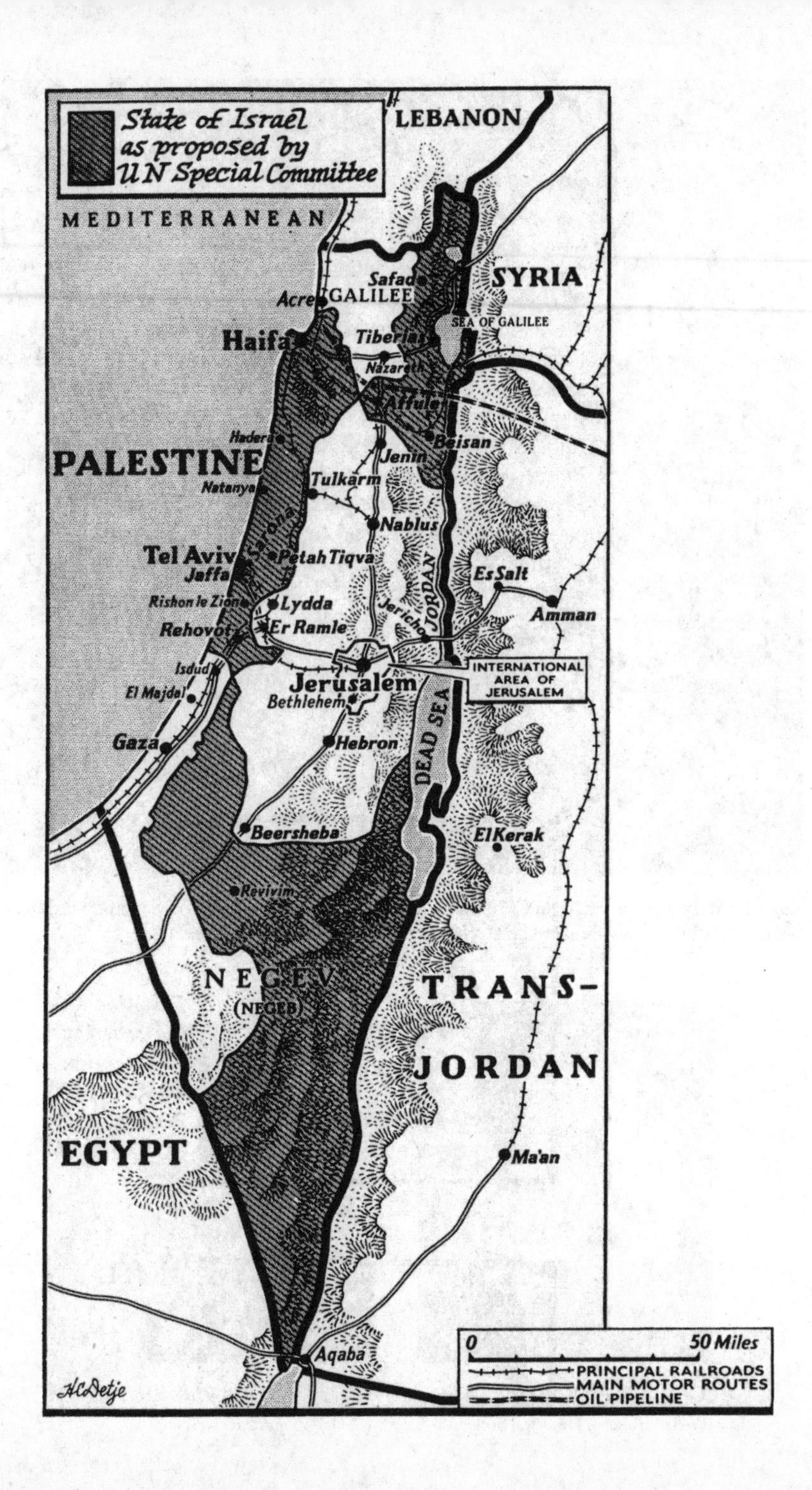
State of Israel as proposed by UN Special Committee
LEBANON
MEDITERRANEAN
SYRIA
Safad
Acre
GALILEE
SEA OF GALILEE
Haifa
Tiberias
Nazareth
Afula
Hadera
Beisan
PALESTINE
Jenin
Tulkarm
Natanya
Nablus
Tel Aviv
Petah Tiqva
Jaffa
JORDAN
Es Salt
Rishon le Zion
Lydda
Amman
Jericho
Rehovot
Er Ramle
Isdud
INTERNATIONAL AREA OF JERUSALEM
Jerusalem
El Majdal
Bethlehem
Gaza
Hebron
DEAD SEA
Beersheba
El Kerak
Revivim
NEGEV
(NEGEB)
TRANS-
JORDAN
EGYPT
Ma'an
0
50 Miles
Aqaba
PRINCIPAL RAILROADS
MAIN MOTOR ROUTES
OIL PIPELINE
H.C.Detje

ABOVE: At the end of the Yalta Conference, Roosevelt met with King Ibn Saud of Saudi Arabia on board the Navy cruiser U.S.S. *Quincy.* (Franklin D. Roosevelt Presidential Library)

Rabbi Abba Hillel Silver.
(Western Reserve Historical Society)

Rabbi Stephen S. Wise.
(Central Zionist Archives)

The New York Times, Monday, February 8, 1943 - Page 8

ACTION—NOT PITY

CAN SAVE MILLIONS NOW!

EXTINCTION OR HOPE FOR THE REMNANTS OF EUROPEAN JEWRY?—IT IS FOR US TO GIVE THE ANSWER

Daily, hourly, the greatest crime of all time is being committed: a defenseless and innocent people is being slaughtered in a wholesale massacre of millions. What is more tragic—they are dying for no reason or purpose.

The Jewish people in Europe is not just another victim in the array of other peoples that fell prey to Hitler's aggression. The Jews have been singled out not to be conquered, but to be exterminated. To them Hitler has promised . . . and is bringing . . . DEATH.

It is a satanic program beyond the grasp of the decent human mind. Yet, it is being carried out. Already 2,000,000 of the Jews in German-occupied Europe have been murdered. The evidence is in the files of our own State Department.

The Germans dared to undertake this process of annihilation because they knew that the Jews are defenseless; that the Jews are forgotten and deserted even by the democratic powers.

The Germans believe that the Inited Nations, indoctrinated by wenty years of anti-Jewish propa-anda, are to a great extent apathetic nd indifferent to the sufferings of he Jews. They believe that for rimes committed against the Jews, o retaliation on behalf of the Gov-rnments or armed forces of the Inited Nations will be carried out. 'hey know that there is no instru-ent of power and force on this earth ith which the Jews can fight back, o avenge their dead and save the emaining millions.

Of what avail are the statements f sympathy and pity and promises f punishment after the war? Since he perpetrators of these slaughters re to be punished for the murders hey have already committed, then hey can lose no more by further ourder.

Such mere statements of sym-athy and pity are to the Germans roof that their judgment of Democ-acy's attitude toward the Jews is ustified, and in their criminal minds hey understand them as "carte lanche" to go on with the slaughter.

What can be done?

What is necessary is to impress he Germans that the Governments f the United Nations have decided o change their present policy of assive sympathy and pity to one f stern and immediate action; that hey consider the cessation of atro-ities against the Jews as an imme-liate aim of their military and po-itical operations. Under this premise igorous United Nations' interven-ion to save European Jewry would ecome a matter of course. Exactly s it would be if it were American r British civilians who were being illed in a systematic campaign by he Nazis, the whole of the forces of these great democracies would be utilized to find an immediate and effective solution.

The inauguration of such a new policy on behalf of the United Nations would logically result in enabling all those Jews who have managed to escape the European-German hell to fight back. The first dictate, therefore, would be the immediate approval of the demand for a Jewish Army of the Stateless and Palestinian Jews—an army 200,000 strong.

Suicide squads of the Jewish Army would engage in desperate commando raids deep into the heart of Germany. Jewish pilots would bomb German cities in reprisal.

A Jewish Army would imply a call to arms of all Stateless Jews living in North Africa so that they may participate in the imminent invasion of the European continent.

A Jewish Army would immediately give a decisive moral relief to the agonized Jews of Europe. Their psychology of despair and helplessness would be transformed into one of hope for revenge and survival. A Jewish Army will give a meaning to their sufferings—to their death.

They will then realize that they cease being helpless victims and become partners in the global struggle for a better world, in which their survivors will live in freedom and equality as all other human beings.

The Jews of *Palestine* and the *Stateless* Jews want to fight . . . AS JEWS. They want to prove to Hitler, and to the world, that the Jews can be more than "the persecuted people" . . . that Jews can die in other ways than through murder. They want the right to fight for the world's freedom, under their own banner. To die, if needs be, but to die *fighting*.

Of course, these are not all the practical proposals which the human mind is capable of conceiving. It is unfair to ask for a single solution to such a disastrous problem. What we must realize is that it is our duty not to resign ourselves to the idea that our brains are powerless to find any solution; not to resign ourselves to the idea that the forces of Democracy are too weak to enforce such a solution.

Remember when a few thousand British soldiers were put in chains by the Germans? How swift the retaliation? . . And how practical . . .

The Germans chained no more British soldiers.

Remember when a tiny town in Czechoslovakia was horribly "punished"? How swift the hurricane of world-indignation that answered . . .

There have been no more Lidices.

Remember when small and encircled Sweden opposed vigorously and stubbornly the expulsion of Norwegian Jews. The Germans abandoned their plan.

The Jews of Norway are still there.

The American sense of justice and decency and American ingenuity must also find ways to overpower the diabolical plan to exterminate the Jewish people. It must find a way now, before millions more perish.

It is, therefore, our primordial demand that an inter-governmental commission of military experts be appointed with the task of elaborating ways and means to stop the wholesale slaughter of the Jews in Europe. This must be done now—before the greatest homicidal maniac extends his policy of extermination to other peoples; before he dares introducing poison gas and bacteriological warfare.

Remember that for years the Germans rehearsed on the Jews what they later practiced on other peoples.

Therefore we have decided to launch an all-out campaign to save European Jewry. We will spare no efforts and have no rest until the American public will be fully informed of the facts and aroused to its responsibilities.

We believe in the overwhelming power of public opinion, as the greatest, if not the only, power in democracy. Governments in democratic countries like the United States and Great Britain can act only when they feel sure that they are backed by a powerful movement of public opinion. We plead with everyone to help and to co-operate in this sacred campaign we have launched. Join in this fight—write to your Congressmen, contribute to our work, so that this message may be carried to every city and hamlet in the United States as is being done in Great Britain. You are part of the collective conscience of America; this conscience has never been found wanting.

NO AMERICAN CITIZENS are wanted as soldiers in the Jewish Army. They fight in the Armed forces of the United States, where millions of men *of all* creeds have joined the struggle.

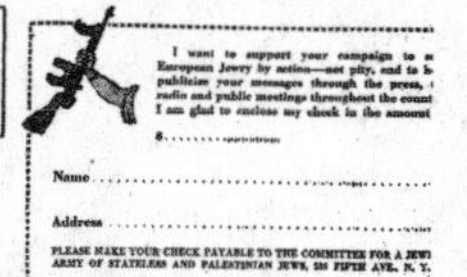
I want to support your campaign to s
European Jewry by action—not pity, and to h
publicize your messages through the press,
radio and public meetings throughout the count
I am glad to enclose my check in the amount
$............
Name ....
Address ....
PLEASE MAKE YOUR CHECK PAYABLE TO THE COMMITTEE FOR A JEWI
ARMY OF STATELESS AND PALESTINIAN JEWS, 535 FIFTH AVE., N. Y.

Committee for a Jewish Army of Stateless and Palestinian Jews

NATIONAL HEADQUARTERS • NEW YORK, 535 Fifth Ave. • MUrray Hill 2-7237

TOP: In 1942, Truman signed the petition for the Committee for a Jewish Army of Stateless and Palestinian Jews. (The David S. Wyman Institute for Holocaust Studies)

LEFT: Peter Bergson (aka Hillel Kook). (The David S. Wyman Institute for Holocaust Studies)

Shortly after becoming president, Truman stopped by to see Eddie Jacobson's new Kansas City store, Westport Menswear. (Courtesy Kansas City Star)

David K. Niles. (Harry S. Truman Library)

Samuel I. Rosenman. (Harry S. Truman Library)

Jewish DPs expressed their sentiments to the Anglo-American Committee in the winter of 1946. (*Witness: One of the Great Correspondents of the Twentieth Century Tells Her Story*, by Ruth Gruber, copyright © 2007 by Ruth Gruber. Used by permission of Schocken Books, a division of Random House, Inc.)

Refugee ship *Ben Hecht* before leaving Port-de-Bouc, France, for Palestine. (The David S. Wyman Institute for Holocaust Studies)

Truman with Britain's new Labour prime minister, Clement Attlee, and Joseph Stalin at Potsdam. (Harry S. Truman Library)

Loy Henderson.
(Harry S. Truman Library)

General George C. Marshall.
(Harry S. Truman Library)

Truman with his special counsel Clark Clifford at Truman's Little White House in Key West, 1949. (Harry S. Truman Library)

LEFT: Truman with United States ambassador to the UN Warren R. Austin and Assistant Secretary of State Dean Acheson, March 25, 1947. (Harry S. Truman Library)

RIGHT: On May 25, 1948, Dr. Chaim Weizmann, the first president of Israel, presented Truman with a Torah. Truman later said it was one of his most prized possessions. (Harry S. Truman Library)

BELOW: Eliahu Elath (Epstein), the Jewish Agency's representative in Washington and later Israel's first ambassador to the United States, presenting an ark to President Truman, October 26, 1949. (Harry S. Truman Library)

TOP LEFT: Truman with Israeli prime minister David Ben-Gurion and Israeli ambassador Abba Eban, 1951. (Harry S. Truman Library)

TOP RIGHT: Commemorative Israeli stamp. (Harry S. Truman Library)

BOTTOM LEFT: When Truman's presidency ended, he returned home to Independence, Missouri, and was often sighted in nearby Kansas City taking walks and having lunch with his friend Eddie Jacobson. (Harry S. Truman Library)

NINE

UNSCOP: PRELUDE TO PARTITION

On April 2, 1947, the British, admitting that they could find no solution to the Palestine quagmire, formally requested that Palestine be put on the U.N. General Assembly's agenda for the fall session. Before the meeting took place, however, the British representative, Sir Alexander Cadogan, asked that a special session of the General Assembly be convened to create a committee to investigate the entire question of Palestine. Things moved along rapidly. On April 13, a majority of the United Nations' members agreed to call the special session.[1] Truman hoped that the United Nations as a representative world body would have the authority and wisdom to come up with a workable solution. He was also relieved that this perplexing issue would be out of his hands for the time being.

A debate now took place on the question of which countries would be chosen to serve on the committee and by what criteria. The committee's recommendations would certainly influence any final settlement in Palestine. Secretary of State George C. Marshall told Truman that the State Department thought it should be a small committee made up of "comparatively disinterested states, excluding the Big Five and the Arab States." Truman liked the idea of using small, neutral nations.[2]

The Zionists took the opposite tack. Meeting with Dean Acheson and Loy Henderson, Moshe Shertok, the Jewish Agency's representative to the United Nations, argued that small nations would only turn the issue over to the larger and powerful nations and ask them what to do. Instead, the Jewish Agency preferred presenting a resolution to the General Assembly asking that Britain administer the Mandate faithfully–in other words, permit Jewish immigration to Palestine and lift the ban on land sales. If nothing was done, he was afraid, two forces operating in Palestine could easily destroy the area: the British occupation and the actions of the Irgun and Stern Gang terrorists. The cycle of "repression breeding terror" and "terror leading to new repression," he argued, had to be broken. If it were not, handing the issue over to the United Nations would be nothing but an academic question. The question Shertok most wanted an answer to: What was U.S. policy toward Palestine? If the United States had no policy, Shertok thought, "the situation of the Jews is almost hopeless since they have no other powerful friend on whom they can depend."[3]

Acheson disagreed. The United Nations, he told Shertok, might be the only institution left that could solve the issue in a fair and peaceful manner. If it was to succeed, it had to have the support of world public opinion, and that could develop only after careful study in an atmosphere free from pressure from the big powers and intrigues behind the scene. Again, he reminded Shertok that the United States had been trying to get the British to allow immigration, without any results. He did not tell Shertok that the U.S. government had not yet decided on their policy. (Truman's Yom Kippur statement that "the creation of a viable Jewish state" in part of Palestine was a solution that "would command the support of public opinion in the United States" had not been an official policy statement. It had been meant by Truman only to give the public his personal opinion.) Now Marshall and Truman, busy with the equally pressing developments in Europe and the new Cold War, were improvising when it came to the Palestine issue. In any event, they would have breathing room while the United Nations conducted its study and made its recommendations, arguing that the United States should not influence the outcome by coming out in support of any particular solution.

This stance was unacceptable to the Zionists and their supporters in the United States. Thirty members of Congress released a public letter addressed to Secretary of State Marshall and Warren R. Austin, the U.S. representative at the United Nations. They were "deeply concerned," they wrote, about the inability of Britain to fulfill the terms of the League of Nations' Mandate for a Jewish national home in Palestine. The congressmen wanted clarification. They wanted to know from Marshall and Austin whether the United States had formulated policy with respect to matters put before the United Nations. Noting that the U.N. Charter contained a clause saying that trusteeship had to be agreed upon by states "directly concerned," they asked, "Will the United States take the position in the United Nations Assembly . . . that it is one of the states 'directly concerned' in the disposition of Palestine?"[4] Marshall replied that the government felt it was not time to announce a completed Palestine policy that might hinder the operation of the new U.N. committee. The United States would make its views known after the report was in.[5]

On April 28, the U.N. special session met at the new U.N. headquarters at Lake Success on Long Island, about an hour from New York City. A proposal made by the United States that a committee be formed to investigate the situation in Palestine and that it be composed of neutral nations was adopted by a majority vote, with the Arab states voting against it. The United Nations Special Committee on Palestine (UNSCOP) was then set up and given the power to investigate the current situation as it related to the status and future of Palestine. The countries were chosen on a territorial basis. Western Europe was represented by Judge Emil Sandström of Sweden and Dr. Nicholas Blom of the Netherlands; the British Commonwealth by Justice Ivan Rand of Canada and John D. L. Hood of Australia; Latin America by Dr. Jorge García Granados of Guatemala, Professor Enrique R. Fabregat of Uruguay, and Dr. Artur García Salazar of Peru; Eastern Europe was represented by Dr. Karl Lisicky of Czechoslovakia and Vladimir Simić of Yugoslavia. Asia was represented by Nasrullah Entezam of Iran and Sir Abdur Rahman of India. Victor Hoo of China, the United Nations' assistant secretary-general for trusteeship, was picked by Secretary-

General Trygve Lie as his representative. The African-American intellectual Ralph Bunche was chosen by Lie to be a special assistant to Victor Hoo. The committee, overwhelmed by the tasks facing it, appointed Bunche as chairman of its subcommittee on arrangements. In this capacity, he would make all the logistical arrangements and write all of UNSCOP's reports.[6] UNSCOP was to submit its proposals to the General Assembly by September 1.

The attitude of Britain, the mandatory power in Palestine, toward UNSCOP and the General Assembly's decisions were crucial. Would the British cooperate, and if they did, to what extent? There was an inkling of things to come when the British representative, Cadogan, announced that the British did not intend to carry out any decision of which they did not approve and that, as he put it, they "could not reconcile with our conscience." Furthermore, the British government would not participate in enforcing any solution in Palestine that was not acceptable to both the Arabs and the Jews.

The Arab states requested that a motion calling for the immediate "termination of the Mandate over Palestine and the Declaration of its independence" be placed on the agenda of the special session. If adopted, it would make Palestine into an Arab state. The United States opposed it, claiming it was premature to consider such a proposition before UNSCOP conducted its study and made its recommendations. The Jewish Agency also put in a request to present the Zionist case before the committee. The State Department at first opposed this on the grounds that it would receive a negative response from the Arab states and that the United Nations had no provision for representation from nongovernmental groups. As a result, the White House and State Department were inundated with protests from Jewish, Zionist, and liberal organizations. The U.S. delegation to the United Nations then shifted its stance and agreed to allow the Jewish Agency representation as a nonvoting member.[7]

This was very good news for the Zionists. The Jews had an "unanswerable case," Chaim Weizmann believed, but their success would depend on the manner in which they presented it. If it was done "without too much bitterness and too many recriminations" against the British, they could prevail. "It is a great chance," he wrote to a colleague, "and I am afraid it is our last chance." There would be "no appeal from this tribunal."[8] On May 8, Moshe Shertok and David Ben-Gurion took their seats among the world's nations to make the first presentation before the United Nations of the Jewish case for a homeland in Palestine. The British then argued that if the Jewish Agency were allowed to represent the Jews of Palestine, the Palestinian Arabs should also be represented. Both the U.S. delegation and the State Department agreed and supported the British stand that UNSCOP hear a delegation of the Arab Higher Committee of Palestine.

Support for the inclusion of the Jewish Agency and opposition to the appearance of the Arab Higher Committee came from an unusual but influential source, Freda Kirchwey, the editor of *The Nation*. Long associated with liberal causes, Kirchwey was best known for her positions in favor of the Republicans during the Spanish Civil War, antifascism during World War II, and, later in the Cold War era, support of the positions of the Soviet Union and opposition to the policies of various American presidents. Running the magazine was not sufficient to attain political clout and influence. For that purpose, Kirchwey had created the Nation Associates as a mechanism to provide funding for the poorly financed magazine, as well as to have her own organization that could influence policy. Its board included Philip Murray, president of the Congress of Industrial Organizations, the industrywide body representing organized labor; the theologian Reinhold Niebuhr; James G. Patton, the president of the Farmers Educational and Cooperative Union; the left-wing radio commentators Frank Kingdon and Raymond Graham Swing; the playwrights Lillian Hellman and Eugene O'Neill; Frank P. Graham, the president of the University of North Carolina; and most significant of all, the former first lady, Eleanor Roosevelt.

Under Kirchwey's leadership, aided by Lillie Shultz, the director of the Nation Associates and a former assistant of Stephen Wise, the Nation Associates became an important research arm of the Zionist movement and one of the most active supporters of the Jewish state in the media. Kirchwey counted among her friends and associates both Eliahu Epstein and Chaim Weizmann, as well as most of the American

Zionist leaders, who greatly appreciated her devotion to their cause. Unrelenting, she carried on a one-woman campaign on behalf of the Jewish homeland, sending her reports, letters, and requests to virtually every senator and congressman and scores of administration members, including the president, as well as internationally renowned and influential individuals. Kirchwey, her biographer writes, "fueled the battle for a Jewish homeland with speeches, Nation Associates' studies, and *Nation* articles . . . at the expense of other news. To Kirchwey . . . special interest in the issue merited the stress."[9]

When Kirchwey learned that the U.S. delegation to the United Nations backed the British on giving a platform to the Arab Higher Committee, she immediately went into action. First, Kirchwey and the Nation Associates gave out a lengthy report on the pro-Axis activities of the Grand Mufti of Jerusalem to all fifty-five delegations to the United Nations. Most striking was the fact that it was based on classified U.S. government files that had been leaked to Kirchwey by someone in the government. The Arab Higher Committee, the report charged, was the "creature of the Arab League" and was run from Egypt by the Mufti, Haj Amin al-Husseini. The committee's members were not elected representatives but "nothing more . . . than a deal among leaders of the various Arab factions in Palestine–and the will of the grand Mufti." Three members of the Palestine Arab Higher Committee, who were delegates to the U.N. General Assembly, were called, in the Nation Associates report, men who ranked with the "worst of the Axis war criminals." Moreover, its leaders were part "of that cabal that ruled Hitler's Germany." The account of the Mufti, Kirchwey explained, was documented from captured files belonging to the mufti and the German High Command, all found by American military authorities in Germany.[10]

Along with the report were documents and photos that substantiated the charge that the Mufti controlled and directed the Arab Higher Committee. Photos used showed the Mufti and other Arab leaders with Adolf Hitler, Heinrich Himmler, and Dino Alfieri, Mussolini's ambassador to Berlin. The originals, the report noted, were "now in the possession of the State Department." The Nation Associates named

Emil Ghouri, the head of the Arab delegation to UNSCOP, and delegates Wasef Kamal and Rasem Khalidi as "notorious for their long-time association with the Mufti and his Axis activities." Ghouri, the delegation's chief, was accused of being partially responsible for "internal terror against Arab opponents of the Mufti and Arabs who sell land to the Jews." Kamal, the report claimed, had left his native Iraq after the 1941 rebellion against the British and escaped to Turkey, where he had become a "paid agent of the German secret service." The Mufti's relative Jamal Husseini, who was also a delegate, had joined the Mufti in Iraq in 1939. While there, he had organized a pro-Axis fifth column that led to the anti-British rebellion.

As for the Mufti himself, the report was scathing and thorough. It traced the Mufti's actions back to 1920, revealing a history of violent organized attacks he had planned against the Jews of Palestine. Captured German files revealed that he had planned the Palestine riots of 1936, using funds supplied by the Nazis to carry out the action. The U.S. investigator had concluded, "Only through funds made available by Germany to the Grand Mufti of Jerusalem was it possible to carry out the revolt in Palestine." After escaping to Iraq, the Mufti was directly responsible for the death of four hundred Jewish men, women, and children, who were murdered on Baghdad's streets. The Mufti then fled to Italy and later to Nazi Germany. Once he was there, the Nazis had established an office in which the Mufti was put in charge of propaganda, espionage, Muslim military units, and Arab legions, which were ordered to carry out fifth-column activities in the Middle East, including sabotage. In the Hague, Arab students were trained in explosives and parachuted into Turkey, Syria, and Iraq. Captured records also revealed that the Mufti had accompanied Adolf Eichmann to visit the gas chambers at Auschwitz. Moreover, the Mufti had put an end to negotiations being carried out to ransom Jews in Bratislava, insisting that they all be liquidated. A letter the Mufti had written to Heinrich Himmler revealed the Mufti complaining that Joachim Ribbentrop and Himmler had been too lenient, since they had let some Jews leave Germany. "If such practices continue," the Mufti was quoted as saying, "it would be incomprehensible to Arabs and Moslems and provoke a feeling of disappointment."[11] Refusing to comment on the Nation Associates' report, members of the Arab Higher Committee told the press only that "the Axis issue should be forgotten."[12]

Next, Kirchwey presented to every delegate to the U.N. General Assembly a 133-page investigative report, "The Palestine Problem and Proposals for Its Solution." It became the major media brief on behalf of the position of the Jewish Agency. The report's headings covered such subjects as "The Jewish Claim to Palestine," "British Labor's Pledges on Palestine," and "Oil, Communications, and Bases," and concluded with an explanation of why a Jewish state could and should be created in Palestine. The Nation Associates favored the partition of Palestine into two states and provided evidence for why American oil interests would not be harmed by its establishment, while minimizing the possibility of an Arab revolt.[13]

Kirchwey made sure that the Nation Associates' reports reached the White House. Writing to Truman about their report on the Mufti and the Arab Higher Committee, David Niles explained to the president just how important the material was:

> You have received a copy of the Documentary Record submitted to the United Nations. This contains very confidential material that is in the files of the State Department.
>
> I think it is important to find out how it got out. It is very damaging evidence that the Arab representatives now at UNO were allies of Hitler. There is also included in this material the diary of the Grand Mufti, which Justice Jackson found at Nuremberg. Copies of this document have already gone to all the Members of Congress.[14]

It is hard to believe that Niles was outraged that this "damaging evidence" about the Mufti and the Arab Higher Committee had been leaked. He most likely was using the leak to make sure that Truman saw the report. "Thanks, glad you sent it," Truman replied in handwritten notes he scribbled on the memo. "I knew all about the purported facts mentioned and, *of course,* I don't like it." He went on to again note his frustration that neither the Anglo-American Committee nor the Morrison-Grady recommendations had worked out. "We could have settled this Palestine thing if U.S. politics had been kept out of it," Truman wrote. "Terror and [Rabbi Abba Hillel] Silver are the contributing causes of *some,* not all, of our troubles." As to the document itself, Truman added that it "could have been used by us for the welfare of the world had not our own political situation come into the picture. I surely wish God Almight[y] would give the Children of Israel an Isaah [*sic*], the Christians a St. Paul and the Sons of Ishmeal [*sic*] a peep at the Golden Rule. Maybe he will decide to do that."[15]

The Soviet Union now had a surprise. Stalin had always been opposed to Zionism, which the Communist leadership viewed as a reactionary form of "bourgeois nationalism," an ideology that pushed Jews away from the universalist doctrine of Marxism-Leninism. In addition, the Soviet leadership's views were influenced by the continuation of traditional Great Russian anti-Semitism. Most observers, including the U.S. State Department, believed that the Soviets would adhere to a pro-Arab policy and support Arab nationalists as a counterpoint to the Zionists and the British. The earliest stages of the Cold War had already begun, and it was clear that the Soviets intended to play a greater role in the Middle East. But the Soviets were playing a dual role. Although their diplomats in Arab capitals regularly backed the Arab states, the Soviet government was permitting the emigration of Jews from Eastern Europe and Poland to the Western occupation zones in Germany and Austria, knowing full well that most of them planned to go to Palestine.[16]

It came as a shock when, on May 14, the Soviet ambassador to the United Nations, Andrei Gromyko, proclaimed the Soviet Union's belief that the Mandate was bankrupt and said that the Soviet Union supported the "aspirations of the Jews to establish their own State." No Western European nation, he said, "was able to provide adequate assistance to the Jewish people in defending . . . its very existence." That failure explained the desire of the Jews to create their own state. His government's preference was for a unitary democratic Arab-Jewish state in Palestine. But if that was impossible to implement, the United

Nations had to consider "the partition of Palestine into two independent autonomous States, one Jewish and one Arab."[17]

What had caused the Soviets to change their position? No one was sure. Just a few days before Gromyko delivered his speech, the American charge d'affaires in the Soviet Union had noted that the Soviets regarded creation of a Jewish state as a "Zionist tool of the West" that would be hostile to the Soviet Union and hence supported the Arab side in the Palestine controversy.[18] The American ambassador to Moscow, Walter Bedell Smith, explained that the USSR was now changing its position in order to gain a foothold in the Middle East, at a time when the Soviets saw that the Western nations were opposing Soviet pressure in Iran and Turkey. With Britain giving up the Mandate, a vacuum would be created in the region. They hoped that a new Jewish state in Palestine would be the vehicle by which they could eliminate British influence in the Middle East.[19] Yet other diplomats, such as Dean Rusk, the director of the Office of Special Political Affairs at State, argued that the Soviets were playing both ends against the middle in order to gain support from both Jews and Arabs. Since they were not as yet ready to give complete support to the Arab side, they concentrated on attacking Britain, the mandatory power. But Rusk was certain that at a later date they would shift to the Arabs, when they could "reap the greatest benefits in the Moslem world."[20]

Contemporary historians argue much as did Bedell Smith and Rusk. "During World War II," Laurent Rucker writes, "the USSR had been reluctant to undermine the empire of its British ally, but the postwar division of the powers into two antagonistic blocs was creating a new situation. In this environment, Moscow's goal became to weaken Great Britain and 'exploit the contradictions between London and Washington.' . . . The Palestine issue thus provided Moscow with an opportunity to strike a blow at a place of strategic importance to the British Empire, and to exacerbate the Anglo/American divide over the Jewish DP issue."[21]

The Zionists were taken by surprise. The speech, David Horowitz remembered, "was like a thunderbolt out of a clear blue sky after so many years during which our cause had been ostracized by the Russians."[22] Now that it looked as if there was a good chance that the Russians would support partition, the Zionists' main concern focused on the failure of the United States to make known its own position. When they asked that the U.S. representatives issue a new statement of the American position, Acheson again explained that the United States did not wish to be viewed as pressuring UNSCOP or trying to influence its decisions.[23]

The UNSCOP investigation in Palestine came at a time when tensions between Jewish Palestine and the British were at the breaking point. The Revisionist Zionists–the Irgun, led by Menachem Begin–was supporting terrorism as a tactic for gaining independence as a Jewish State. They would lay down their arms, they proclaimed, only when the British ceased their suppression of the Yishuv. Both the Irgun and the Fighters for the Freedom of Israel–commonly called the Stern Gang–refused to suspend their operations while UNSCOP was conducting its study, as the United Nations had requested.

In the United States, the playwright Ben Hecht, a supporter of the Bergson group, took out a large incendiary ad in the *New York Herald Tribune* titled "Letter to the Terrorists of Palestine." "Every time you blow up a British arsenal, or wreck a British jail, or send a British railroad train sky high, or rob a British bank or let go with your guns and bombs at the British betrayers and invaders of your homeland," he proclaimed, "the Jews of America make a little holiday in their hearts." The ad infuriated the British government. Lord Inverchapel told the State Department, according to news reports, that the ad was an incitement to murder British soldiers and subjects living in Palestine. The British were particularly annoyed at the ongoing fund-raising in America for such terrorist projects and the tax exemption such American groups evidently enjoyed.[24]

Though only a small minority of the Jews in Palestine advocated the terrorist tactics of the Irgun and the Stern Gang, the entire Yishuv supported bringing the European Jewish refugees to Palestine and rejected the British contention that it was illegal. Recently, Jewish immigrants sailing to Palestine on the *Hatikvah* were forcefully removed from their ship at the Haifa port and taken in British deportation ships to detention camps in Cyprus. The British naval party was allowed to board the ship, and the Jewish immigrants did not resist them. They did so because Jewish Agency leaders believed that any violence or resistance to the British might spread terrorism–and they did not want that to occur while UNSCOP was studying the situation.[25] Moreover, David Ben-Gurion, the chairman of the Agency's executive board, had just announced publicly what he had previously kept private: that a Zionist state throughout all Palestine was not feasible and that the Jewish Agency was ready to accept partition under the right circumstances. "We must not ignore realities," he told a Jerusalem audience. "The United Nations will not acquiesce in turning the whole of Palestine into a Jewish State now." Building a state in the areas where Jews were already prominent, including the desert area of the Negev, would be acceptable.[26]

In the opinion of the Irgun leader, Menachem Begin, this kind of timidity and accommodating behavior would lead nowhere. If not for the resistance shown by the dissident's fighters, said Begin, "there would have been no inquiry by the United Nations." It was their actions that had made "the British position untenable." The problems were caused by Britain alone. "With 120,000 foreign British troops in this country," Begin said, "with foreign British naval units stationed in Palestine's waters to prevent Jews from entering Palestine, it is obvious who is causing insecurity in this country." And unlike Ben-Gurion and the Jewish Agency leadership, Begin and his followers opposed partition and demanded all of Palestine as a Jewish state.[27]

Harry Truman took notice of the ongoing strife and made a statement lending his government's support to a May 15 resolution of the General Assembly requesting all groups to refrain from violence while the United Nations conducted its study. In an obvious reference to the Ben Hecht ad, Truman urged "every citizen . . . of the United States . . . meticulously to refrain, while the United Nations is considering the problem of Palestine, from engaging in . . . any activities which tend to further inflame the passions of the inhabitants of Palestine . . . or to promote violence in that country."[28] He said nothing, however, about taking the kind of action the British requested, such as ending the tax exemption of the Bergson group. The statement had little effect, other than formally putting the administration on the side opposed to violence.

A week later, Secretary Marshall sent out a sweeping statement to twenty major U.S. diplomatic and consular posts, warning them that the United States was not advocating any specific plan for Palestine and that anyone referring to an "American plan for a partition of Palestine" should cease and desist. Those using the phrase may have been doing so for descriptive purposes, but they had to be aware that "others may reiterate such descriptions for propaganda reasons," implying that the United States supported a specific solution. All the United States stood behind, the secretary cautioned, was consideration of the Palestine issue by the United Nations.[29] Moreover, Marshall informed the U.S. ambassador to the United Nations, Warren Austin, the United States should not make any public statement about the administration's views about the future government of Palestine at all. If and when the United States took a position, it would have to be based on a review of the entire situation and on principles that America could defend to the world, "now and in the future."[30]

Why was Marshall so reluctant to come forth with the U.S. policy on Palestine? Writing on behalf of the Jewish Agency, Eliahu Epstein reported to Rabbi Silver about what he thought the new "Marshall Doctrine," as he called it, really meant. Marshall had recently given a speech to Harvard's graduating class announcing the Marshall Plan to aid Western Europe's reconstruction. The foreign policy of the United States was in a fluid phase, Epstein pointed out. It was therefore difficult for Secretary Marshall to fit the Palestine question into the new overall foreign policy goals. When Marshall told Silver personally that he was not ready to announce the U.S. position, he was not engaging in a cover-up. The problem was the following, Epstein wrote:

> While Truman remains a potential supporter of our cause–for party reasons–the State Department, and especially Henderson, are apparently doing their best to counter-balance this by overemphasizing, as usual, the Arab angle. This was proved recently on an occasion where Loy Henderson presented the

> State Department's views on Palestine. Speaking before a small body . . . Henderson's anti-Zionist views were expressed in no ambiguous terms when he spoke on American policy in the Middle East.

As Epstein saw things, the problem was that Marshall was new to the job of secretary of state and was being briefed by Henderson, whose role had increased in importance since Acheson had left the department and been replaced by Robert Lovett. To compound the problem, Marshall got reports and data from both the War and Navy Departments, whose spokesman agreed with the British and the American oil companies. Felix Frankfurter, whom he had recently seen, had advised Epstein not to press Marshall for an immediate statement but instead to concentrate on trying to achieve the best decision possible when a future policy statement was issued. Epstein had just learned, moreover, of how Truman and Marshall "were able to check the efforts of some of our friends" in the Senate and House who had been prepared to present a statement on behalf of partition to UNSCOP. Nevertheless, Epstein was hopeful that State had not made up its mind and was not necessarily motivated to argue against partition.[31]

Meeting with a group of Democratic members of Congress, Secretary Marshall assured them that the United States had not changed its established Palestine policy, including unlimited Jewish immigration and the eventual creation of a Jewish commonwealth; it just did not want to prejudice the U.N. investigation. *The New York Times* reported the story under the headline "Marshall Reaffirms Policy on Palestine." Senator James E. Murray (D.–Montana) led the group, arguing that the United States had to take the leadership or the United Nations would not take a stand. For the time being, although they favored a restatement of U.S. policy, the group agreed to allow time for Marshall to develop his own restatement.[32]

Less than one week later, the congressmen went public with a statement against the slow progress. Senator Owen Brewster (R.–Maine) and Senator Murray charged that the State Department was not acting constructively. Speaking before the convention of the Zionist Organization of America, they accused State of not carrying out the announced U.S. policy toward Palestine. "If American policy has any sense left," Senator Brewster said, it would see to it that "a substantial portion of Palestine at the very least is now made into a Jewish commonwealth." As to Secretary Marshall's argument that it was premature to formulate a new, firm policy, Brewster charged, "the Secretary is twenty-five years behind the times." Invoking the Truman Doctrine, he noted that the United States had openly bypassed the United Nations to take unilateral action on Greece and Turkey. If the United States permitted the United Nations to stymie taking action, he predicted, the United Nations might not "endure more similar shocks to its prestige without following the League of Nations into innocuous desuetude." Senator Murray added that the State Department was postponing taking a clear and definite position on Palestine, and he thought that executive branch officials should quickly implement policies laid down by Congress.[33]

Opposition to the State Department's position also came from other Republicans. Thirty Republican members of Congress, led by New York Representative Jacob K. Javits, wrote to Marshall, asking whether Warren Austin, the U.S. delegate to the United Nations, was going to appear before UNSCOP and declare that the United States would adhere to the goal of creating a Jewish national home in Palestine. By not informing UNSCOP of the United States' willingness to aid in the implementation of a new policy, Javits argued, the United States would be leaving the United Nations in the dark as to what recommendations it could make that would be acceptable to the United States.[34]

This same week, Moshe Shertok returned to Palestine to observe the situation. Reporting to his Agency colleagues there, he said that he had found "a wave of shallow sympathy" in America for the Irgun–Stern Gang terrorist actions. Some news reports, he feared, glamorized the terrorism and made it harder for American Jews to resist their appeal. To combat the proterrorist propaganda of the Bergson group in America, Shertok said, the Zionist movement was "trying to organize a campaign of enlightenment among Jews to open their eyes to the freak charlatan aspect of that group." Most disturbing was that Bergson was managing to move "a fair amount of Jewish money" from legitimate groups to bodies favoring "useless and destructive ends."[35]

Shertok's position was supported by Rudolf G. Sonneborn, the American businessman who was the chief fund-raiser in America for the Haganah, as well as cochairman of the United Jewish Appeal and a member of the ZOA executive board. Sonneborn issued a blistering attack on both the Irgun and Stern Gang. Both "underground groups in Palestine," he said publicly, "were endangering the efforts and sacrifices of three generations 'by their insane actions.' "[36] Sonneborn's own fund-raising body, the Sonneborn Institute, had by April 1947 raised half a million dollars for that year. All of the money would be used explicitly for the purpose of helping Jewish illegal immigration into Palestine. Sonneborn's major concern in 1947 was that the Irgun terrorists in Palestine were confusing Americans who were sympathetic to the cause of Palestine's Jews. As a result, thousands of Americans were contributing funds in the hope that they would be put to good use, not realizing to whom their money was going. Bergson's American League for a Free Palestine, he feared, was particularly successful. Ben Hecht had managed to get support for their cause from the world of show business, and drew huge crowds throughout the country to his play *A Flag Is Born*, produced by the impresario Billy Rose and written by Hecht. Their production raised what was then the staggering sum of more than $1 million, and the contributors did not realize their money was going to the Irgun. As a result, Sonneborn formed Americans for Haganah, which tried to counter the Bergson group activity with their own rallies and advertisements.[37]

In Palestine, the split between the Haganah, the Yishuv's official defense force, and the Irgun and Stern Gang had widened. The Haganah's leadership realized the necessity of keeping the peace while UNSCOP carried out its investigation. In November 1946, they had begun to turn against the terrorists. Now, on June 18, they foiled a plot by the Irgun to blow up Citrus House, the British military headquarters in Tel Aviv. The Irgun had dug a forty-five-foot tunnel from a basement opposite the headquarters to the British site, where it planned to place explosives. The Haganah had a commando detachment seal the tunnel, leaving a note announcing its opposition to "dissident groups . . . planning a lunatic act which would have resulted in unnecessary slaughter." The Irgun's retort was to accuse the Haganah of being "collaborators" working with the British.[38]

While the Jewish Agency and the Haganah were bending over backward to oppose terrorism and show their willingness to compromise, the Arab states remained steadfast in their position: Palestine must be an Arab state in which Jews could live as a minority.

Jamal Husseini, vice chairman of the Arab Higher Committee, informed the secretary-general of the United Nations that the Palestinian Arabs would not be "collaborating" with UNSCOP and its representatives would not testify before it.[39] Any U.N. decision that did not grant the Arabs an independent Arab state in Palestine, he threatened, would have immediate repercussions throughout the Arab world. Fifty million Arabs, he said, were ready to back the Palestinian Arabs by armed force. Partition would not work, since "Palestine is too small," and if the United Nations failed them, they would be prepared.[40]

The United Nations mandated that both the Arabs and Jews could appoint two liaison officers to UNSCOP. Though the Arabs declined, the Jewish Agency appointed the Jewish Agency's chief economist, David Horowitz, and a thirty-two-year-old British Zionist, Aubrey (later to change his name to Abba) Eban, who were given instructions "to work for the creation of a Jewish state in a suitable area of Palestine." Eban, unlike most of the Jewish Agency leadership, was not from Eastern Europe, nor had he gone to Palestine as a pioneer to develop the Yishuv. He had been born in South Africa and raised in Great Britain. Seeing his intellectual potential, Eban's grandfather Elaihu Sacks hoped that his grandson would become a major Hebrew scholar. For six years, Aubrey spent weekends with his grandfather memorizing, studying, and learning the great arguments of the Talmud. All this was in addition to his rigorous education in private British schools. He graduated with high honors from Cambridge, where he had joined the Cambridge Zionist Society, as well as socialist organizations at the university. By the time he was in his mid-twenties, Eban's articles and essays appeared in the major English Zionist publications. The Zionist elders, especially Chaim Weizmann, took notice. Before long, Weizmann became his mentor.[41]

Horowitz and Eban couldn't understand why the Arabs would miss an opportunity to interact with the committee, something they fully intended to do, with good effect. For this and what he thought were other Arab misjudgments, Eban reflected, "we live on the mistakes of the Arabs."[42] Even Harold Beeley in the British Foreign Office bemoaned the Arabs' "exceedingly inept" diplomacy and failure to "recognize that their unyielding and dogmatic stand would create sympathy for the Jews rather than support for Arab nationalism."

With the Arabs boycotting UNSCOP, it fell to Eban, Horowitz, and the British liaison officer, Donald C. MacGillivray, to establish the committee's itinerary in Palestine. The committee had three months to conduct its investigation. That included travel as well as substantive hearings with both Arabs and Jews. At the end, they were expected to prepare a report. Eban and Horowitz took them throughout Palestine–to Haifa, the Dead Sea, the Arab town of Ramallah, Mount Carmel, Tel Aviv, and down to the Negev–visiting factories, canneries, orange groves, and harbor installations. The men took their assignment very seriously. "The enlightening, the pointing out, the persuading," Eban's biographer wrote, "went on at breakfast, lunch and dinner, as they drove from place to place; sometimes in French, sometimes in English or German; unceasingly but subtly."[43]

From the start, UNSCOP members had a very different experience with the Arab community. When the committee started its work on June 16, it was greeted by an Arab general strike, held in protest at their presence. UNSCOP's head, the Swedish Chief Justice Emil Sandström, went on the radio to plead with the Arab Higher Committee to give the U.N. group its cooperation. "I cannot put it too strongly," he told them, "that this Committee has come to Palestine with a completely open mind. . . . We are impartial. . . . We have reached no conclusions."

The same week, the Palestinian press gave publicity to the Haganah's operation against the Irgun, a step that could not but impress UNSCOP's investigators. They had to deal, however, with the request made by the parents of three convicted Jewish terrorists, members of the Irgun, who were scheduled to be hung by the British. The three teenagers had carried out a raid on the Acre prison that had enabled about 250 prisoners to escape. UNSCOP members expressed concern as to what they feared might be unfavorable repercussions were the hangings to be carried out. They explained to the parents that they could not directly intervene with the Palestine governing authority, it being beyond their instructions to do so.[44]

The British decision to impose the death sentence on Irgun youths, whose apparent crime was posting illegal leaflets and participating in a jailbreak, horrified one of UNSCOP's members, the delegate from Guatemala, Jorge García Granados. Granados was from a family of political reformers and had himself been imprisoned and exiled for his political activities. Learning of the harsh sentence, he wrote, "every fibre of my being protested against the principle of capital punishment for political acts," something he recalled his own people had "so fiercely fought in Guatemala." Granados saw the British actions as a purposeful act meant to create difficulties for UNSCOP, which the British did not want there. Approaching a British general, Granados confronted him: "These youths have been sentenced to death because they helped break open the doors of a prison. They killed no one. They are innocent of bloodshed. I find the death penalty an extremely drastic one." The general responded that "British officials are bound to respect the law," and anyone breaking it was a criminal and must take the consequences.[45]

The military regulations that the British military had imposed in Palestine were indeed harsh, although the British argued that they could have been much tougher in light of the dissidents' terrorism. As it was, a person could be arrested without a warrant and imprisoned indefinitely for years, without a trial or being informed of any charges. Asher Levitsky, a Palestinian Jewish attorney, gave Granados a complete account of Britain's emergency regulations for the Yishuv. The British could order deportation, confiscate a home, or sentence an arrested person to death without showing due cause. Anyone could be detained by a soldier and held in prison pending a decision of the military commander. When trials took place, it was before a British military court and the defendant had no access to a judge-advocate. One could even be arrested for an act that had taken place years earlier and had been legal at the time. "I had crossed half the world," Granados wrote, "to find myself in the one truly police state remaining in the 20th century."[46]

The British did not make the committee's work easy. They insisted on giving their testimony in private and wanted to be informed in advance about whom the committee was interviewing. This would be impossible to do for groups such as the Irgun, which was outlawed but with whom the committee might like to speak.

The committee began its hearings in a hail of publicity. "In the glare of movie floodlights," Clifton Daniel reported for *The New York Times,* "the United Nations committee met in the ornate cavern of the Y.M.C.A. auditorium. Its eleven members and two principal secretaries were arranged around the table." Covering their meeting were 80 journalists and 125 spectators.[47] At the opening, Shertok spoke for an hour, giving some background to the Jewish case, and Horowitz spoke on the economy. The delegates then visited Jewish and Arab areas of Palestine, although the Arab boycott and general strike made the latter very difficult. When the delegates entered Arab areas, people lined up as they passed, greeting them with hostile stares and complete silence. Passing through one Arab village, the committee found all the buildings deserted because the inhabitants had been given the signal to leave.[48] When the committee members toured an Arab tobacco factory, they saw the poor conditions and the child laborers. The Arab owners would not allow any Jewish visitors, so Horowitz was forced to stay behind, which made a statement in itself. From the reports he got, he thought the rehearsed answers of the children, all under fourteen years of age, had worked to "heighten the bad impression."[49]

The U.S. consul general at Jerusalem, Robert Macatee, wrote to Marshall that he thought the Arabs were self-defeating. In Ramle, a town near Jaffa, the Municipal Council declared that representatives from the Jewish press would not be allowed to accompany the committee on its visit. In Jaffa, when journalists, including Jewish journalists, entered an Arab factory, the management "stalked out in protest." Finding an English-speaking Arab at the door of the town hall, the UNSCOP group asked him what he thought of the committee coming to Jaffa. "I will tell you this," he answered, "if it was up to me . . . I would hang everyone who came here today." The reason was the presence of Jews on the UNSCOP mission. "They have no right to be here," the young man continued, asserting that he knew three quarters of the committee members were Jewish.[50] "The Arab Higher Committee's uncompromising attitude," Granados explained, "its refusal to consider the possibility of any conciliatory course, was to prove a convincing argument for partition."[51] As the evidence of racial discrimination mounted, many of the delegates wondered what it said about the fate of a Jewish minority in an Arab Palestinian state.

The Jews of Palestine gave the committee a strikingly different reception. In Tel Aviv, they were flooded with a "staggering volume of information, and crowds clapped and sang for the delegates and pressed around their cars to shake their hands." An enormous crowd gathered in the main square to cheer and greet them. At a major synagogue, the Chief Rabbi of Palestine called upon the Almighty "to instill in the hearts of the United Nations Committee knowledge, wisdom and intelligence, to judge honestly and to gather the people of Israel in their Holy Land to revive and rebuild it." In an understatement, Horowitz noted that "the warm reception by the Yishuv, in contrast with the cold malevolence shown by the Arabs, did not pass unnoticed by UNSCOP."[52] It was in Tel Aviv, Granados wrote, "that I first really understood what the coming of our Committee meant to the Jewish people. We held in our hands life or death for all those men, women and children who were gazing at us so eagerly, so hopefully. We could give them peace and happiness, or we could plunge them into the depths of suffering and sorrow."[53] This reception and good impression were marred, however, by the continuing attempts of the Jewish terrorists to kidnap British soldiers and by several murders. So far, two things were apparent, Macatee informed Marshall: "Jewish terrorism is as rampant as ever," and "the Arab boycott is as firm as ever."[54]

The group then traveled to the Negev in "suffocating heat." When they saw looming before them the "green-clad bluffs, plantations, lawns, fields and water-tower" of the Jewish settlement of Ruhama, Horowitz thought it was one of the most impressive moments of the entire tour. He pointed out the sections of the Negev pipeline to the committee, and they met with a group of irrigation engineers. There, according to Horowitz, "we sat down in a room with maps of the country on its walls, and the experts described the great hydrological plans that were designed to refresh the vast, arid Negev."[55]

Traveling to Ramallah and then Haifa, Horowitz worked at winning over Canada's chief justice, Ivan Rand. Rand wanted to know what solution was possible and was especially interested in the possibilities of partition. Horowitz told him that under existing conditions the proposal offered the "sole possibility of extricating the country from its political dilemma."[56] As evening approached, Horowitz and Rand went to the top of Mount Carmel to view the town and harbor beneath. Horowitz pointed out the "rusting hulks of the Jewish 'refugee fleet' " and described the "epic drama of the fight for free immigration, its grandeur and suffering, its anguish and heroism." The judge stood quietly, "his blue eyes gazing out over the vista of the harbor and rickety vessels illumined by the glow of the setting sun." In the eleven hours he had spent with Horowitz, Rand had seen evidence of the Arab attitude toward the Jews and the plight of the Jewish refugees fighting their way to their haven, and had learned about the "broad panorama of the Jewish struggle." Horowitz believed Rand's reservations about the Zionists were slowly disappearing and sympathy was emerging. As they parted, Rand told him, "I fully appreciate that you're fighting with your backs to the wall."[57]

As UNSCOP's third week in Palestine approached, the group was scheduled to hear from the Zionists' top leaders, including David Ben-Gurion and Chaim Weizmann. Ben-Gurion started with a two-hour speech laying out the case for ending the Mandate and establishing Palestine as a Jewish state. Calling the Palestine issue "the supreme test of the United Nations," Ben-Gurion told UNSCOP of how, despite all attempts to crush the Jewish people's identity, it had been preserved.

Addressing the issue of how Arabs would fare in a Jewish state, Ben-Gurion pledged that they would remain safe, whereas a Jewish minority in an Arab Palestine "would mean the final extinction of hope for the entire Jewish people."[58] Hitler, Ben-Gurion said, had been able to carry out his policy of extermination because the Jews were the only people without a state of their own, which had left them "unable to protect, to intervene, to save and to fight." Commenting on the contrast continually made between Jews and Arabs, the Jewish Agency's director explained:

> There are some 600,000 Jews in Palestine and some 1,100,000 Arabs. The Arabs own 94% of the land, the Jews only 6%; the Arabs have seven States, the Jews none; the Arabs have vast underdeveloped territories . . . the Jews have only a tiny beginning of a National Home and even that is begrudged them by the Palestine Administration. The most glaring disparity of all is that the Arabs have no problem of homelessness and immigration, while for the Jews homelessness is the root cause of all their suffering for centuries past.[59]

Under questioning by the Czech delegate, Ben-Gurion publicly announced that the Jewish Agency felt that *"we are entitled to Palestine as a whole, but we will be ready to consider the question of a Jewish State in an adequate area of Palestine"*[60] (our emphasis). David Horowitz followed Ben-Gurion, speaking for an hour and a half as he had done before an Anglo-American committee. Using an array of statistics and charts, he laid out the case that no Arabs had been displaced by the Jews and that because of Jewish advances, the Arabs in Palestine were better off economically and had better living conditions, as evidenced by lower infant mortality and increased longevity, than the Arabs in surrounding countries.

The fourth week began with the testimony of Chaim Weizmann. Although he had lost the support of the international movement, everyone recognized his years of leadership and his stature. He came to the hearing as a man who had more years in the Zionist movement than anyone in Palestine. For the first time, the YMCA auditorium was packed, since everyone knew that, as the American consul put it, Weizmann was the "star performer."[61] Because of his impaired eyesight, Weizmann was helped to the witness stand by an assistant. His presentation had been typed in large letters, but this was still a strain on his eyes, and he read slowly. Despite his maladies, Macatee observed, Weizmann spoke in a "well modulated tone which contrasted pleasantly with [the] intense pitch often attained by Mr. Ben-Gurion." He took the delegates through the entire history of the Jews and the Zionist movement.[62]

Weizmann avoided directly attacking the British government as the other witnesses had done. Instead, he praised it and its efforts on behalf of Zionism in years past, especially during the years of the Peel Commission of 1936, which had called for a Jewish state and partition of Palestine, and the Balfour Declaration, promising the Jews a national home. His words were a "vivid contrast," Consul Macatee reported, to the "torrential flow of denunciation of that Government which had been poured forth by the Agency." Weizmann made a deep impression as he spoke of the historical roots of the Jews in Palestine. He told them that as soon as a Jew came into contact with Palestine, he began "to feel as if he has returned. The country releases energies . . . in the Jewish people which are not released anywhere else." Everything that UNSCOP had seen, from the drained marshes to the homes and plants to the great institutions like Hebrew University, were "the work of Jewish planning, of Jewish genius, of Jewish hands and muscles."

Concluding his comments, Weizmann pleaded with UNSCOP to decide in favor of partition. He read to them a letter he had just received from Jan Smuts, the South African premier, who had been among those who had drafted the Balfour Declaration. "I see now, at this sad stage," Smuts wrote, "no escape except by way of partition." Weizmann thought the release of this letter would help persuade the delegates. "I am not so foolish to think that if you proclaim partition, passions will die out." But, Weizmann told them, with the United Nations proclaiming it and appealing to Jews and Arabs, "it will prevail."

Previous spokesmen for the Jewish Agency had referred to partition in passing, but Weizmann, no longer an official of the agency, did not face their constraints. His emphasis on partition as the best solution to an intractable situation made an impression on the UNSCOP delegates.[63]

Weizmann's testimony was reinforced when two successive groups of the UNSCOP delegates visited him at his home in Rehovoth. As they sat for drinks and dinner, looking out over the Judean hills and the beautiful sculpted gardens, Weizmann regaled them, as he had the Anglo-American Committee members, with stories of his years in the Zionist movement.[64] Horowitz, who had heard Weizmann speak many times, was still moved as he "interwove the story of his own early years with the broad narrative of the Jewish people's past and destiny." Weizmann captivated his guests with a "wonderful synthesis of Jewish wit and delicate irony, . . . his tales about his father, the timber-merchant of Pinsk, and of the school in which he studied in his youth; and thrown in casually, his references to meetings with the great ones of the earth." Ralph Bunche declared his emotional identification as an African American with the feelings of Jewish destiny that Weizmann spoke of. The group was so in awe that few spoke or asked any questions. On the drive back, Horowitz noticed that Sandström and Rand were unusually quiet. After a while, one of them murmured, "Well, that's really a great man."[65]

Having heard the testimony of the Jews, UNSCOP decided that it would try one more time to get the Arab Higher Committee to cooperate. However, its members repeated that they would not recognize or testify before the committee. Meanwhile, the British authorities reaffirmed their intention to execute the young Jewish terrorists who had carried out the prison break at Acre. The Irgun announced that its Committee of Retribution would take action if the executions were carried out. It succeeded in capturing two British sergeants on July 11, which led the British to send thousands of troops to comb townships around Netanya, where the sergeants had been kidnapped, and to threaten to again impose martial law.[66]

As the committee's time in Palestine was coming to an end, its members heard that a ship carrying 4,500 Jewish refugees had been intercepted by British destroyers. The British had been trying to stop the ships from leaving European ports, and Bevin had been in touch with Marshall, pleading for the United States to support a crackdown on the boats, which were funded largely by American Jews. The British had managed to slow the traffic down, but the boats kept coming.[67] In the United States, the public already had a romance with the flight of the Jewish refugees as they undertook the dangerous voyage to Palestine on the old, crowded ships. Back in April 1946, the American journalist I. F. Stone had been asked by an American he knew who was close to the Haganah if he would be interested in meeting people involved in the Aliyah Beth, the underground railroad to Palestine. Of course, he replied. A meeting was quickly set up with twenty-four young American Jews who were preparing to sail for Europe the next day. Stone asked if he could accompany them.

When Stone got to Europe, he was assigned to the *Beria,* a ship manned entirely by American Jews, most of whom had served during the war in the Army, Navy, and merchant marine. The ship, which carried 1,500 refugees, was one of the last the British administration allowed to land in Palestine and disembark. "The exodus of the Jews from Europe," Stone wrote in the conclusion to his best-selling book about the trip, *Underground to Palestine,* "is the greatest in the history of the Jewish people, greater than the migrations of the past out of Egypt and Spain. . . . Full support of the so-called illegal immigration is a moral obligation for world Jewry and a Christian duty for its friends." As for the ships and the immigrants being illegal, Stone wrote, "so was the Boston Tea Party."[68] Stone's reportage created a vast reservoir of sympathy for the new Jewish exodus. His series in the New York newspaper *P.M.* shot up its circulation, and when his book was published in 1946, he was a favorite on the lecture circuit.

But these were different times and the British were now in no mood to put up with the boats and their cargo. It was time to teach them a lesson. The American journalist Ruth Gruber, who had covered the Anglo-American Committee, was back in Palestine reporting on UNSCOP for the *Herald Tribune*. As she drove to Haifa to cover the developing story of a ship called the *Exodus 1947,* she listened to the radio. The former Chesapeake Bay excursion steamer, formerly called the *President Warfield* and used as a troopship in the Allied landing at Normandy, had a final mission to make. The ship, purchased by the Friends of the Haganah, had been refitted and brought up to par in Baltimore. It left America in late March, boarding Jewish DPs at the French port of Sètt, near Marseille. Over the radio station Kol Israel, the Voice of Israel, Gruber heard John Stanley Grauel give his account of how the boat had been attacked by five British destroyers and one cruiser seventeen miles from the shore of Palestine, in international waters. He said that the British had "immediately opened fire, threw gas bombs, and rammed our ship from three directions. On our deck there are one dead, five dying, and one hundred and twenty wounded."

By the time Gruber arrived at the Haifa docks, they looked like a war zone "with coils of rusted barbed wire, British Army tanks and trucks and some five hundred troops of the 6th Airborne Division." As the battered ship came into the harbor, she wrote, "It looked like a matchbox splintered by a nutcracker. One whole deck had been blasted open. I could see plumbing, broken staircases, and children running, their faces tormented with fear. Thousands of people crowded on the uppermost deck." The blue-and-white flag flew and a sign on the ship read, "HAGANAH Ship EXODUS 1947." The people were all singing "Hatikvah" (The Hope), the song, she noted, that they sang at every crisis. As the refugees came off the boat, the soldiers smashed their green water bottles. "Next their tattered bags and bundles were taken away" and they were told they would get everything back in Cyprus. The men were separated from the women, and some of them began screaming because in the concentration camps separation had meant death. Then they were frisked and sprayed with DDT.[69]

Witnessing this scene were UNSCOP members Judge Emil Sandström, its chairman, and Vladimir Simić of Yugoslavia, who had come to Haifa with Aubrey Eban. What they witnessed upset them and would prove to be a public relations disaster for the British, who, instead of taking the refugees to Cyprus as they said they would do, took them back to Germany on prison ships. "I had a feeling that the British Mandate died that day," thought Eban. "A regime that could maintain itself only by such . . . squalid acts was clearly on its way out. What they were doing was against the whole temperament and structure of the British character. It was both cruel and ridiculous. . . . I was sure that seeing with their own eyes this British behavior, the committee members would realize that the British simply had to go." And indeed, it had that effect.[70] After watching the scene, Sandström announced, "Britain must no longer have the mandate over Palestine."[71]

Later that day, two American journalists rang the bell of Granados's apartment in Palestine. With them was "a tall, blond-haired, blue-eyed man of about thirty, in a pitiable state of nervousness and exhaustion, his hair unkempt and obviously uncut for weeks, and his clothes apparently borrowed, for they were too small for him." He was introduced to Granados as John Stanley Grauel, a Haganah volunteer aboard the *Exodus*. The reporters told Granados that they had rushed him here from Haifa–the British were trying to jail him, and he wanted to tell Granados his story. Grauel told him that the British had taken all of his identification and he hadn't slept in sixty hours. Grauel, a former Methodist minister, told them he had been deeply disturbed by the plight of the Jewish displaced persons. Helping them get to Palestine, he felt, was his true calling in life. When he heard about the plans for the *Exodus,* he wanted to be part of it. Officially, Grauel had come on board the *Exodus* as an observer for the American Christian Palestine Committee and a reporter for the Episcopal Church publication *The Churchman*. In reality, Grauel had signed up with the Haganah and considered himself a militant fighting for a Jewish state. On board, not content just to function as a journalist, he served as a cook and able-bodied seaman. He told Granados that there were forty-three crew members, all Jews except him. Most were in their twenties, about a third war veterans, and the chief mate was a twenty-four-year-old naval reserve officer from Los Angeles, Bill Bernstein, whom Grauel considered his best friend. He cried as he told Granados that "the British clubbed him to death in the fight this morning." The next morning Grauel went to Sandström's apartment and recounted his story to

UNSCOP members Rand, Fabregat, and Blom, and the three members of the secretariat, Dr. Hoo, Dr. Bunche, and Dr. García Robles. He ended by telling them, "the Jews in the European Displaced Persons camps insist on coming to Palestine, they will come to Palestine, and nothing short of open warfare and complete destruction will halt them."[72]

A few days after this, Sandström, Dr. Hoo, and Dr. Bunche sat down with five members of the Haganah's High Command, military leaders who preferred to remain anonymous. The next day they gave the other UNSCOP members a report of what had been said. The military leaders told them that the Haganah was organized as an army with a branch devoted to immigration. Its High Command was appointed by a committee, which in turn, was elected by the entire Jewish community. It was not "a band of conspirators, nor a private army, nor a political faction." Rather, it claimed to be "a national volunteer army in whose ranks could be found practically every Jewish young man and woman capable of bearing arms." The Haganah leaders said that they had joined with the Irgun and Stern Gang after November 1945, when they realized that Britain had begun its anti-Zionist policy but had broken with them over their "insane actions" when it became clear that the two groups "could not be depended on to accept our authority and to work under our plans." Its current trained and armed strength was about 55,000 with about 35,000 reserves. Then Sandström asked, would the Haganah with its own means be able to defend itself against Arab attacks? The High Command responded that it could repulse any attack by the Arab population of Palestine. They also believed they could meet the situation if the other Arab states came to fight them. The Haganah was training its forces underground; its morale was high; and Jewish Palestine had an ammunitions industry that would put it far ahead of the Arab countries in the next few years. All of this, assumes, however, they told the committee, that UNSCOP's decision "will be in favor of the Jews and will give the Jews a legal basis for arming and defending themselves."[73]

In Washington, Secretary of the Treasury Henry Morgenthau went to the White House to speak to Truman about what the president called "that Jewish ship in Palistine [*sic*]." Truman was annoyed but told him he would talk to Marshall about it. When Morgenthau, whom Truman did not care for, left the Oval Office, Truman wrote in his diary, "He had no business, whatever to call me. The Jews have no sense of proportion nor do they have any judgement [*sic*] on world affairs." Then Truman, referring to the small number of Jewish refugees let into the country during the war to settle in Oswego, New York, wrote, "Henry brought a thousand Jews to New York on a supposedly temporary basis and they stayed." When the Republicans won Congress in 1946–"When the country went backward," as Truman put it–the Oswego refugees "loomed large on the D.P. program." Then Truman took his complaints about the Jews still further:

> The Jews, I find are very, very selfish. They care not how many Estonians, Latvians, Finns, Poles, Yugoslavs or Greeks get murdered or mistreated as DPs. as long as the Jews get special treatment. Yet when they have some physical, financial or political power neither Hitler nor Stalin has any thing on them for cruelty or mistreatment to the under dog. Put an underdog on top and it makes no difference whether his name is Russian, Jewish, Negro, Management, Labor, Mormon, Baptist he goes haywire. I've found very, very [few] Jews who remember their past condition when prosperity comes.[74]

A few weeks later Truman wrote a toned-down version of his mantra to Eleanor Roosevelt. The former first lady had written the president to protest the British handling of the *Exodus 1947.* In her syndicated newspaper column, she publicly expressed her strong reaction to what the British had done. "The thought of what it must mean to those poor human beings seems almost unbearable," Mrs. Roosevelt wrote. "They have gone through so much hardship and had thought themselves free forever from Germany, the country they associate with concentration camps and crematories. Now they are back there again. Somehow it is too horrible for any of us in this country to understand."[75]

The situation was "most embarrassing . . . all the way around," Truman wrote back to her. The problem was that the refugee ships "were loaded and started to Palestine with American funds and American backing–they were loaded knowing that they were trying to do an illegal act." He continued, "The action of some of our United States Zionists will eventually prejudice everyone against what they are trying to get done. I fear very much that the Jews are like all under dogs–when they get on top they are just as intolerant and as cruel as the people were to them when they were underneath."

"I regret this situation," he concluded his letter, *"because my sympathy has always been on their side"*[76] (our emphasis).

By the end of August, Truman realized that the *Exodus* had to be paid attention to. At a cabinet meeting on August 22, Truman asked Undersecretary of State Lovett to stay and discuss problems arising out of the "British ultimatum to the Jews" to disembark in France, as the French government had offered, or to be taken by force to a British-controlled port in Germany and forcibly disembarked. As Lovett noted, the harsh action by Britain had "caused a storm of protest" in the United States. State, he reported, had told the British that they should alter their plans and avoid having Jewish refugees land in Germany.[77] Lovett wired the U.S. Embassy in Britain, informing it that transferring the refugees to Germany would "arouse bitterness and aggravate the situation." He wired in a second message, "protests to White House and [State] Dept are piling up." To carry out such a step would harm the British image in America. Lovett urged that the embassy threaten to take up the message with the British government at a high level and in a formal manner.[78] American Ambassador Douglas then reported that he had gone ahead and spoken with Roger Makins, the assistant undersecretary of state in the British Foreign Office, to lodge the U.S. government's complaint. If there was a problem, Makins argued, it was due to the Jews, who would not disembark in France and insisted upon going to Palestine. By refusing to disembark in France, he said, the Jews had "made their own choice."[79]

Meanwhile, the UNSCOP delegates had proceeded to Sofar, Lebanon, where they were scheduled to hold informal meetings with delegates from the Arab states. At a cocktail party hosted by the Ministry of Foreign Affairs, Granados got a chance to speak with Camille Chamoun, the chief Lebanese delegate to the United Nations. Granados had met Chamoun at Lake Success and thought of him as a "cultured and cultivated man." When Granados raised the issue of partition as a possible solution, Chamoun was candid. "It will be very difficult to reach an understanding," he pointed out; ". . . the Arabs will never accept it. They will fight it." What about cantonization, which had been favored by the authors of Morrison-Grady? "It would be difficult to have even that accepted," Chamoun replied. Granados hoped that with cantons, Jewish immigration could take place in those areas assigned to the Jewish areas. "Let the other countries of the world . . . accept their share" of the Jews, said Chamoun; "the Arabs will never accept any further Jews."

Soon afterward, the committee met with the heads of the Arab states, including Hamid Frangie, the foreign minister of Lebanon, Emir Adel Arslan of Syria, and Fouad Hamza of Saudi Arabia. "It was a colorful meeting," Granados recalled, "the Saudi-Arabians in their white and brown robes, and white, gold-braided headdress; the Yemenites in their picturesque garb carrying curved daggers in their belts; the others wearing Western clothes, their only concession to the East being the bright red tarbooshes they wore instead of hats." Again the Arabs would not budge from their position; Palestine must be an Arab state where the Jews could live as a minority.

Some took an even harder line. Frangie told them that all of the Jews who had come to Palestine since Balfour–November 1917–were as far as they were concerned, illegal immigrants. It meant, in effect, that the Arabs would deport more than 400,000 Jews if an Arab Palestine was created. What about the Arab Higher Committee being led by supporters of the Nazis? It was immediately denied. "The members of the Arab Higher Committee had nothing to do with the Nazis," the UNSCOP delegates were told. The only reason the Mufti had taken refuge in Nazi Germany was because it too was "fighting against the Jews." Sandström tried to reason with them about the necessity of some kind of compromise, going through the different alternatives that had been put forth–a binational state, a federal state, partition. The Arabs were not interested and would not consider any of them. UNSCOP's Polish delegate, Karl Lisicky, summed up the Arab position: "We ask for 100% of our claims and the others can share the rest."[80]

In Lebanon, the only group the delegates met with who were sympathetic to the idea of a Jewish state were the Lebanese Christians. The Maronite Christians were the oldest Christian community in the world, they were told. Their orientation was Western and European, and they felt that Lebanon should be an independent state and not be forced into the Arab orbit. They also told the delegates that the Jews were justified in asking for their state and would be a positive force in the region.[81]

The next stop for UNSCOP and its final destination was Geneva. There they had a respite from the heat and sand of Palestine and were given offices in the Palais des Nations, the home of the now defunct League of Nations. They had been placed, Granados quipped, in "monumental halls haunted by the ghosts of international futility." The delegates had to decide whether they should visit the refugee camps in Germany and Austria. After a vote, it was decided that a subcommittee of alternatives would go to the camps for seven days. Wanting to see the camps for themselves, the two Latin American delegates, Granados and Fabregat, with Hood of Australia, decided to join them. While they were getting ready to leave, horrifying news came from Palestine. The British had hung the three Irgun members as promised, and in retaliation the Irgun had hung the kidnapped British sergeants. To Granados, it was a "vicious circle of repression and retaliation," which only a strong UNSCOP report could stop.[82]

The delegates who proceeded to the camps came back with stories similar to those of the Anglo-American Committee. Any doubts about where the refugees wanted to go disappeared as it became plain that almost all of the detained Jews would give up everything to reach Eretz Israel, the Land of Israel.[83] The subcommittee traveled to Munich and met with Rabbi Philip Bernstein, the adviser on Jewish affairs of the U.S. Army, who told them that the U.S. Army could not cope with the problem either materially or psychologically. The DPs had now grown from 100,000 to 225,000. The Army had done magnificent work, he told them, but it had insufficient resources to carry on. The rabbi had been searching for countries that would take in DPs. "I have peddled my wares through many capitals of the world, and while I have not found a lack of sympathy, I have gotten no results."

All of the Jewish DPs they talked to in the camps had been touched by violent death and had lost their families. One woman Granados could not forget was seventy-three-year-old Branda Kalk, who had been born in Poland, where she had lived all her life. The Germans had killed her husband in 1942, and she had escaped to Russia with the rest of her family. After the war, they had returned to Poland, and eight months after the liberation there had been a pogrom and her whole family–eight children and eighteen grandchildren–had been killed by the Poles. She alone had escaped but had been shot in the eye. She couldn't go back to Poland, she told the subcommittee's members, and didn't want to stay in Germany. "I want to go to Palestine," she said. "I know the conditions there. But where in the world is it good for the Jew? Sooner or later he is made to suffer. In Palestine, at least, the Jews fight together for their life and their country."

Their next stop was Vienna, into which more than a thousand "panic stricken Jews from Rumania and the Balkans" were pouring every week. On April 18, the United States declared that no further refugees would be admitted into the Army camps. As a result, facilities such as Vienna's Rothschild Hospital were packed with refugees. Conditions there were so horrible, the Iranian alternate, Dr. Ali Akdalan, told Granados, "This is a crime against humanity. I never imagined I would witness anything like this." They spoke with Lieutenant Colonel Henry C. McFeeley, chief of displaced persons in the U.S. zone in Austria. When asked for his opinion for a solution to this tragedy, McFeeley told them, "When I was in Palestine I saw a large area there that was deserted." Traveling to the British zone, the delegates and alternates went to the railway terminal, where they encountered Jews who were going to Palestine as legal immigrants. Granados observed that here were people who thought they were going to their "promised land." "And I thought to myself looking about me at these smiling faces, at these really joyous men and women, 'These are the only happy Jews I have seen in Europe.' That picture–of the sheer ecstasy which transfigured the faces of the Displaced Persons knowing they were going at last to Palestine–was the picture we took back with us to Geneva."

In Geneva rumors swirled as UNSCOP deliberated over its recommendations, which were due to be completed by September 1. Eban and Horowitz heard that the conflicts within the committee seemed to focus on three points: "the question of the form of the future government, the question of area . . . and the question of implementing a solution." They also received reports from both American and French statesmen that if the report was negative, there wouldn't be "a single country ready to give serious support to Jewish demands after the considered report by the representatives of eleven disinterested powers."[84] Weizmann had been right; this might be their last chance.

After much debate, the delegates finally agreed on eleven points, including an end to the British Mandate; that independence be given to Palestine under the auspices of the United Nations; that the sacred nature of all holy places be safeguarded; that international agreements for 250,000 Jewish DPs in Europe be worked out; and that there be some kind of economic union among the groups of Palestine. They could not agree, however, on what sort of entity or entities would be created in Palestine.

Finally, on August 31, UNSCOP presented two plans, a majority report and minority report. Both reports were written by Ralph Bunche, who worked without a break from 6:00 P.M. on August 27 to 6:00 A.M. on August 28.[85] He was not satisfied with the plans but wrote to his wife, "this is the sort of problem for which no really satisfactory solution is possible."[86] The majority report proposed the creation of Arab and Jewish states, which would become independent after a two-year transitional period, during which time the United Nations would supervise the Mandate under a trusteeship. The two states would have an economic union. In addition, 150,000 immigrants would be allowed to move to Palestine, and Jerusalem would be established as an international zone. Seven of the members–Canada, Czechoslovakia, Guatemala, Netherlands, Peru, Sweden, and Uruguay–supported this plan. Three member states–India, Iran, and Yugoslavia–voted for the minority report, which called for a federal state to be created within three years, of which the Arab and Jewish states would become provinces with a common capital in Jerusalem. The Australian delegate abstained from supporting either plan.

Liaison Officers David Horowitz and Aubrey Eban, representing the Jewish Agency, and Donald MacGillivray, representing the mandatory government, were instructed to appear by 9:00 P.M. to receive copies of the signed report. Eban and Horowitz knew what was in the reports, but until they were signed, nothing was final. They paced up and down the halls as the hours passed, and there was no sign of a report. A few minutes after eleven, a member darted out for a moment and, seeing the three men, exclaimed, "Oh, here are the expectant fathers!" But still they waited. Finally, at midnight on September 1, the eleven members of UNSCOP emerged and officially handed the report to them. Several had tears in their eyes, and Enrique Fabregat came over to Horowitz and embraced him. "It's the greatest moment in my life," he said.[87]

UNSCOP's report was the major story in the American press that day. Most editorials were favorable. Typical was that of *The Christian Science Monitor*, which said the "majority report–political partition and economic unification offers the best basis for a settlement that we are now likely to see."[88] The report, *The Atlanta Constitution* proclaimed, "was a statement which neither Zionist nor non-Zionist could quarrel."[89] The *Chicago Sun* saw it as "a workable and fair proposal." Partition, its editors thought, "appears to offer the best hope of a peaceful solution with the least injustice to both sides."[90] The *St. Louis Post-Dispatch* noted that "the State Department has been highly reticent about our policy toward Palestine," and that now "the time has come for a clear statement of our government's views."[91] The liberal and pro-Zionist *New York Post* called on Truman to "insist that our representatives in U.N. back the new recommendations to the utmost. . . . President Truman and the Congress must keep faith."[92]

The New York Times was more sanguine. In an editorial it wrote, "some of us have long had doubts as to the wisdom of erecting a political state on a basis of religious faith." It would, however, be willing to accept any favorable U.N. decision. But it warned that "no one can make up his mind in an hour or two as to the practicability and justice of the proposed partition lines."[93] *The Times*'s commentator, Arthur Krock, saw both UNSCOP reports as "conceding great merit to the case for a Jewish homeland." Noting the broad international support they received, Krock quoted an unnamed Washington official: "If this doesn't constitute world opinion, what would?" The message to the fifty-five U.N. members was clear: "The case for a Jewish homeland in Palestine has undeniable merit."[94]

Even I. F. Stone, who had once favored a binational state and who had first stirred the conscience of America with his account of travel on a refugee ship, now supported UNSCOP's conclusions. The majority proposal, he wrote, would "provide within a few years for the displaced Jews of Europe" as well as giving the Jews a "livable territory of their own." The Arabs' rights should not be ignored. But, he wrote, "the bulk of the land which would be given [the Jews] is theirs already, on empty desert on which only the stout-hearted would propose to make a home. The cities and fields they have already won were largely swamp, sand and rock before their coming." As Stone saw it, "a Jewish nation already exists in Palestine." The Arabs were asked only to accept "an inescapable fact."[95]

It was now up to the U.N. General Assembly to consider the UNSCOP recommendations and make a final decision about Palestine.

TEN

THE FIGHT OVER PARTITION: "A LINE OF FIRE AND BLOOD"

The fate of Palestine would soon be in the hands of the U.N. General Assembly, where the delegates would consider the UNSCOP's reports. As *New York Times* reporter Clifton Daniel wrote, the expectation was that the Arabs would "fight to the last diplomatic ditch to prevent the implementation of the United Nations committee proposals for partitioning the country into Jewish and Arab states." The Arabs were taking no chances, he wrote from Jerusalem, and were sending their most forceful personalities to represent them at the U.N. General Assembly to make their case.[1] Among them were the strident Jamal Husseini, representing the Palestinian Arabs for the Arab Higher Committee, and the Arab League's Faris al-Khouri of Syria.[2] On September 8, when the text of the reports was released, the Arab Higher Committee attacked both the majority and minority proposals, proclaiming them "absurd, impractical and unjust." The Arabs, its representative said, "would never allow a Jewish state to be established in one inch of Palestine," and he warned that attempts "to impose any solution contrary to the Arabs' birthright will only lead to trouble and bloodshed and probably to a third World War."[3]

Soon after, the Political Committee of the Arab League Council unanimously approved a statement denouncing partition and Zionism as totally illegitimate. Samir Rifai, the minister for foreign affairs for Trans-Jordan, further elaborated the Arabs' position in a letter he sent to Bevin and to British representatives in Arab countries. In their eyes, wrote Rifai, once the League of Nations was dissolved, everything that the Great Powers had accepted, including the Balfour Declaration, was null and void. UNSCOP's majority report, he wrote to Bevin, would "destroy the independence of Palestine as an Arab state." If it were instituted, he threatened, "the whole of the Middle East [would] flare up in disastrous and widespread disturbances." The Arabs would never accept partition and would take up arms in defense of their country with the support of the entire Arab world. In addition to supplying arms and money, he predicted, the Arab governments themselves would take action.[4]

In contrast to the Arabs, the Jewish Agency, the U.S. consul in Jerusalem, Robert Macatee, reported to Marshall, seemed "very satisfied," although some officials thought the report had two serious drawbacks, the inclusion of Jerusalem in the Free Zone and the failure to include western Galilee in the Jewish state. Nevertheless, a Zionist correspondent of a large American newspaper had confided to him, "To say the Jews are pleased with the report is an understatement, they are elated."[5] Golda Meyerson, head of the Political Department of the Jewish Agency and the only authorized Zionist spokesperson left in Palestine, the others being at the United Nations, told the press that the agency was relieved but also skeptical because there had been so many commissions and nothing had come of them. It was also aware that a combination of the British, Soviet, and Arab blocs might prevent the two-thirds General Assembly majority necessary to adopt partition. She was troubled as well by the proposed internationalization of Jerusalem, the exclusion of western Galilee from the proposed Jewish state, and the indefiniteness of the transition period. To Meyerson, "The words Palestine and Jerusalem were almost synonymous to Jews . . . we can hardly imagine a Jewish state without Jerusalem."[6]

But not all of Jewish Palestine was happy with the partition plan. Newspapers of the left-wing Socialist Party and Orthodox and Revisionist parties proclaimed their dismay; the Holy Land in its entirety belonged to the Jews; it was their partrimony and should not be divided. Clashes between the Haganah and the Irgun and Stern Gang intensified as the Haganah tried to maintain the peace during the U.N. deliberations, even to the extent of protecting British installations against attacks. On September 7, the Irgun made a secret broadcast that partition would be "a national disaster." Denouncing the campaign of the Haganah to clamp down on them, it stated, "We shall give the terrorists in the camp of the defeatists their due. We shall repulse their attacks by continuing to fight the Nazi-British enemy."[7]

By all indications, however, the majority of American and Palestinian Jews supported the UNSCOP partition plan. Now their leaders were focused on how to get it implemented. To this end, Emanuel Neumann, Nahum Goldmann, and David Horowitz approached the U.S. Embassy in Britain asking to set up meetings. The chargé d'affaires at the embassy reported to Marshall that their primary concern was the implementation of the UNSCOP report. All seemed to take it for granted that the General Assembly would approve the majority plan since it had been created by an impartial group appointed by the General Assembly itself. And even though they considered Bevin to be the "number one enemy of Zionism," the Jewish leaders believed that the British government would either stand aside or accept the final decision of the General Assembly. That being the case, the Jewish leaders believed that the success or failure of the UNSCOP majority plan would depend on the position taken by the U.S. government. Therefore, the Jewish Agency was going to mount an intensive campaign in the United States in favor of the majority plan until the final vote in the General Assembly took place.[8]

In the United States, the Zionists had not only public opinion on their side but important leaders of liberal and progressive thought. Observing the scene, David Horowitz thought that the cultural, economic, and intellectual centers of America backed partition and had thrown the full weight of their influence on its behalf. The entire Jewish and liberal community, Horowitz wrote, "from coast to coast, was aflame with zeal and ardor of the battle; its heart beat in unison with ours."[9] Horowitz's estimate was reinforced by Chaim Weizmann, who was happy to find support for partition coming from Joseph Proskauer of the American Jewish Committee, who had once been an anti-Zionist foe; the banker Bernard Baruch; and the business leader Herbert Bayard Swope. Within the administration, he was pleased to find support from Henry Morgenthau and others.[10]

Arthur Lourie of the Agency quickly went to work lining up congressional support. First he got in touch with Congressman Emanuel Celler. Could he organize members of both the House and Senate to write Truman in support of partition? Enclosed were three sample letters that Lourie had prepared, each with a slightly different slant. They were worded so as to not offend Truman, who was already angry at the immense lobbying and pressure. One of the drafts, for example, would say to Truman that he must have read the majority report "with great satisfaction, for they completely vindicate your judgment in the matter, both in respect of the admission of Jewish refugees to Palestine and the establishment of a viable Jewish state in an adequate area of Palestine." Noting that the report was supported by the American press, Lourie's draft went on to say, "there is every likelihood of a solution along the lines you have indicated in the past."[11] Whether or not the president would bite, however, was not at all certain. He was still smarting at his inability to achieve a solution to the Palestine issue, for which, in part, he blamed the Jews as well as the British and the Arabs. Sympathetic congressmen immediately sent letters to the president.

The Nation Associates then went into action organizing a mid-October dinner at the Waldorf-Astoria hotel on "The Palestine Solution and its Relationship to World Peace." Pointing out suspected State Department plans to prevent the establishment of a Jewish state, *The Nation* publisher Freda Kirchwey wrote the invitation letter, signed by the Nation Associates, warning their constituency that "there is a gigantic double-cross in the offing at the United Nations." Not only was that an actual possibility, Kirchwey argued, but "President Truman is reported as capitulating to the Arabs," which meant that "there will be no hope for settling the Jews of Europe in Palestine." The letter announced that the scheduled dinner would feature some of the most prominent American and world figures who supported a Jewish state, including two members of the Anglo-American Committee, Bartley Crum and Richard Crossman; former Undersecretary of State Sumner Welles; and Kirchwey herself. The dinner would be a "public demonstration" timed for precisely when the United Nations would be dealing with Palestine and could "play a decisive role in making clear to the United States delegates that American public opinion will not accept the planned betrayal."[12] Eleanor Roosevelt, who was one of the U.S. delegates, took the letter as an insult. She promptly withdrew as cochairperson of the dinner, asked that her name be removed from the letterhead, and resigned from the Nation Associates. "I do not feel I can be affiliated with a group which does such irresponsible things," she said.[13]

Before the dinner took place, the Nation Associates submitted another detailed report to the U.N. General Assembly, "Could the Arabs Stage an Armed Revolt Against the United Nations?" The seventy-seven-page report argued that the Arabs were ill trained, ill equipped, ill disciplined, and undernourished. Their armies had few modern weapons, hardly any air force, and no navy. The only decently trained forces were those of the Trans-Jordan Arab Legion of 24,000 men, which was in effect controlled by Britain. If they nonetheless chose to attack, the report predicted, the Jewish resistance would amount to a force they "could not overcome." The Jewish community's forces of the Haganah consisted of almost 70,000 troops, who were well trained, disciplined, and led by experienced officers.[14]

The report was based on material gathered in Palestine by the Nation Associates' director, Lillie Shultz, and the writer Saul K. Padover, and was, as they called it, "the first factual presentation of the social, economic, political and military aspects of life in the Arab countries," which, its editors thought, "exposed the inability of the Arabs to wage a military war or to stage any long-term revolt." They believed their report influenced the delegates. The report, along with the magazine's special October 4 issue, was "the only comprehensive documents coming from any non-Jewish source and presented to the U.N." As the magazine noted, its reports and future ones–including one called "Police State: Nazi Model"–were distributed not only to the U.N. delegates and Secretariat but to the Democratic National Committee and major American intellectual and cultural figures. It was also sent to liberal senators and congressmen, to every governor, and to all college and university presidents, union leaders, and other organizations.[15]

Most important, Shultz believed that the magazine had direct access to the ear of the president through David Niles and other advisers. It also was now, she thought, "the only direct line to the Democratic Party and the Administration" because Truman refused to meet with officials of American Zionist groups, especially Rabbi Silver, who had become persona non grata at the White House. Shultz thought that this left the Nation Associates as one of the few pro-Zionist organizations that could present the Zionist case to the administration without receiving a hostile response.

At the United Nations, Truman appointed Senator Warren R. Austin of Vermont to lead the U.S. delegation, with Herschel V. Johnson, the former minister to Sweden, as his deputy. The U.S. delegation included some of the most prominent Americans: John Foster Dulles, Eleanor Roosevelt, Major General Matthew B. Ridgway, Philip Jessup, and General John Hilldring, who was recommended by David Niles to be a liaison with the White House.[16] The group would regularly be in touch with Secretary of State Marshall, working through the undersecretary of state, Robert Lovett. At times Lovett would give the delegates direct instructions; at other time he would travel to New York to meet with them in person. Truman would then consult with both Marshall and Lovett.

To deal with the Palestine issue, U.N. Secretary-General Trygve Lie recommended that an ad hoc Palestine committee be set up, with a representative from each of the member states in the General Assembly to study the majority and minority proposals. Herschel Johnson was appointed the chief U.S. member of the committee, and General Hilldring was appointed to assist him. Diplomats George Wadsworth and Paul Alling were made consultants to the committee, which was charged with considering the minority and majority proposals of UNSCOP and presenting its conclusions to the General Assembly, which would then vote on it.[17]

A key to the U.S. position was the question of what stance Secretary of State Marshall would take on Palestine. Eleanor Roosevelt, a supporter of partition, said that Marshall at first had thought that America was obligated to support the establishment of a Jewish state as a matter of principle. But after the General Assembly began, the State Department people told him he was wrong and, according to Mrs. Roosevelt, put it in such a way that "to stand by his support of the UNSCOP recommendations [would be] to go against the advice of all the qualified experts in the Department."[18]

The influence of the department's qualified experts was apparent when, on September 15, Marshall met with the U.S. delegates to discuss the U.S. position and what he would say in his upcoming speech at the United Nations. Marshall first wanted to know the delegates' views–should they take a stand in favor of partition and back the UNSCOP majority report? In a statement Loy Henderson and Undersecretary Lovett had prepared for him to consider using as the text for his speech to the General Assembly, they advised stressing that not all UNSCOP members supported partition, and they hoped that the General Assembly would study the report and then reach a general agreement.

Hilldring immediately answered that such a statement by the secretary "would certainly be a disappointment to American Jews and Jews everywhere, who hoped the United States would take a favorable position on Palestine . . . in favor of the majority report." He thought they should accept the UNSCOP majority report, while telling the General Assembly that the United States would be willing to amend it as a result of General Assembly debate. Marshall said that the Zionists were pressing him for a decision on Palestine, but he was concerned that if they adopted the majority report it would provoke a violent Arab reaction. He thought the United States should avoid "arousing the Arabs and precipitating their *rapprochement* with the Soviet Union in the first week or ten days of the General Assembly." This, he thought, would happen if the delegation took a clear stand in favor of partition when he delivered his speech. But then again, if he didn't take a clear stand, he would be accused of "pussyfooting." At this point Mrs.

Roosevelt interrupted and asked if their concern shouldn't be more about the effect on the United Nations if the United States did not back the majority report.

Henderson then spoke up and warned that if partition were accepted, it would have to be implemented by force. The United States would be fighting Jewish terrorists at the same time as Arab terrorists. UNSCOP said it was impartial, but, as Henderson saw things, although the United States could use force on behalf of principle, it shouldn't do so when there was no principle but "only expediency," which he thought was the case here. Austin added that the delegation had to decide whether to take a position at present or "wait until the row got hot." He was having a hard time seeing how partition could work because Palestine was "already too small for a still smaller state." Each side would have to defend itself with "bayonets forever, until extinguished in blood." That meant if the United States supported partition, it would also have to agree to defend it with American troops.[19]

Two days after this meeting, Marshall gave his speech to the General Assembly. Trying to take a middle course, he told the anxious delegates, "while the final decision of this Assembly must properly await the detailed consideration of the report, the Government of the United States gives great weight not only to the recommendations which have met with the unanimous approval of the Special Committee, but also to those which have been approved by the majority of that Committee." This noncommittal statement did not make anyone happy. The Zionists were disappointed by the lack of outright support for partition, while the Arabs took it to mean that the United States intended to support it.[20]

Meanwhile, Henderson was disturbed by the drift of the September 15 meeting and by his unpersuasive performance. He knew he could have made a stronger case. He felt sideswiped; he had thought he was going to be meeting with Marshall alone and had been surprised to find the delegates there. When he got back to his office, he prepared a lengthy memo for Marshall.[21]

In the memorandum, Henderson let out all the stops. He told the secretary that he felt it was his duty to tell him that "it would not be in the national interest of the United States for it to advocate any kind of a plan at this time for the partitioning of Palestine or for the setting up of a Jewish State in Palestine." He thought the best course of action for the United States to take during the U.N. debates was to remain "strictly impartial."

To back up his rejection of partition, Henderson argued several points, most of them familiar: support would undermine U.S. relations with the Arab world, causing the Arab nations to join with the Soviet Union against the United States; the United States would be expected to carry the major burden in providing force, materials, and money for the implementation of partition. He then offered the tactics he thought the United States should follow at the General Assembly. The United States should say publicly that it would give "due weight" to the views of other nations and all other interested parties. Then there must be discussion and agreement among moderate Jews and moderate Arabs, not by an outside international organization. The goal should be to get the moderates on both sides to "enable the setting up of a trusteeship" that would function "in a neutral manner so as not to favor either partition or a single state." That meant, in effect, continuation of the status quo, with control falling either to one of the big powers or to the United Nations itself.[22]

After Marshall received the memo and obviously had given it to Truman, Henderson was summoned to the White House by Truman's counsel, Clark Clifford. Clifford, who had grown up in St. Louis, had attended both college and law school at Washington University. He had come to Washington as an assistant to James K. Vardaman, who had been appointed by his friend Harry Truman to be his naval aide, a largely ceremonial position. Samuel Rosenman, noticing that Clifford had little to do, had asked the young attorney if he wanted to work with him. During the time they spent together, Rosenman and Clifford had discussed the situation of the Jews and their need for a homeland. "I learned a good deal from Judge Rosenman," Clifford later wrote; ". . . he felt strongly about it." By the time Truman consulted with Clifford on the issue, Clifford wrote, "I was sympathetic" and had become "an advocate of the Jewish State."[23]

When Rosenman left, Truman did not think he would fill the position. But now there was no one to handle the speeches, to act as a liaison with the cabinet, or to work with the Justice Department. He reconsidered, and Clifford was officially brought on board as his special counsel on June 27, 1946. He would handle these tasks, which evolved, as Clifford put it, into a "forerunner of a National Security Assistant." His other responsibilities included acting as a liaison with State and Defense and serving as one of Truman's chief political advisers.[24] "Tall, handsome and friendly," Clifford was immediately popular. Truman also found him "energetic and highly capable."[25]

Clifford's attitude was apparent when Loy Henderson arrived at the White House to discuss his memorandum. When he got there, he found Truman with Clifford, David Niles, and a few other White House personnel. Clifford, according to Henderson, "took the lead in the conversation." It was the president's understanding that Henderson "was opposed to the United States adopting a position of supporting the establishment of Jewish State in Palestine." What, Clifford asked, were the sources of his views? Responding by reiterating his arguments, Henderson proceeded to endure what he saw as a cross-examination. "Were they merely my opinions" or were they "based on prejudice or bias?" "Did I think that my judgment and that of members of my office were superior to that of the intelligent group that the United Nations had selected to study and report on the Palestine problem?"

Henderson thought "the group was trying to humiliate and break me down in the presence of the President." Trying as best as he could to hold his own ground, he replied that the views he expressed were not only his but those of the U.S. legations and consular officials in the Middle East, as well as the views of those in the State Department who had responsibility for the region. Henderson and the Office of Near Eastern and African Affairs were doing their job, which was to acquaint the secretary of state with their conclusions, which he would pass on to Truman. In particular, they had the responsibility of letting the president know the ramifications of what support of a Jewish state would mean to the United States' international position and interests.

As the questioning became "more and more rough," Truman got up and muttered, "Oh hell. I'm leaving." Henderson took this to mean that despite the late date, Truman had not as yet "made the final and definite decision to go all out for the establishment of the Jewish State." He conjectured that Truman was desperately hoping "that the Department of State would tell him that the setting up in Palestine of Arab and Jewish states . . . would be in the interest of the United States," something Henderson and the Division of Near Eastern Affairs would never do.[26]

Despite Henderson's efforts, Truman and Secretary of State Marshall did not accept the policy proposals of the Division of Near Eastern Affairs. On September 24, Marshall met with Eleanor Roosevelt, Dean Rusk, Charles Bohlen, and General John Hilldring. They agreed that U.S. Representative to the United Nations Warren Austin would not make an opening statement at the start of the Ad Hoc Committee on Palestine but would call for a discussion of the Palestine question among the British government, the Arab Higher Committee, and the Jewish Agency, after which Herschel Johnson would present the U.S. view for the first time. The U.S. position would take into account previous American commitments and the UNSCOP majority report, as well as the U.N. delegates' views. The United States would also reserve the right to make amendments to the reports. Most important, it would make it clear that it would retain "in the plan the provisions for partition and large-scale immigration."

However, if two thirds of the General Assembly failed to give it their support, the U.S. delegation would consider two options: to force a vote in the Ad Hoc Committee to reveal that the U.N. majority did not back the UNSCOP majority report's conclusion, even though it had been prepared by the United Nations' own committee; or decide to propose an alternate solution that would be able to gain a two-thirds majority. The group then pledged itself to "utmost secrecy" about the U.S. position until the time for it to be made public.[27]

A week later, the State Department prepared a memorandum with suggestions for fleshing out the general outlines of a U.S. plan, which began (to their chagrin) to be referred as the "American Plan." A lengthy working paper, the memorandum was meant to offer the U.S. delegation a series of goals and recommendations. Its working premise was that the United States "should give support to the majority plan in principle with a view to perfecting the plan in certain of its features." The "perfection" would lie in its making it more palatable to the Arabs by giving them more land. Its recommendations included giving Jaffa—a city with 70,000 Arab residents and only 10,000 Jewish ones—to the Arab state as an enclave and not to the Jewish state, as suggested by the UNSCOP majority report. The eastern boundary of western Galilee should go to the Arab state as well and redrawn to include Safad, whose population was 9,500 Arabs and 2,500 Jews. This meant that the "southern portion of the Negev, allocated to the Jewish State by the majority plan, should be included in the Arab State." An area useful only for seasonal grazing, the memorandum argued, it was inhabited by 60,000 Arabs and had no Jewish settlements. This last point is one that would be of most concern to the Jewish Agency. The largely unsettled area was deemed necessary for a Jewish state, since it could provide space for new immigrants and only the Jews would have the incentive and resources to develop its economic potential. Despite these proposed changes benefiting the Arabs, the memorandum acknowledged that the Arab states would most likely reject any U.N.-imposed solution, outside of all of Palestine becoming an Arab state. "It is difficult," the memorandum put it, "to predict whether any solution short of immediate independence would obtain even the reluctant acquiescence of the Arab States."[28]

Herbert Vere Evatt, the Australian minister for external affairs, headed up the Ad Hoc Committee on Palestine, which began its proceedings on September 27. Evatt had only a few weeks to make the committee's recommendations to the General Assembly. He decided to deal with the majority and minority reports by setting up two subcommittees to consider each of them, placing proponents of each plan on each committee. For Subcommittee 1, representing partition, he chose Canada, Czechoslovakia, Guatemala, Poland, South Africa, the United States, the Soviet Union, Uruguay, and Venezuela. Subcommittee 2, representing the unitary government minority proposal, was made up of Afghanistan, Colombia (which dropped out), Egypt, Iraq, Lebanon, Pakistan, Saudi Arabia, Syria, and Yemen. The committees had until October 29 to prepare and present their reports with recommendations.[29]

Evatt also suggested that the Jews and the Arabs be given a hearing. Consequently, Jamal Husseini gave an 8,000-word statement announcing that the Arab Higher Committee would not accept either UNSCOP's majority or minority report. Instead, it promised to drench the soil of the Holy Land "with the last drop of our blood in the lawful defense of all and every inch of it." There were three simple points to Arab policy: "no partition, no further Jewish immigration and no Jewish State." It would accept only an "independent, democratic Arab state, embracing all of Palestine."

Husseini said the Arabs' case was "simple and self-evident," based on international justice: "It is that of a people who desire to live in undisturbed possession of their country, in which they have continually existed and with which they have become inextricably interwoven." Going back to ancient history, Husseini claimed that the European Jews were not descendants of Palestine's ancient inhabitants; he also said that that the Balfour Declaration violated the League of Nations Covenant and was illegitimate, and that the Zionists were preparing a Jewish invasion "to take Palestine from the Arab inhabitants."[30]

Testifying for the Zionists were Rabbi Silver and Moshe Shertok. Silver addressed the Jews' historical claims to Palestine and disputed those of the Arabs. In 1917, when the Allies had liberated Palestine along with other provinces of the Ottoman Empire, Silver argued, Palestine had been just a "segment of a Turkish province. . . . There was no politically or culturally distinct or distinguishable Arab nation in that province." When the Arabs had conquered Palestine in A.D. 634, the country had contained a heterogeneous population, and then after A.D. 1071 it had been conquered by various non-Arab peoples, including the Kurds, the Crusaders, and finally the Ottoman Turks. Most important, by the time the Arabs had conquered Palestine in A.D. 634, Silver continued, the Jews had "already completed nearly 2,000 years of national history in that country during which they created a civilization which decidedly influenced the course of mankind." When the Palestinian Mandate recognized the historical connection between the Jewish people and Palestine, it was, according to Silver, "only stating a fact that was universally acknowledged through the ages."

Silver also laid to rest the Biltmore Declaration calling for a Jewish state in all of Palestine and publicly affirmed his support for partition. The Jewish Agency, he told the delegates, would "assume this burden as one of the sacrifices designed to find a way out of the present intolerable impasse." This was not an easy thing for Silver to support; it was done, he said, "in sadness, and most reluctantly." It was a "sacrifice." Silver also voiced the American Zionists' opposition to the minority plan, which, he said, would bring "all the disadvantages of partition–and a very bad partition geographically–without the compensating advantages of a real partition: statehood, independence and free immigration."[31]

During the proceedings, Weizmann was also called upon to give a statement. He asked Aubrey Eban to help him prepare a draft. Weizmann decided to put in a little humor to deal with the Arabs' arguments that the Jews were not descendants of the ancient Hebrews but came from the Khazar tribes of southern Russia. Apparently, it went over well with the audience, who thought it was hilarious when the aging and most eminent of Zionist leaders said, "It is strange, very strange, but all my life I have been a Jew. I have felt like a Jew. I have suffered like a Jew. So now it is fascinating to learn that I am a Khazar!" Weizmann wanted a biblical verse for the ending. Searching the room, Eban and Weizmann found a Bible in the hotel table and spent a half hour looking for an appropriate "Return to Zion" passage. They found one in Isaiah. "Well, this is it," said Weizmann, "over the top for the last time." He ended his speech with words appropriate to the occasion: "The Lord shall set his hand the second time to recover the remnants of his people. And he shall set up an ensign for the nations, and shall assemble the outcast of Israel and gather together the dispersed of Judah from the four corners of the earth."[32]

Kirchwey's predictions of an impending double-cross were not off the mark when, on October 3, the American delegates met in Marshall's Washington office, along with Dean Rusk, Charles Bohlen, Ambassador Alling, and others. Marshall told them that although the U.S. position in principle was to support the UNSCOP majority report, they now favored "certain modifications" that would be proposed as amendments. These were candidly, Marshall told them, "modifications . . . of a pro-Arab nature," concerning boundary issues and the plans for an economic union of Jewish and Arab states. Moreover, Marshall predicted that at present the majority report would probably not get a two-thirds majority of the General Assembly. Should that occur, Marshall continued, "some form of Trusteeship for Palestine might be desirable." And, he emphasized, although the United States was committed to Jewish immigration into Palestine, it was *"not committed to support the creation of a Sovereign Jewish State"* (our emphasis). And, finally, Marshall told them that the United States "should not attempt to persuade members of the General Assembly to vote for the majority plan."[33]

Pressures were now mounting on Truman to come out in favor of partition. Eddie Jacobson wrote to him, "Again I am appealing to you on behalf of my People." He continued:

> The future of one and one-half million Jews in Europe depends on what happens at the present meeting of the United Nations. With winter coming on with its attendant hardships, time is short for action by this meeting to alleviate further suffering by these helpless people.
>
> How they will be able to survive another winter in concentration camps and the hell holes in which they live, is beyond my imagination. In all this World, there is only one place where they can go–and that is Palestine. You and I know only too well this is the only answer.
>
> I have read Secretary Marshall's recent statement that the U.S. would give great weight to UNSOP's [*sic*] recommendation; that was a great deal to be thankful for. Now, if it were possible for you, as leader and spokesman for our country, to express your support of this action, I think we can accomplish our aims before the United Nations Assembly.

Continuing, Jacobson told his friend that he trusted "to God that he give you strength and guidance to act immediately." He concluded:

> I think I am one of a few who actually knows and realizes what terrible heavy burdens you are carrying on your shoulders during these hectic days. I should, therefore, be the last man to add to them; but I feel you will forgive me for doing so, because tens of thousands of lives depend on words from your mouth and heart. Harry, my people need help and I am appealing to you to help them.[34]

Asking that he not be quoted, Truman responded. Since the matter was at the U.N. General Assembly, he said, "I don't think it would be right or proper for me to interfere at this stage, particularly as it requires a two-third vote to accomplish the purpose sought. Marshall was handling the issue," he told him, "I think, as it should be and I hope it will work out all right."[35]

Pressure on the administration to support the majority report was also coming from both Republicans and Democrats. Republican Senator Arthur Vandenberg, the chairman of the Senate Foreign Relations Committee, met with Rabbi Silver in New York City and told him that he favored the majority report as "the only hopeful basis for a settlement of the Palestine conundrum." But, he added, he could not accept imposing it by force, which might lead the United States to take responsibility and send in troops. Vandenberg then wrote to his Senate colleague Robert A. Taft that he wanted to develop a Republican position, in which they would "unequivocally continue to assert our belief that viable partition is the sound and hopeful answer." His hope was that the General Assembly would vote for partition and that a formula would "be found which will produce the acceptance by both Arabs and Jews without too much grief or bloodshed."[36]

At the same time as Vandenberg expressed what he saw as the right Republican position, leading New York Democrats, along with the governors of twenty-three states, publicly urged that the United States come out in support of the majority report and partition.[37]

As much as he wished the Palestine problem to be out of his hands, Truman was going to have to make a decision. Was the United States going to support partition, and if so, how strongly? He consistently argued that the proper place for the settlement of such disputes was the United Nations. Now that it had come up with a plan, he had to consider Eleanor Roosevelt's point; if he wanted the infant United Nations to succeed, how could he not support the UNSCOP majority report recommending partition? The Arabs' intransigence helped him to commit himself. On October 9, Clifton Daniel of *The New York Times* reported that the Grand Mufti had arrived in Lebanon for a meeting of the Arab League, to discuss military measures should partition be attempted. The League announced that they intended to occupy Palestine if the British forces withdrew and to resist by force any effort to create a Jewish state.[38] When Truman was given the report in the Oval Office, he thought the League's position was "belligerent and defiant." Therefore, he claimed in his memoir, he then *"instructed the State Department to support the partition plan"*[39] (our emphasis).

On October 11, Herschel Johnson, U.S. representative on the Ad Hoc Committee on Palestine, gave a speech publicly endorsing partition. Though rejecting the use of U.S. troops in Palestine, he called on the United Nations to establish a volunteer police force to help keep the peace in there, if needed. Until the future status of Palestine was decided, he said, it should remain the responsibility of Britain, the mandatory power, to administer the area in its transition to independence. *The New York Times* called Johnson's statement "clear and explicit" and remarked that it gave "considerably stronger support to partition than had been expected."[40]

Supporting partition was now the official U.S. policy, but the State Department set about adjusting the borders of the proposed two states. The territorial modifications it suggested when meeting with the U.S. delegation included taking Safed and the southern Negev away from the Jews and placing them in the new Arab state.[41] Marshall explained the decision of State and the U.N. delegation to have the Negev go to the Arabs by pointing out that it was "overwhelmingly Arab" and a "barren, arid and topographically inhospitable area suitable only for marginal cultivation and seasonal grazing," a task carried out only by its present inhabitants, "semi-nomadic Arabs." Experts, he incorrectly wrote, "admit that there is extremely slight chance of any large-scale development in the area." If it went to the Jews, "it would create a wedge in an Arab area inhabited by traditionally truculent and fanatical Moslems" and would increase the difficulty of creating the two states. As for the Jews' argument for the desirability of their state having an outlet to the Red Sea and the port of Aqaba, Marshall argued that "Aqaba is not in Palestine," and any plan to have any part of Palestine bordering on the Red Sea as a port "is open to serious doubt."[42]

Not everyone on the U.S. delegation agreed. The two American representatives on the Ad Hoc Committee, Herschel Johnson and General Hilldring, wrote Austin a memo that Austin forwarded to Marshall, pointing out that the United States was the only country asking for the Negev to be reassigned to the Arab state. Both of them felt strongly that it would be a mistake to make this change. If the United States introduced this publicly, the Jewish Agency would vigorously oppose it, as would "all the friends of partition in the 37-nation committee," of which there were many. The only support would come from the Arab states. That meant that if the vote for partition failed, the blame would be put on the United States for "raising this major doubt as to the justice of the partition plan."[43]

The Jewish Agency was becoming increasingly alarmed about the United States' attempt to give the Negev to the Arabs. Not only was it backtracking on previous promises, but it was dangerous because the Yishuv's enemies might use it as a weapon against the Ad Hoc Committee's overall support of UNSCOP's majority plan. They would try to use it to change U.N. delegates' minds about partition; they would now think it had no chance of succeeding since the U.S. delegation was vacillating. Ironically, the Soviet delegate, S. L. Tsarapkin, had made a statement before the Ad Hoc Committee endorsing partition and opposing any changes to the UNSCOP recommendations. The Soviets, who had long opposed Zionism, were being more steadfast in their support than their wavering American friends.[44]

How could the State Department be stopped from giving away the

Negev? "The situation," Epstein wrote, "called for immediate and drastic action on our part to prevent that calamity." There was only one possible action that could help them: *"Direct and instant intervention by the President of the United States himself."* (our emphasis). And the only person who would be able to influence Truman was Chaim Weizmann. Weizmann had just arrived in New York, where he was to address the United Nations, when Epstein called him at the Waldorf-Astoria telling him that it was crucial that he see Truman as soon as possible and convince him not to take away the Negev from the Jewish state. Weizmann agreed to try if Epstein could set up an appointment. Epstein then called Niles, who was able to set one up with the help of the British ambassador, Lord Inverchapel, for November 19.[45] Just as the former British ambassador, Lord Halifax, had arranged for the British subject Weizmann to see Truman, so Inverchapel cooperated in arranging a meeting. At times, Inverchapel expressed strong Zionist sympathies, but Acheson found him to be so eccentric that he did not know what he really believed.[46]

Epstein was passionate about the Negev and especially the inclusion of Eilat–then called Elath–within the proposed borders of the Jewish state. If there were any chance at all of cultivating the desert, it could be realized only by Jewish pioneers, who had already succeeded in regions in Palestine that had previously been dismissed as uncultivable. The Negev would increase the absorptive capacity of the Jewish state and create new opportunities for large-scale immigration and settlement. It was crucial because it had outlets to both the Mediterranean and the Red Sea, giving the country access to both the Atlantic and Indian Oceans. Given Arab opposition, the Jewish state probably would not be able to use the Suez Canal. The only alternative was to use the port on the Gulf of Aqaba.[47]

Weizmann was buoyant and optimistic about what might transpire. The Harrison Report had launched the events that led to UNSCOP's majority report and he attributed all this to Truman's actions as president. He compared Truman favorably to Lord Balfour, whom all knew to be "a friend and sympathizer of the Zionist cause."

Now he wanted to believe that Zionism had once again found a "powerful champion."

On the nineteenth, the day he was to meet Truman, Epstein joined Weizmann and Justice Felix Frankfurter for breakfast. Epstein gave Weizmann a memo and a map locating the Gulf of Elath and demonstrating its importance to the future Jewish state. The memo, which Weizmann used as the basis of his discussion with Truman, stated:

> For the Jewish State, this outlet will be one of the most important routes for commercial relations with that part of the world. The Jewish State, to absorb the refugees coming from Europe, will have to do its utmost to develop its industrial and commercial capacities, and in this connection the importance of Elath is much greater than just a piece of land on the Red Sea.
>
> In view of present strained relations with the Arabs, and the existing Arab boycott, there is a permanent threat that, by the cutting off of the Persian Gulf by the Iraq Government, or through the Suez Canal by the Egyptian Government, the Jewish State can be isolated from some part of the world which may be of great importance. . . .
>
> Elath has played a major role in Jewish history, from the early days of the Jewish Kingdom, and the UNSCOP Majority Report, giving this place to the Jewish State, has recognized the historic connection of the Jews with this part of the Red Sea.
>
> Elath, in the hands of the Arabs, may be a constant menace in the rear of the Jewish State. The Arab States have an outlet to the Red Sea and the Gulf of Aqaba through Jordan, Egypt, and Saudi Arabia.[48]

Before he left for the White House, Epstein gave Weizmann a piece of advice: just focus on one issue–the necessity of including the Negev and Elath in the Jewish state. Weizmann would not have enough time with Truman to cover other points, and he would have a better chance of getting Truman's support if he kept his request to the single issue that was of the greatest importance. If that issue were postponed, it might prove the last chance to make them part of the Jewish state.

The group moved on to the White House, where Weizmann and Epstein were joined by Clark Clifford, David Niles, and the chief of protocol, Stanley Woodward. Epstein remained outside the Oval Office as Weizmann went in alone to join the president. A half hour later, Weizmann emerged, looking pleased.[49] As suggested, Weizmann had contained his conversation with Truman to the issue of the Negev and diligently kept the president's focus on it when he began to wander. The chemist Weizmann had painted a picture for Truman that appealed to the farmer he once had been. By using desalted water, the Jews would make the desert bloom. Their experiments with desalination were already producing carrots, bananas, and potatoes in areas where nothing had grown for hundreds of years. If the Negev were taken from the Jews, it would remain a desert. Aqaba too was crucial, Weizmann told the president. It was now a useless bay, which had to be dredged, deepened, and made into a waterway that could accommodate ships of sizable dimensions. If it were part of a Jewish state, Weizmann told Truman, "it will very quickly become an object of development, and would make a real contribution to trade and commerce by opening up a new route." It would be a parallel highway to the Suez Canal, shortening the route from Europe to India by a day or more. "I was extremely happy," Weizmann later wrote, "to find that the President read the map very quickly and very clearly. He promised me that he would communicate at once with the American delegation at Lake Success."[50]

Two hours later, at 3 P.M., Herschel Johnson was standing in the lobby of the building used by the General Assembly. He was about to talk to Moshe Shertok about the change regarding the Negev when he was summoned to the phone. The president was calling. Truman told Johnson that he was not happy with the instructions given to the U.S. delegates by the State Department. "Nothing should be done," Truman instructed Johnson, "to upset the apple-cart";[51] the Negev should go to the Jews. Truman told the same thing to General Hildring: that he personally agreed with Weizmann's views and that "he wished the Delegation to go along with the majority report on the Negeb."[52] The State Department plan had been foiled. "Obviously the President," Weizmann wrote, "had been as good as his word."[53]

On November 22, Ambassador Johnson delivered a speech to the United Nations Ad Hoc Committee on Palestine in which he formally withdrew the previously stated U.S. reservations about the Negev and introduced a new resolution to make Beersheba and the surrounding territory included as part of an Arab state and not part of a Jewish state. The Jewish Agency had already indicated that it would be willing to give this area up and make some other modifications. Including the Negev in a Jewish state was now the official U.S. position.

Johnson, who had by this time become sympathetic to the Jewish position, then criticized the British for not giving the Ad Hoc Committee their full cooperation and for creating a very difficult situation by declaring they could not take part in the implementation of a plan if they did not have the approval of both peoples of Palestine. This condition, as the British knew very well, was impossible and was the very reason they had referred the problem to the United Nations in the first place. And how, he asked, did they expect to contribute to a solution if they abstained from voting, which is what they intended to do?[54]

Britain was in fact being most uncooperative. Sir Alexander Cadogan, its representative to the United Nations, told the delegates that British troops would not be available to enforce any settlement against either Arabs or Jews and that Britain planned to evacuate the last of its troops by August 1948. He warned that this did not mean that Britain would continue to maintain a civil administration in Palestine until that date. It intended to lay down its Mandate anytime it wanted after the General Assembly took a position not accepted by both Arabs and Jews. He concluded with what the press called a "cryptic" statement. "It follows," he said, that if the United Nations were in Palestine taking preparatory steps for a settlement, "it must not expect British Authorities either to exercise administrative responsibilities or to maintain law and order." They would do so only in the areas they still occupied. Confusing as his statement was, some delegates who favored partition saw this as grounds for optimism. If the General Assembly approved any solution by a two-thirds majority, he added, the British government "would not take any action contrary to it."[55]

Johnson's speech angered Henderson, who sat down and wrote another memo to Undersecretary Lovett. Henderson wondered if President Truman realized that the U.S. partition plan would leave only local law enforcement organizations to preserve the peace in Palestine. There would be wide-scale violence on all sides, he predicted, and the local police would not be able to cope with it. As he saw it, the United States was fostering a plan that would draw in U.S., Soviet, and perhaps other troops to Palestine, and he urged that the administration think twice before sending in U.S. troops.[56] Lovett immediately took the Henderson memo to Truman and read it to him at their daily 12:30 P.M. meeting. "I explained to him," he noted, "that the Department thought the situation was serious and that he should know of the probable attempts to get us committed militarily."[57]

Lovett also discussed with Truman the question of what role the U.S. delegates should play, if any, in ensuring that partition passed the General Assembly. Should the United States actively lobby other nations' delegates? *"The President,"* Lovett stressed, *"did not wish the United States Delegation to use threats or improper pressure of any kind on other Delegations to vote for the majority report favoring partition of Palestine. We were willing to vote for that report ourselves because it was a majority report but we were in no sense of the word to coerce other Delegations to follow our lead"* (our emphasis). That instruction would turn out to be critical. At the present moment—a few days before the vote was scheduled to take place—it was not certain that a majority of the U.N. General Assembly would in fact vote for partition. Such a passive posture, although Truman might not have been aware of its implications, would be interpreted as indifference by the smaller states.[58]

On November 24, Moshe Shertok and Jamal Husseini gave their concluding remarks to the Ad Hoc Committee. Shertok kept his short and to the point, arguing that the Jews' right to immigrate and settle in part of Palestine was no less legitimate than the Arabs' claim. He ended by telling the delegates, "A nation that has in a period of six years, lost six million will not be deterred from its attempt to base itself in the only spot on earth that it views as belonging to itself." Jamal Husseini summed up the Arabs' position by declaring that they absolutely rejected partition and demanded an independent state in all of Palestine.[59] Then, after hearing weeks of testimony, the committee voted, rejecting the proposal of Committee 2, set up to handle the minority report, by 29 to 12. The next day it approved the majority report by a vote of 25 to 13 with 17 abstentions. The matter would now go back to the General Assembly, where a two-thirds vote of the countries actually voting (as opposed to the entire U.N. membership) would be required for passage. By that measure, the votes for partition in the Ad Hoc Committee would have been short by one vote to be passed in the General Assembly.

The vote in the General Assembly was scheduled for the next day. Everyone was on pins and needles. Emanuel Neumann, Rabbi Silver, Moshe Shertok, and some others arrived early at Flushing Meadows, milling around and exchanging predictions about how the votes would go. They were nervous but feeling optimistic, as they had a lot of supporters at the United Nations, including the president of the General Assembly, Dr. Oswaldo Aranha, and Secretary-General Trygve Lie. But now Aranha walked by and mentioned that things did not look so good for partition since three or four of the small countries that had been positive or uncertain the day before had now decided to vote against it.[60] This was confirmed during the session when the delegates from the Philippines and Haiti both spoke against partition. The Greek representative also declared that Greece would vote against it. Rumors began to circulate that other countries were considering changing their votes.

During a break, as the Jewish Agency representatives sat in a corner of the corridor looking dejected, one of its South American supporters came over and said, "Go home! The sight of your faces is demoralizing your friends."[61] During the session, Shertok had such a worried and downcast expression that Trygve Lie took Goldmann aside and asked him to say something to Shertok because he was sending a message that he thought "the partition plan was done for." Lie himself favored partition because it was the result of UNSCOP, and he believed that if one wanted the new world body to succeed it was important to support its committees' recommendations.[62] The Arabs, David Horowitz observed, were "smiling broadly and crowing, their heads held proudly." He wasn't surprised to see Britain's Harold Beeley "radiant with joy."[63]

Given these alarming developments, the Jewish Agency decided that the vote had to be delayed at all costs. The next day was Thanksgiving, when no session would be held. If they could postpone the vote until after the holiday, it would give them more time to make a last-ditch effort to marshal the necessary votes. The Zionists and their allies told their friends in the General Assembly to "deliver full-length addresses in support of the resolution and to keep talking as long as possible." Aubrey Eban approached UNSCOP member Professor Rodríguez Fabregat of Uruguay and asked him if he thought he could make such a speech. Fabregat was happy to help, as was his friend García Granados, replying, "For me an hour-long speech is not a filibuster. It is a brief observation." Others followed suit, provoking the Arabs to make their own long speeches.[64]

At the lunch break, Neumann took General Hilldring aside and explained the situation. Could he get the U.S. delegation to request that the evening session be canceled on the ground that it was the eve of Thanksgiving Day? Hilldring said he would try.[65] It was getting late, and if they were to continue they would be there all night. Finally, Aranha asked for a vote to adjourn until Friday. It passed 24 to 21.[66] On Friday, Alexandre Parodi, the French representative, moved that another twenty-four hours pass before a vote to see if the two sides might make some progress in reconciling their differences. This request seemed reasonable, and it passed.[67]

Now the lobbying intensified. Of particular interest were Haiti and Liberia, which had abstained during the Ad Hoc Committee vote, and the Philippines, which had not participated. The Arabs had made inroads with these countries and wanted to hold on to their gains. Prince Faisal of Saudi Arabia reported to the British representatives that delegates from the Philippines, Liberia, and Greece were "deliberately lying low to avoid pressure," since they had now promised their votes to the Arab states. The Philippines delegate had just pronounced that he would not back any dismemberment of Palestine. The Greek delegate, despite the fact that Greece was a major economic and political beneficiary of the Truman Doctrine, also announced his opposition to partition.[68] The Greek ambassador later explained that Greece had voted against partition "as part of a deal which the Greek representatives had made with the representatives of the Moslem states. In return for Greek support on the Palestine issue, the Islamic states agreed to give full support to Greece in the future on Greek questions arising before any UN organ."[69] Camille Chamoun of Lebanon, Faris al-Khouri of Syria, and other Arab delegates told the Guatemalan delegate, Granados, that they were putting pressure on the president of Guatemala in his nation's capital. The Arabs had offered Costa Rica support for a seat on the U.N. Trusteeship Council if it changed its declared intention to vote for partition.[70] Most of all, the Arabs were banking on what U.S. State Department officials had told them: that the United States would vote for partition but would not use pressure to coerce other delegations to follow its lead.[71]

It was becoming obvious to the pro-partition forces that the U.S. delegation's hands-off policy was having disastrous effects and was giving the impression to the other delegations that the creation of the Jewish state was not especially important to the U.S. government. This view was now being reinforced since the Philippines, Haiti, and Greece were going to vote against partition.[72] The fate of the partition vote now lay in the hands of those nations, as well as Liberia and Ethiopia. Now that they had a little reprieve, the Jewish Agency and its allies rallied their forces, ready to do whatever it took. "Cablegrams sped to all parts of the world," David Horowitz recalled. "People were dragged from their beds at midnight and sent on peculiar errands. . . . not an influential Jew, Zionist or non-Zionist, refused to give us his assistance. . . . Everyone pulled his weight . . . in the despairing effort to balance the scales to our favor."[73]

Horowitz was not exaggerating. On November 26, when Philippine Ambassador Carlos Romulo addressed the General Assembly, he indicated that the Philippines would be voting against partition. There had to be some way to change this vote. The agency leaders learned that an obscure civil servant named Julius Edelstein had once worked in the Philippines and was a personal friend of the president of the Philippines. They called the State Department and finally tracked Edelstein down in London through the U.S. Embassy. Joseph Linton, the agency representative in London, called him in the middle of the night and asked him to immediately call President Manuel Roxas and convince him to have his delegation vote for partition. Sleepy and complaining, Edelstein nevertheless followed through. He woke Roxas up from his afternoon nap and tried to persuade him to have his delegates vote favorably. Even more persuasive was a call from "a United States Representative intimating that failure to support the United States position on the Palestine question might have an adverse effect upon Philippine-American relations." Two Supreme Court justices, Felix Frankfurter and Frank Murphy, wired Roxas, telling him "the Philippines will isolate millions and millions of American friends and supporters if they continue in their effort to vote against partition."[74] Roxas received a telegram from ten U.S. senators and then a cable from Senator Robert F. Wagner, Jr., of New York, with an additional twenty-five senatorial signatures.[75] In a similar fashion, Clark Clifford lobbied the Philippine ambassador to the United States in Washington.[76] All of these interventions led the Philippines' delegates to eventually vote for partition.[77]

The Agency people met with scores of delegations. There were three groups they thought they could count on: Russia and its satellites, the British Dominions, and the Latin American republics.[78] But to win the vote, they had to go beyond this group. Moshe Shertok and David Horowitz, for example, talked to the Ethiopian delegate, whom they expected to vote against it. They reminded him of the days of the Ethiopian war with Italy in the mid-1930s and the fact that the Italian consul in Jerusalem had asked the Jews to have their press stop backing Ethiopia, which they had refused to do. Shertok had answered that the Jews "would not sell the conscience of our people for a mess of pottage." The Ethiopian ambassador decided to abstain, which meant one less vote against partition.[79]

Liberia was another potential negative vote. Former Secretary of State Edward Stettinius was asked to help by using his business connections. Stettinius called Harvey Firestone, whose rubber plantations produced a large part of that nation's export trade. Firestone phoned Liberian President William Tubman, telling him that if his government voted against partition, as it had in one of the subcommittees, Firestone would not go ahead with plans for expansion in Liberia. Liberia ended up voting for partition. Pro-partition forces had less success with Greece. David Niles spoke to Thomas Pappas, a successful Boston businessman of Greek origin. Pappas got back to him on November 26, reporting on cables he, Spyros Skouras (a film executive), and others had sent to the Greek government. Pappas complained that Niles should have contacted him a week earlier, when there would have "more time to do something and we could have put on a lot more pressure." Under the circumstances, "we did everything humanly possible."[80]

The biggest problem for the propartition forces remained the United States' policy of neutrality. In his memoirs Truman wrote that he had "never approved of the practice of the strong imposing their will on the weak, whether among men or among nations." This stance, recommended by the State Department, appealed to his sense of fair play. However, now an intense public outcry for partition hit the White House. Truman wrote that the barrage "was unlike anything that had been seen there before.[81] Truman was also lobbied by members of Congress. Emanuel Celler cabled Truman from New York to let him know that he was "seriously disturbed that the vote for partition of Palestine . . . may fail by one or two votes." All of his efforts, he told Truman, "will be frustrated at the eleventh hour." Celler wanted pressure brought against "recalcitrant countries like Greece" that were "immeasurably and morally indebted to us." It was not too late to get them to change their vote, Celler argued. "It may also be necessary," he added, "that Greece be spoken to on a higher level than our delegation at the United Nations" by men such as Undersecretary Lovett.

The same pressure, he suggested, should be placed on Haiti, China, Ecuador, Liberia, Honduras, and Paraguay.[82] Each time he got a message like that, Truman fumed. He had read Celler's telegram with a "lot of interest," Truman replied, but it had "no foundational facts in the long run." On the bottom Truman later scribbled, "The pressure boys almost beat themselves. I did not like it."[83]

Despite his announced hands-off policy, there is much circumstantial evidence that the president did act, if indirectly through others, in his administration and at the midnight hour. Former Undersecretary of State Sumner Welles, who had many contacts in the Truman administration, claimed that "by direct order of the White House every form of pressure, indirect and direct, was brought to bear by American officials upon those countries outside of the Moslem world that were known to be uncertain or opposed to partition. Representatives or intermediaries were employed by the White House to make sure that the necessary majority would at length be secured."[84]

On November 11, eighteen days before the final vote was taken, there had been a cabinet meeting, and attached to a memorandum about the meeting was an unsigned and undated note in what appears to be Truman's press secretary Charlie Ross's hand:

Palestine votes look a little better.

1. We have been in touch with Liberian minister to try to get the Government's instructions to support us.
2. I think we have Haiti.
3. We may get Philippines out of No and into abstention, or with luck, yes.
4. Cuba still won't play.
5. Greece is uncertain but has the excuse of the Balkan Commission vote trade with Moslems.[85]

A few days after the final vote had been taken, Michael Comay, the director of the Jewish Agency's office in New York, wrote an account of what went on behind the scenes at the United Nations to Barney Gering of the South African Zionist Council. "The key to the struggle," Comay wrote, was the Agency's relations with the United States, which "were never satisfactory until right at the end, and were at times disheartening." The other delegations reported that the United States, usually active to make its wishes known, was unusually quiet on the issue of partition. It was so bad that some friendly delegates told the Agency that "Washington did not insist on their support on this particular issue," and that the American delegates "were not anxious to go out in order to line up votes." It amounted to what Comay called a "policy of indifference." He thought that as a result of America's attitude three countries "completely dependent upon Washington"–Greece, the Philippines, and Haiti–came out against partition."

The Jewish Agency then took advantage of the Thanksgiving holiday to create an "avalanche" to descend upon the White House. Truman, Comay continued, *"became very upset and threw his personal weight behind the effort to get a decision.* From then on Washington exerted itself to rally support and the situation improved" (our emphasis). Beforehand, sympathetic men such as Johnson and Hilldring had done their best but had been hampered by the State Department. *"It was only in the last 48 hours; i.e., on Friday and Saturday, that we really got the full backing of the United States"* (our emphasis). By the end, wrote Comay, "we had got the United States to take a firm stand." Finally, at Weizmann's urging, Prime Minister Jan Smuts of South Africa, who had been one of the original architects of the Balfour Declaration, sought to pressure Attlee and Bevin.[86]

Meanwhile, Loy Henderson was horrified when he heard reports about the change in the behavior of the U.S. delegates. They were ignoring the State Department's instructions and were obviously following White House orders.[87] He phoned Herschel Johnson at the United Nations and asked him bluntly, *"What's going on up there? We are being told here in Washington that the Americans at the United Nations are engaging in a lot of arm-twisting in order to get votes for the Majority Plan."* At that, Johnson *"burst into tears. He said, 'Loy, forgive me for breaking down like that, but Dave Niles called us up here a couple of days ago and said that the President had instructed him to tell us that, by God, he wanted us to get busy and get all the votes that we possibly could; that there would be hell if the voting went the wrong way. We are, working, therefore, under terrific strain trying to carry out the President's orders'* " (our emphasis).

Henderson had called Johnson, but it is hard to believe Henderson's account that Johnson burst into tears. By this time, Johnson supported partition, and it looked as if it could go down to defeat if something weren't done. When Johnson asked Niles what to do about the State Department, Niles told him, "Never mind the State Department! This is an order from the President!" Johnson and Eleanor Roosevelt, then went from delegation to delegation, telling them that they had been instructed by Truman to seek their support for partition.[88]

The British were quick to pick up on the new U.S. course. British delegates reported to the Foreign Ministry that the U.S. delegates were now "putting pressure on other delegations to support partition" and were "trying to rush matters to a final vote with as little discussion as possible."[89] The British continued to make things difficult. On November 28, in the General Assembly, Bevin handed Marshall a summary of a revised British plan for withdrawal. The British would withdraw their military in four phases and would end their civilian administration and their mandatory responsibilities for Palestine on May 15, 1948. And if partition was voted in, they did not want anyone from the U.N. commission to arrive in Palestine before May 1 because it would not fit in with their withdrawal plans.[90] This would leave a transition period of only two weeks.

Finally the day of the vote arrived. On November 29, tension built as the world waited to hear the General Assembly's decision. A small number of votes would decide the future of Palestine. When the French delegate voted yes, cheers broke out in the Assembly Hall from the audience who had packed into every seat. The final vote was announced by Secretary-General Trygve Lie: 33 in favor of partition, 13 opposed, and 11 abstentions. An overwhelming majority of the General Assembly had done what a few days earlier had seemed impossible. A combination of pressure, reasoning, and the joint unity of the Soviet and American delegations on the issue had all contributed to getting other originally undecided or opposed delegations to change their vote to favor partition.

As the votes were being counted, a shiver went down Emanuel Neumann's spine. He felt the spirit of Theodor Herzl haunting the proceedings. In 1897, Herzl, the father of modern Zionism, had made the incredible prediction that a Jewish state would be founded within fifty years. Here it was, fifty years later.[91]

As the Arab delegates heard the vote, many stormed out of the hall. Abdul Rahman Azzam and Prince Faisal stood up, "their faces twisted with anger," the journalist Ruth Gruber wrote, and "their white robes and headdresses shook as they motioned to the delegates from the other Arab states to follow them. At the door in Arabic-accented English, Azzam Bey shouted, 'Any line of partition drawn in Palestine will be a line of fire and blood.' "[92]

The British had abstained, although the world knew that Bevin and Attlee were hoping that partition would never be put into place. The British strategy had backfired. "Mr. Bevin is our sworn enemy," Weizmann noted. ". . . He has twice tried to break us: first through the Anglo-American Committee, which produced a verdict quite contrary to his expectations; and a second time by handing us over to the UN, which he hoped might bring about the liquidation of the whole affair. But in fact it has done just the opposite."[93]

In the streets of New York, pandemonium reigned. A few days later, crowds overflowed a theater booked for the celebration, taking their jubilation into the street. They were addressed by Mayor William O'Dwyer of New York and the major American Zionist leaders. Outside, five thousand people packed the adjacent blocks, as the speeches were broadcast over loudspeakers to those who could not get into the hall. The police estimated that twenty thousand people had sought admission into a venue that held only a quarter of that amount. Speaking to the crowd, Emanuel Neumann credited Harry Truman for the achievement, telling his audience, "If we now have this decision of the United Nations, it is due in very large measure, perhaps the largest measure, to the sustained interest and the unflagging efforts of President Truman."[94]

Despite his ambivalence and anger about the pressure coming from the "extreme Zionists" and New York's Jewish community, Truman was glad then and in the future to take the credit. Congressman Emanuel Celler sent him a telegram of praise, thanking him for "the effective work you did" in getting partition passed. Celler added, "I shall make it my business to emphasize the wonderful work you did when I address New York audiences as well as other audiences."[95]

A few days later, Eddie Jacobson and Abe Granoff paid Truman a visit at the White House. They just wanted to thank him. After Granoff said, "Thank you and God bless you," it was Jacobson's turn to speak. "Thank you," he told his old friend, "for the wonderful efforts to establish a Jewish State."[96] Truman then said, Jacobson wrote in his diary upon leaving the White House, that "he and he alone was responsible for swinging the votes of several delegations."[97] When he returned home, Jacobson wrote the president that he and Granoff were writing an unsigned editorial for a Jewish paper. It would, he assured the president, inform the world of his role in getting a positive vote on partition. The editorial appeared in the B'nai B'rith magazine, *National Jewish Monthly*.[98]

Truman had waited until the last possible moment to bring the full weight of the White House to secure the votes for partition. He had hoped that a simple announcement that the United States intended to vote for it would be enough. When it became obvious that it would not, he moved quickly and gave permission and encouragement for direct pressure to be applied to assure its passage. Partition, which had been sanctioned by the United Nations, was apparently the only answer for the Jews still in DP camps, American public opinion supported it, and so did his close advisers in the White House. He felt the pressure to act both politically and morally. And he realized that if partition went down, there was one person who would be blamed for it: Harry Truman.

ELEVEN

A "BIG CONSPIRACY" BREWS IN WASHINGTON: THE STATE DEPARTMENT FIGHTS PARTITION

Partition had won, but the representatives of the Jewish Agency were apprehensive about the roadblocks that lay ahead. Aubrey Eban observed that unfortunately, in the Jewish experience, "to regard worst-case contingencies as more real than successful outcomes is our legacy."[1] He was right to be wary. Almost from the first day, the U.S. State and Defense establishment waged a relentless campaign to undermine partition and President Truman's support for it. Returning to Washington after the vote, Epstein wrote to Lillie Shultz that he could "smell that official Washington is not at all jubilant about the decision taken by the U.N. and backed by the White House." He saw "a great deal of disappointment coupled with bitterness among many people who did not like the idea of a Jewish State and who will not so easily digest it."[2]

His observations were confirmed on December 5, when a story appeared in *The New York Times* reporting that the State Department had announced an embargo on the sale of arms to the Middle East, extending to "arms and ammunition of the US outside this country, such as surplus war property." They also announced that, although passports were still being issued to Americans who wanted to go to the Middle East for "valid reasons," they would not be issued to persons who wish "to serve with armed forces not under the United States Government."

Loy Henderson had suggested the embargo to Marshall on November 10, noting that the Arabs and the Jews would be seeking to buy arms from the United States. "I am of the opinion," said Henderson, "that in view of the tense situation in Palestine and on its frontiers, we should not permit the export of any material of this nature to Palestine or neighboring states so long as the tension continues." They would use these arms against one another, and the United States would be subject to "bitter recriminations." Henderson recommended, effective immediately, that "we suspend authorization for the export from the United States of arms, ammunition and other war material intended for use in Palestine or in neighboring countries, until the situation in that area has become somewhat more clarified." He assured Marshall that his colleagues in the Office of Near Eastern and African Affairs concurred.[3] Marshall agreed to it, and it was made public on December 5.[4]

Truman and the White House staff were taken by surprise, learning about the embargo at the same time as the public. Now Truman was in a bind. He couldn't contradict his own State Department, because doing so would make him appear to not be in command of his own foreign policy. Moreover, State's position sounded reasonable, and Truman did want to prevent more bloodshed. In any event, he had little choice but to accept the State's Department's new policy.[5] Just like State's recommendation to desist in using U.S. influence for the partition vote, the embargo ultimately worked to the advantage of the Arabs. The British government had no intention of canceling its existing arms contracts with various Arab states, and the Arab armies were already well equipped, having bought more than $37 million of surplus wartime arms from the former Allies.[6]

The Jews were at a disadvantage. They were raising money and buying arms overseas, but British arms searches and naval blockades made it difficult to get them into Palestine. At the moment, they had a few thousand guns that were scattered throughout the Yishuv, but no heavy armaments, machine guns, artillery, antitank or antiaircraft guns, or armored cars. In such a situation, the arms embargo felt like a nail in their coffin.[7] Some made the analogy to FDR's ban on arms sales to Republican Spain, which had put the republic on the defensive while Italy and Germany gave the Nationalists troops, advisers, and pilots.

In light of these developments, Epstein wrote to Freda Kirchwey that there were many people "in official Washington" who would stop at nothing to sabotage the partition decision. This meant they still had a lot of hard work ahead of them, including the mounting of a massive propaganda effort to popularize the idea of a Jewish state and to show that it would be an asset to both the Western democracies and the United States.

The UN vote for partition had provoked clashes between the Palestinian Arabs and Jews, which resulted in considerable Jewish casualties. The large number of Jewish victims was the result of the "policy of restraint adopted by the Haganah, which is to neither retaliate nor to attack the Arabs, but to stick to the policy of self-defense." They did not want to give the foes of partition more ammunition by helping the Mufti create major clashes between the Jews and Arabs. The other item on the Agency's agenda, he told Kirchwey, was even more important: to provide the Yishuv with the necessary arms and equipment for self-defense. It needed arms, not foreign troops. That would prove the best way to "avoid international complications" that might occur if foreign troops moved into Palestine after the British terminated their Mandate, which they had announced they would do on May 15. In light of the situation, Epstein told Kirchwey that it would be useful if she could raise some of these points in a *Nation* editorial.[8]

There was no rejoicing for the Arabs after the partition vote, only a resolution to resist. The day after the vote, the Arab Higher Committee called for a three-day general strike of Arabs in Palestine followed by mass demonstrations. On December 5, when the three-day strike was over, the Arabs sporadically began attacking the Jews using Sten and machine guns, focusing on buses, which were now forced to travel in convoys with armored escorts.[9] They called upon the Arab world to maintain a complete boycott of all Jews and Jewish products and threatened that "any contact will be regarded as the greatest crime and highest treason against their nation and religion." The committee's acting chairman said the Arabs would wage a holy war "if an attempt is made to enforce the partition plan." In Damascus an Arab mob killed all the members of the Soviet Union's legation staff. There was unchecked rioting in Syria and Lebanon, and thousands of students demonstrated in Cairo.[10]

In late December, American Consul Robert Macatee reported from Jerusalem that "terror is prevalent and normal life is disappearing." One example was particularly chilling. Jews, he wrote, "are picked off while riding in buses, walking along the streets, and stray shots even find them while asleep in their beds. A Jewish woman, mother of five children, was shot in Jerusalem while hanging out clothes on the roof. The ambulance rushing her to the hospital was machine-gunned, and finally the mourners following her to the funeral were attacked and one of them was stabbed to death." Even so, he reported, the Yishuv was remarkably quiet. Even the Stern Gang waged limited attacks on the British, and the Haganah took only punitive reprisals for especially harmful Arab attacks. The Jewish Agency accused the British of showing partiality to the Arabs, which Macatee thought was accurate. Jews' requests to organize their own protection were always turned down, he said, and police concentrated on searching Jewish personnel and settlements while ignoring the Arabs. Looking ahead, he warned of what he thought was a "serious preparation" for war by the Arab states.[11]

Immediately after the U.N. vote on partition, the General Assembly created a five-nation Commission on Palestine, which would assume temporary authority over Palestine and help with the implementation of partition after the British Mandate ended. The commission was composed of Bolivia, Czechoslovakia, Denmark, Panama, and the Philippines, with Ralph Bunche working as the chief of its secretariat. Heading the commission was the Czech representative, Karl Lisicky, who soon dubbed himself and his colleagues the "five lonely pilgrims." They were indeed lonely. The committee had planned to go to Palestine in two or three weeks. The Arab Higher Committee responded by announcing that it would boycott the commission. The British then made their task next to impossible by proclaiming that they did not want them there until May 1, only two weeks before Britain was to end its Mandate. Added to this was the escalating violence, which made them fear for their safety.[12] David Horowitz observed that "the commissioners were conscious of their helplessness and had no confidence in their ability to cope with the massive forces and situations pitch forked around them. They groped and blundered along in bewilderment."[13]

The Jewish Agency was strenuously trying to downplay the violence in Palestine in order to bolster their argument that Jewish Palestine could take care of itself. Truman had been clear that he did not want to send U.S. troops to Palestine and would do so only if they were part of a U.N. police force created by the Security Council.[14] But the Jewish Agency's contention that it would be able to deal with Arab opposition was challenged when its headquarters in Tel Aviv was blown up by explosives planted in the building's garden. Other explosives had been set off in Haifa and Jerusalem. Armed Arabs now began to cut off Jewish communications and raid Jewish suburbs and settlements. They attacked Jewish convoys going from Tel Aviv to Jerusalem with crucial supplies by putting up barricades and ambushing them.[15]

To deal with the problems of arms, the Jewish Agency sent Golda Meyerson (Meir) to the United States on an emergency mission to raise money for the Haganah. Meyerson could be counted on to do the most difficult tasks. She had emigrated to Milwaukee from Russia as a young girl and had worked as a teacher before moving to Palestine in 1920. "Stout and striking with black hair knotted in the back and eyes that perpetually reflected an inner sadness," she was familiar with American Jews and claimed that she would get through to them about the urgency of her request.[16] When she arrived, she was immediately squeezed into the program already taking place in Chicago, at a conference of the Council of Jewish Federations. "I have come to try to impress Jews in the United States," she told the council, "with this fact, that within a very short period, a couple of weeks, we must have in cash between twenty-five and thirty million dollars." It was a staggering amount. Meyerson traveled throughout the United States, giving the same pitch for money to one Jewish group after another. After two and a half months of constant fund-raising, she returned to Palestine with $50 million.[17]

Meanwhile, the State Department was busy developing plans and strategies to kill partition in favor of some form of trusteeship. In mid-November, Loy Henderson and George F. Kennan of the State Department got together to prepare a report, "U.S. Security Interests in the Mediterranean and Near East Areas," which would have widespread circulation. Though Henderson was arguably the most prominent and committed Arabist at State, Kennan was famous in Washington for his "Long Telegram" of 1946, sent by cable from Moscow while he was attached to the U.S. Embassy there. Later made public in an article in *Foreign Affairs* (in which Kennan wrote under the name "X"), his memo spelled out how the United States should confront Soviet totalitarianism in the new Cold War. Back in Washington, Kennan now turned his sights to the Middle East. Their paper was meant to be, Kennan told Marshall, State's initial position, which would be presented to the NSC.[18]

It would also serve as the State Department's guideline for changing course on partition. Warning that "U.S. support of partition has already brought about loss of U.S. prestige and disillusionment among the Arabs," they accused partition of being a violation of the principle of self-determination forged by the League of Nations, when many Arab states had been carved out of the Ottoman Empire. Even Prince Faisal of Saudi Arabia, they argued, had left the United Nations "in a bitterly anti-American mood" and might now support extremist elements unfriendly to America. The end result would be the endangerment of air bases and U.S. oil concessions. If the United Nations sought to actually implement partition, they warned, moderate Arab leaders would be swept out of power by extremists. Men such as Azzam Bey of the Arab League would be replaced by the likes of the Grand Mufti of Jerusalem. Their hatred of Zionists would then be extended to the entire West "in direct proportion to the latter's support of Zionist armies . . . and of partition in particular."

Assistance by the United States might even be a threat to the success of the Marshall Plan, since the Arabs could cut oil production and such a step would leave little oil for use in Europe. The Soviets could also gain entrée to the Middle East under the pretext of helping maintain order, a step that would give "Communist agents . . . an excellent base from which to extend their subversive activities." Even if this did not happen, Henderson and Kennan argued, "the UN decision is favorable to Soviet objectives of sowing dissention and discord in non-Communist countries." The Soviets would use partition of Palestine as an excuse to partition areas in which they had troops, such as Iraq, Iran, Turkey, and Greece, and then set up separate pro-Soviet states. Indeed, they predicted, "the whole structure of peace and security in the Near East and Mediterranean" would be adversely affected.

Partition, they argued, was supposed to lead immediately to a just and workable plan that Arabs and Jews would both support, but since the Arabs would not cooperate, one of the major premises in favor of partition had collapsed. If partition were not given up, anti-Jewish protests in Arab countries would soon extend to the West, providing a place for new "anti-Jewish agitation," which would interfere with the assimilation of individual Jews into the lives of their respective countries. Jews might then be seen as "an alien political factor." They explained, "In the U.S., the position of Jews would be gravely undermined as it becomes evident to the public that in supporting a Jewish state in Palestine we were in fact supporting the extreme objectives of political Zionism, to the detriment of overall U.S. security interests."

The United States would have to face the fact that partition could not take place without outside assistance, and if the United States persisted in supporting it, the government would have to provide economic aid as well as arms and troops. Moreover, the Arab states would view the United States as having announced "a virtual declaration of war" against the Arab world. It would all be for naught. They predicted that "it is improbable that the Jewish state could survive over any considerable period of time in the face of the combined assistance . . . for the Arabs in Palestine from the Arab states." Thus, they concluded, the United States *"should take no further initiative in implementing or aiding partition"*[19] (our emphasis). When it became clear to all that partition had no chance of success, the United States should then shift to a new position–that, in the absence of agreement between Jews and Arabs, the matter "should go back to the General Assembly." The United States could then switch its support to either a federal state or trusteeship.

The paper now circulated at a rapid pace. Dean Rusk made a point that Henderson and Kennan seemed to have overlooked: "A major change in our Palestine policy would require the approval of the President as well as of leading Members of Congress." To achieve their aims, Rusk argued, they would need to show that *"a new situation"* (our emphasis) had arisen in Palestine that supported a policy change. In Rusk's opinion, this new situation had not yet arisen, although he thought it could develop in April as the end of the British Mandate in Palestine grew closer and fighting between Arabs and Jews broke out. This would be the time to suggest alternative lines of action, mainly the "establishment of a United Nations trusteeship for the whole of Palestine."[20] Undersecretary Lovett sent it to James Forrestal, the secretary of defense. Forrestal agreed that the United States was not committed to supporting partition if it meant the use of force and that it was against the country's interests to give arms to the Jews while an embargo was in effect against supplying arms to the Arabs. Forrestal reached the same conclusion as Henderson and Kennan: "The United States should attempt to have the plan withdrawn as soon as possible."[21]

State's plans to roll back U.S. policy were revealed when *The New York Times'* chief diplomatic correspondent in Washington, James Reston, reported that the Defense and State Departments were concerned that President Truman's endorsement of partition was "influenced by the political strength of pro-Zionist organizations in key political centers of this country." Now they were worried that the same Zionist groups might convince the administration to send troops to Palestine to help implement partition. Policy in Palestine, Reston was told, "should not be allowed to jeopardize United States strategic interests in the Middle East." This included the question of oil, which was in short supply and which James Forrestal had told the House Armed Services Subcommittee was essential to America's security. The United States currently had to ship oil to Europe through Middle East pipelines, and that was threatened by "the hostile Moslem response to the United Nations partition plan."

Reston also reported that at present, the administration was not backing away from partition, and some in government even thought the United States should send troops to help implement it. But State and Defense were both strongly opposed to any action that could disturb U.S. relations in the Middle East.[22]

Among those who read Reston's article with great concern was Eleanor Roosevelt. "It seems to me," she wrote Truman, "that if the UN does not pull through and enforce the partition and protection of people in general in Palestine, we are now facing a very serious situation in which its position for the future is at stake." Mrs. Roosevelt pointed out that the United States had "led in the acceptance of the UN majority report on Palestine," and thus she favored creation of a U.N. police force that could act when needed. She also told Truman that she thought the United States should end the arms embargo to the Middle East and "provide such things as are essential to the control of Arabs, namely, modern implements of war such as tanks, airplanes, etc." Britain sought only "to please the Arabs," she thought, and now was the time to strengthen the United Nations as an instrument of peace.[23] Truman did not comment about the first part of her letter. But her statements about the British, Truman told her, "were correct as they can be. Britain's role in the Near East . . . has not changed in a hundred years."[24]

Truman also received a lengthy report and a visit on February 4 from the ambassador to Iraq, George Wadsworth. Wadsworth had also been a consultant to the U.N. Ad Hoc Committee on Palestine and shared Loy Henderson's opinions about Palestine. The regent of Iraq, Prince Abdul Illah, had asked Wadsworth to convey his views to Truman, which Wadsworth wrote up in a report sent to Truman in advance of their meeting. The prince was convinced that the United States was responsible for the United Nations' decision on partition. But he believed that "in spearheading that action the United States was, under Zionist pressure, unfaithful to its own principles . . . of self-determination and majority rule." Wadsworth let Truman know he agreed with the prince. As he saw it, the "Arab fear" had to be addressed and removed. "May I not," Wadsworth asked, "take back with me your personal assurance that the American Government will not support or participate in any project to impose partition by force?" This was necessary, he said, because Middle East oil flow could be affected, and that might harm the Marshall Plan. Like the others, he ended his report to Truman by supporting a U.N. trusteeship over Palestine.[25]

When Wadsworth sat down with Truman, the president agreed with him that the situation in Palestine warranted concern but, he said, he, Bob Lovett, and General Marshall were watching events closely. The problem, he told Wadsworth, went all the way back to the British double-dealing after the Anglo-American Committee report, which Bevin had pledged to support if a unanimous decision was reached. That had not occurred, for two reasons: it "had failed," Truman stated, "because of British bullheadedness and the fanaticism of our New York Jews." One thing he was adamant about, Truman assured him, was that no U.S. troops would be sent to Palestine and that the United States would work through and with the United Nations. That was not enough, Wadsworth replied. To those actually working in the field, partition seemed unworkable and, he said, it should be reconsidered.

To that suggestion, Truman replied that it was for the United Nations to decide. Then he changed the subject. He was most excited about scheduled development projects in the region, such as the Tigris-Euphrates Valley project, in Iraq, which would help get "the world back on its feet." The United States, Truman assured him, wanted to work with other countries to aid economic development. "For the first time in history," Truman said, "the conqueror's policy was to reconstruct the conquered." But, he added, the "Arab leaders should realize that they have to play their parts to make this possible. There was nothing much constructive anyone could do if they . . . started sending their armies into Palestine." The Arabs were worried, Wadsworth pointed out, that the United States would act unilaterally because of Zionist pressure and send in troops. "We won't," Truman assured him. "But they [the Arabs] must first assure me, before I can give them any categoric [*sic*] promises that they won't either."[26]

Truman confided to his friend Oscar R. Ewing, a former vice chairman of the Democratic National Committee who was now serving as administrator of the Federal Security Agency, that he still felt conflicted over what to do in Palestine. "I am in a tough spot," he told Ewing. "The Jews are bringing all kinds of pressure on me to support the partition of Palestine and the establishment of a Jewish state. On the other hand the State Department is adamantly opposed to this. I have two Jewish assistants on my staff, Dave Niles and Max Lowenthal. Whenever I try to talk to them about Palestine they soon burst into tears because they are so emotionally involved. . . . So far I have not known what to do." He wanted some impartial advice.[27]

Ewing offered to help. Truman answered, "Oh, I wish you would, I wish you would. It would help a lot." Ewing was an attorney who claimed not to know much about the situation in Palestine outside what he was reading in the newspapers. As a lawyer, he decided the best approach would be for him to investigate the legal claims that the Arabs and Jews, respectively, had to the land in question. Wadsworth and others had argued with Truman that the Zionist goals might be contrary to international law and that the partition resolution had no constitutional authority to commence with a geographical change in the region. Ewing's conclusion was that in international law, when land is taken by conquest, the conqueror can dispose of it as he desires. This meant that the sovereignty given by the Allies to the Jews of land taken from Turkey after World War I had as much validity as grants of sovereignty that had been made to the Arab countries. Most important, he concluded, "the claim of the Arabs that Palestine had been their land for thousands of years was untrue." For hundreds of years it had been Turkish, and after the First World War, the Allies had given part of their conquered land to the Jews, and "their title to it became indisputable."[28]

The volume of mail to the White House in support of partition continued to swell. Its sheer volume had ceased to have much effect. One survey showed that in the third quarter of 1947, the White House received 65,000 telegrams, postcards, letters, and petitions. This was exceeded in the next quarter, when it received 70,000.[29] By another count, between 1947 and 1948, Truman received 48,600 telegrams, more than 790,000 cards, and 81,000 other pieces of mail on the subject.[30] Before the November 29 vote in favor of partition, Truman wrote Senator Claude Pepper (D.–Florida), "I received about thirty-five thousand pieces of mail and propaganda from the Jews in this country while the matter was pending. I put it all in a pile and struck a match to it–I never looked at a single one of the letters because I felt the United Nations Committee was acting in a judicial capacity and should not be interfered with."[31] Truman, obviously, was speaking metaphorically.

On the home front, politics was suddenly up front. In the 24th Congressional District in the Bronx, New York, the representative who held the seat had suddenly died, and a special election for his replacement took place. The administration and the nation were stunned when the results of the mid-February election were announced: the Democratic nominee for the office, Karl Propper, lost by more than 10,000 votes to a third-party candidate, Leo Isacson, a member of the State Assembly. Isacson had run on the ticket of the American Labor Party, a group that had been formed to provide a separate ticket for left-wing Democrats to vote on for FDR, even though it was far to the left of the official Democratic program. By the start of the Cold War, the group had split up, and the ALP was now made up of only Communists and fellow travelers. They had been pushing to get Henry A.

Wallace, the former secretary of commerce, to run for president on a nationwide third party ticket.

The press took Isacson's victory as a sign that in the forthcoming 1948 presidential election, a large Wallace vote could give the election to the Republicans. What it ignored was the unusual nature of Isaacson's district. It was made up of working-class Jews who were either Communist or socialist and was often referred to as New York's "red belt." Moreover, only 35 percent of eligible voters had turned out for the special election, mainly those of the pro-Communist faction. Yet almost everyone saw the vote as a good omen for a Wallace candidacy and evidence that Truman would soon be out of office. The race was also seen as a mandate on Palestine. Speaking publicly, Wallace said of the president that Truman "talks Jewish and acts Arab." Some observers attributed the vote for Isaacson to dissatisfaction among Jewish voters with Truman's positions on Palestine to date.[32]

James Reston shared that view. The main reason these Bronx voters had put Isaacson in, he wrote, was dissatisfaction with the administration's Palestine policy, especially the arms embargo. They also worried that Truman would not support the formation of a U.N. armed force to fight the Arabs when partition occurred.[33]

The time was approaching when Senator Austin would have to give his statement on behalf of the U.S. delegation to the Security Council, which would be meeting to consider the majority and minority reports issued by the Commission on Palestine. The State Department sent Truman a working draft that it was proposing for his speech. Truman agreed with most of it, but paragraph 12 caught his eye. It said that if it was impossible to impose partition without force and if the Security Council was unable to work out an agreement between the Arabs and the Jews, the entire matter should be referred back to a special session of the General Assembly. Then it stated, "The Department of State considers that it would then be clear that Palestine is not yet ready for self-government and that some form of UN trusteeship for an additional period of time will be necessary." On February 21, Truman wrote to Marshall that he approved the approach in principle, but he added a proviso: *"I want to make it clear that nothing should be pre-*

sented to the Security Council that could be interpreted as a recession on our part from the position we took in the General Assembly [for partition]. *Send final draft of Austin's remarks for my consideration"* (our emphasis).[34]

On February 24, Austin delivered an ambiguous and confusing speech at the United Nations. He started out by saying that the United States supported partition but went on to cite potential roadblocks that could lead to its failure. Security Council military action, he stated, could be used only to prevent aggression against Palestine from outside and maintain international peace. It could not be used to enforce an internal matter such as partition.[35] The next day, he proposed a draft resolution to the Security Council on the Palestine question to establish a committee made up of the five permanent members of the Council. Their function would be to inform the Council of the situation in Palestine and make recommendations to it.[36]

U.N. Representative Ben Cohen, who along with Mrs. Roosevelt was a supporter of partition, immediately criticized the implications of Austin's speech. "Under clouds of legal sophistry," he wrote, "that pusillanimous address confirmed the fears of many of us that the State Department lacks the will to have our government do its part to make the Palestine settlement work. It confirms the fears of many of us that the State Department is faltering in its support of the United Nations."[37] Senator Francis J. Myers of Pennsylvania was so upset that he wrote Truman a personal letter to "get this off my chest." Austin's speech, he wrote, "gave the impression . . . that our country is 'selling out' the Jewish people and undercutting the UN structure." Some Democratic leaders in his state had even resigned in protest, and Pennsylvania's Jews, he reported, "seem to be as bitter against you and the party . . . as some of our brethren in the South profess to be." After reading Austin's speech, Myers found it guilty of "sloppy draftsmanship, perhaps purposely so." He did not think the United States had sold out but did find in Austin's speech "vague, weasel-worded passages so involved and so legalistic and confusing that it is hard to tell just exactly where we stand." He wanted to support Truman, he said, but he and others were at a great disadvantage "because we just can't put our finger on the facts." He reassured Truman that he did not think he was selling out anyone, "and yet," he concluded, "it is so terribly difficult to find solid facts . . . on just where we do stand that I am most disturbed."[38] Truman answered, "I naturally am not happy over the implication that I might be ducking any issues. That is not my manner of meeting a situation."[39]

Publicly, Truman endorsed Austin's speech, despite the criticism it was receiving. It "accurately" explained this country's position, he announced.[40] Representative Emanuel Celler fired off a telegram to Truman: Austin's statement had been "outrageously hypocritical." He recommended yet another commission, "as if there haven't been enough committees."[41] Freda Kirchwey was more emphatic, warning Truman that State was moving to double-cross him. "Wonder if you are aware," she said in her telegram, "that Undersecretary Lovett has told selected group newspaper people this statement is preparation for American effort to revise partition resolution in which you played a leading role." She could not believe, she told Truman, "that you could conceivably be partner to such action."[42]

The State Department's efforts to stall, if not to kill, partition were proceeding. One of its chief arguments was that partition would not be able to be carried out peacefully. The Commission on Palestine report to the Security Council bolstered its arguments when their report concluded that, given the continuing violence in Palestine, partition could not be implemented without an international police force.

It was only a matter of days before Secretary Marshall approached Truman as a follow-up to Austin's speech. He wanted to inform the president that the U.S. delegation would be introducing a new resolution to the Security Council. It would call for setting up a committee to consider the situation in Palestine and advise how to carry out the U.N. resolution on partition. The Belgians would be introducing their own resolution calling for a subcommittee that would see if conciliation could be attained between Arabs and Jews. Marshall thought these efforts at conciliation would prove fruitless, and then the United States would have to reach a decision whether or not to "attempt to carry out partition." The way things were going, he told the president, it looked as if the Security Council would not be able to proceed with partition and would refer the Palestine problem back to the General Assembly for reconsideration.[43]

The same day, Marshall told Ambassador Austin that he believed the interests of the Jews, Arabs, and British were irreconcilable. Eventually, Austin should give a speech emphasizing the inability of all three to implement partition peacefully. If the British gave up the Mandate on May 15 as scheduled, it was clear that there would be major fighting. The United Nations could not permit this to occur. Therefore a special session of the General Assembly had to be convened to look at Palestine again and, with the failure of conciliation, recommend that Palestine be placed under the trusteeship of the United Nations.[44] A draft of the speech announcing the change in U.S. policy was written for Austin by Robert McClintock, Dean Rusk's special assistant in the Office of United Nations Affairs, and Marshall approved it on March 5.[45]

On March 8, the day that Truman declared that he would be a candidate for the presidency, Marshall sent Warren Austin a note that the president had approved a draft of the next statement Austin would be delivering to the United Nations, which included the proposal of trusteeship as an alternative, *"for use if and when necessary"*[46] (our emphasis). In all likelihood, Truman signed it while he was aboard his presidential yacht, somewhere between St. Croix and Key West, believing that he would have the final say in any policy change as important as this one would be.

The State Department had a different interpretation. Robert McClintock then wrote to Lovett that the General Assembly would be convening a special session, and its probable outcome "will be the establishment of a United Nations trusteeship for Palestine." That would mean a movement away from dealing with Arab aggression to "a new threat of Jewish attempts by violence to establish a *de facto* State in Palestine." But State was prepared for this. It had already drawn up draft trusteeship agreements.[47]

It was now quite clear that the State Department was out to reverse Truman's Palestine policy, although the president still did not want to believe it. Truman's White House aides interceded. In March, Clark Clifford enlisted the help of the attorney Max Lowenthal to research and draft a paper for him critiquing the positions enunciated by the State Department in the previous few months against partition and making a case for it.[48] Niles was also lending a hand.

The cool, restrained Clifford appreciated Niles's and Lowenthal's help, although he admitted that he found them both somewhat odd. "If Niles seemed a little strange," Clifford observed, "that was nothing compared to Max Lowenthal, who was never an official member of the White House staff at all, although he came and went as he pleased. . . . Convinced he was under surveillance by J. Edgar Hoover, when Lowenthal wanted to discuss matters he considered sensitive, he insisted that we conduct some of our discussions on a bench in Lafayette Park."[49]

Clifford's paper began on the one point that most everyone acknowledged: "Unless immediate action is taken to preserve peace in Palestine, chaos and war will follow Great Britain's withdrawal on May 15th." Having been responsible for the U.N. action, Clifford argued that it was "unthinkable" that the United States fail to back the partition resolution. He made the following recommendations: First, the United States should work to get the Arab states to accept partition. Second, the United States should, if the Arabs refused to compromise, brand the Arab states as aggressors. Third, the United Nations should require Britain to comply with the resolution. Next, Clifford argued, the United States should immediately lift its unilateral embargo on arms to the Middle East, which would give the Haganah equal opportunity with the Arabs to arm itself for self-defense. Finally, the United States should help the United Nations establish an international security force in Palestine, made up of volunteer recruits.[50]

Clifford then sent another memo to Truman. He began by telling him that to date Palestine had not been discussed with any attention to fundamental issues. It was not a Jewish issue but one that had to be considered from the standpoint of "what is best for the United States of America." Pointing to the ever-present issue of politics and the vote, Clifford wrote, "In advising as to what is best for America [we] must in no sense be influenced by the election this fall. I know only too well that you would not hesitate to follow a course of action that makes certain the defeat of the Democratic Party if you thought such action were best for America."

Clifford termed the State Department's arguments "completely fallacious." Truman's support of partition, he assured him, "was in complete harmony with the policy of the United States." If he turned against it, he would be departing from already established and accepted policy, which would make him "justifiably subject to criticism." Next, Clifford argued that partition offered the best hope to avoid war and to offer a permanent solution. To drift and delay as others advocated would lead to the military involvement the president claimed he wanted to avoid.

Turning to the new Cold War, Clifford argued that support of partition was the only course that would strengthen the U.S. position vis-à-vis the Soviet Union. The United States' key interest was in the development of Western Europe, so that it could join a community of nations that would resist Soviet aggression and capitalize on the Marshall Plan politically. That could be done by cementing alliances through the United Nations. Now many Americans were in despair as they feared the collapse of the United Nations, which they had hoped would be a forerunner of world peace. Much worse, the American people, he warned Truman, felt that their own nation was "aiding and abetting in the disintegration of the United Nations. . . . *Nothing has contributed so much to this feeling as Senator Austin's recent statement* (our emphasis). . . . it seemed to be the sophistries of a lawyer attempting to tell what we *could not* do to support" the United Nations. U.S. foreign policy was failing, and Clifford warned that, to many, it appeared that the United States was "drifting helplessly."

To bolster confidence and maintain an anti-Soviet alliance, it was necessary for the United States' own "selfish interests" to support the U.N. resolution on Palestine. *"We 'crossed the Rubicon' on this matter,"* Clifford reminded Truman, *"when the partition resolution was adopted . . . largely at your insistence"* (our emphasis). To back away would lead other nations never to trust America's commitments. If the United States retreated, he warned, Russia would move into the Arabian Peninsula, "and this is as certain as the rising of tomorrow's sun." Then the

Soviets would portray themselves as the defenders of world peace, as the only force trying to stop a new Jewish-Arab war. As for those who argued that the United States would lose oil because the Arabs would not sell their supply to America, he answered, "The Arab states must have oil royalties or go broke." The Saudis got 90 percent of their oil revenue from the United States, and they would not jeopardize that. "Their need of the United States is greater than our need of them."

Concluding his argument, Clifford told the president that the people claiming that partition would not work were "those who never wanted partition to succeed and who have been determined to sabotage it." State, he noted, "has made no attempt to conceal their dislike for partition." If the United States were not firm, it would appear to be appeasing a few nomadic tribes, and big powers such as the Soviet Union would treat America with "contempt in light of our shilly-shallying appeasement of the Arabs." Citing the Truman Doctrine, Clifford admonished Truman to take a similar firm stand on partition, no matter what the domestic consequences.[51]

Suspicions were now growing in the United Nations that the United States was changing its position on partition. The Big Five permanent members of the Security Council (the United States, Soviet Union, France, China, and Britain) were charged with holding consultations with all the parties, although Britain declined to participate. The change in the United States' attitude was palpable. Andrei Gromyko asked Austin if all the questions the United States was asking Arab, Jewish, and British representatives "were aimed at changing the recommendations of the G.A.?" Denying that such was the case, Austin answered that they were only trying to find out if there were any modifications that might lead them to find some basis of agreement.[52] By now, U.N. Secretary-General Trygve Lie had concluded that "only the Soviet Union seemed to be seriously intent upon implementing partition; the United States clearly was not."[53] Lie told Austin that although he appreciated the United States' attempt to see if there was any way partition could be implemented without violence, he thought it would be impossible to implement it by peaceful means. Nevertheless, it would have to be enforced. If it were not, Lie feared, "the UN would go downhill rapidly to nothing."[54]

But Marshall was convinced that the partition plan could not go forward. The situation in Palestine, he informed Austin, "grows daily more fraught with danger to international peace." For that reason, he thought, "It is . . . of the first importance that the Council disposes one way or another of the partition issue." All parties agreed that it could happen only with force, and thus "the necessary conclusions can rapidly be drawn,"–i.e., a shift from partition to trusteeship was necessary.[55]

Austin was having qualms about so brazenly overturning Truman's support of partition in the speech he was about to give to the Security Council. However, Rusk and John Ross, Austin's deputy, "had succeeded in persuading Ambassador Austin that the tactics . . . should be adhered to." To placate Austin, they agreed that he could say that trusteeship "would be a temporary measure" until some future settlement could be attained by both Arab and Jewish Palestine.[56] This concession was necessary, Rusk said, because if Austin found himself in "outright disagreement" with his instructions he might be tempted to discuss it with Truman. Rusk was almost 90 percent sure that Austin was "back on the track."[57]

It was also obvious to the Zionists that behind the State Department's maneuvers in the United Nations lay a plan to sabotage partition. The only person who could prevent it was the president, but they did not know where he really stood. Truman was having a very negative reaction to all their lobbying and had even banned them from the White House. In his memoir, Truman was later to recall that "the Jewish pressure on the White House did not diminish in the days following the partition vote in the U.N. Individuals and groups asked me, usually in rather quarrelsome and emotional ways, to stop the Arabs, to keep the British from supporting the Arabs, to furnish American soldiers, to do this, that, and the other. . . . As the pressure mounted, I found it necessary to give instructions that I did not want to be approached by any more spokesmen for the extreme Zionist cause."[58]

They turned once again to Chaim Weizmann, whom they knew Truman liked and respected. After the partition vote, Weizmann had stopped briefly in London and was looking forward to going back to his home and institute in Rehovoth, Palestine. But the American section of the Jewish Agency had different plans for him; it needed him in New York. Weizmann was reluctant. He no longer held any official position in the Zionist movement, and his health was not good. Finally, his protégé Aubrey Eban cabled him with an official invitation in the name of the Jewish Agency: "The most crucial phase of all now approaches here in which we sorely miss your presence, advice, activity, influence." Returning to the United States on the *Queen Mary,* Weizmann and his wife landed in the city in the midst of a blizzard. Weizmann wrote to Truman on February 10, asking for an audience before Truman left on vacation for the Caribbean. That could not be arranged, Matthew Connelly responded. The president's calendar was full.[59]

About a week later, Frank Goldman, B'nai B'rith's national president, called Eddie Jacobson in the middle of the night and told him that Truman was refusing to see any of the New York City political leaders who had been imploring him to see Weizmann. He had even turned down Ed Flynn. Truman was angry "at leading American Zionists who denounced him for refusing to send American troops and supplies to fight the Arabs opposing the partition plan," and they were afraid that the president was washing his hands of the whole matter and would let the United Nations decide what should be done. Goldman wanted Jacobson to charter a plane immediately and see Truman before he left for Key West. Jacobson was their last hope.[60]

Jacobson couldn't make the arrangements but sent a telegram to Connelly asking him to give it to the president immediately. "I know that you have very excellent reasons for not wanting to see Dr. Weizmann," Jacobson wrote. He understood more than anyone else the pressure on the president. "But as you once told me," he wrote, "this gentleman is the greatest statesman and finest leader that my people have. He is very old and heartbroken that he could not get to see you." Noting that he had not asked Truman for "favors during all our years of friendship," he now wrote that I "am begging of you to see Dr. Weizmann. . . . I can assure you I would not plead to you for any other of our leaders."[61]

Truman wrote back nine days later. "There wasn't anything he could say to me that I didn't already know," he explained. "Anyway, I had also made it a policy not to talk with anyone regarding the Palestine situation until the Security Council has had a chance to act on our suggestion for a police force to enforce partitioning." He continued:

> The situation has been a headache to me for two and a half years. The Jews are so emotional, and the Arabs so difficult to talk with that it is almost impossible to get anything done. The British, of course have been exceedingly noncooperative in arriving at a conclusion. The Zionists, of course, have expected a big stick approach on our part, and naturally have been disappointed when we can't do that. I hope it will work out all right, but I have about come to the conclusion that the situation is not solvable as presently set up; but I shall continue to try to get the solution outlined in the United Nations resolution.[62]

But this meeting was so crucial that the Agency asked Jacobson to come to Washington once again and see the president when he got back from Key West. On March 12, Jacobson took a flight from Kansas City, paying his own way as always. He had spent so much of his money traveling back and forth to Washington that he apologized to his daughter Elinor that he would have nothing left to leave his children. It cheered him up when she told him the important thing was that he was leaving them his good name.[63]

Jacobson was very anxious as he prepared to see Truman when he returned to the White House. What could he say that would change his friend's mind? He called Eban for some advice and told him that he was searching for ways to convince Truman to see Weizmann and thought of the president's reverence for Andrew Jackson, whose statue was sitting in Truman's office. Did Eban have any suggestions? Eban told him that although "no two human beings had ever walked on the face of the earth with fewer common attributes than Chaim Weizmann and Andrew Jackson," he should go ahead and try it.[64] As he had done before, Jacobson didn't make an appointment but took his chances. At the White House, he encountered Matt Connelly, who "advised and urged and begged [him] not to discuss Palestine with the President." Jacobson was candid, recalling "I quickly told Matt that that's what I came to Washington for, and that I was determined to discuss this very subject with the President."[65]

As always, Truman was glad to see Jacobson. After a few minutes of personal talk about their families and Jacobson's business, Jacobson brought up the issue of Palestine. Truman "immediately became tense in appearance, abrupt in speech, and very bitter in the words he was throwing my way," recalled Jacobson. "In all the years of our friendship he never talked to me in this manner." He didn't want to talk about Palestine, the Jews, or the Arabs. Jacobson then got up his nerve and began to argue with him "from every possible angle." Reminding Truman that he had always said he revered Chaim Weizmann, he added that he could not understand why he would not give him an audience. He was an "old and sick man," he told Truman, and he had "made his long journey to the United States especially to see the President."

Truman remained unmovable. Referring obviously to his disdain for Rabbi Silver, Truman replied how "disrespectful and how mean certain Jewish leaders had been to him." At that moment, Jacobson was shocked and thought the unthinkable: "I suddenly found myself thinking that my dear friend, the President of the United States, was at that moment as close to being an anti-Semite as a man could possibly be." He was more dismayed that "some of our own Jewish leaders should be responsible for Mr. Truman's attitude." Jacobson was crushed and believed that his mission had been a failure. Then, as his eyes rested on Truman's statue of Andrew Jackson, he decided to make the Weizmann-Jackson connection. He told the president:

> Harry, all your life you have had a hero. You are probably the best read man in America on the life of Andrew Jackson. I re-

member when we had our store together and you were always reading books and papers and pamphlets on this great American. . . . Well, Harry, I too have a hero, a man I never met, but who is, I think, the greatest Jew who ever lived. I too have studied his past and I agree with you, as you have often told me, that he is a gentleman and a great statesman as well. I am talking about Chaim Weizmann; he is a very sick man, almost broken in health, but he travelled thousands and thousands of miles just to see you and plead the cause of my people. Now you refuse to see him because you were insulted by some of our American Jewish leaders, even though you know that Weizmann had absolutely nothing to do with these insults. . . . It doesn't sound like you, Harry, because I thought you could take this stuff they have been handing out to you. I wouldn't be here if I didn't know that. If you will see him, you will be properly and accurately informed on the situation as it exists in Palestine, and yet you refuse to see him.

Truman began to drum on his desk with his fingers. As Jacobson stopped talking, Truman "turned around while still sitting in his swivel chair and started looking out the window into what in the summer is a beautiful rose garden, gazing out the window just over the pictures of his mother, his wife, and his daughter. I knew the sign. I knew that he was changing his mind." Seconds that seemed like many minutes passed in silence. Truman suddenly swiveled around and, facing Jacobson, "looked me straight in the eyes and then said the most endearing words I had ever heard from his lips":

"You win, you baldheaded son of a bitch. I will see him. Tell Matt to arrange this meeting as soon as possible after I return from New York on March 17."

At that moment, Connelly entered the room. Truman immediately told him to schedule in Weizmann. It was to be an off-the-record meeting, the president said, and the press and public were not to know anything about it. Rushing to the Statler Hotel, Jacobson met Frank Goldman and Maurice Bisgyer of B'nai B'rith, who were anxiously waiting to hear Truman's decision. Jacobson was so nervous and excited that he first went to the bar and guzzled down two double bourbons, an unprecedented act for him. The three men then made their arrangements to go to New York, where Jacobson would meet Weizmann for the first time.

Weizmann's meeting with Truman was scheduled for March 18. The White House told Jacobson to make sure that Weizmann entered through the east, rather than the main northwest, gate, so that the press would not notice his arrival. Weizmann spoke to Truman about plans for the economic development of Palestine, the scientific work he and his associates were carrying out in Rehovoth, and the need to attain more land for future Jewish immigrants. Truman was impressed, he recalled, by Weizmann's emphasis on the Jewish state's need to retain the Negev. Truman explained to him why he had put off seeing him. "My primary concern," he told Weizmann, "was to see justice done without bloodshed." By the end of the meeting, Truman thought, Weizmann left his office with a complete understanding of his policy.[66]

Weizmann's wife, Vera, recorded her own account of the meeting. She noted in her diary that after her husband returned from his meeting with the president, he told her that he had raised three points: "lifting of the arms embargo, support for partition, and increased and free immigration into Palestine." Truman told him, in response, that the State Department was considering the embargo question and, on the point they were most concerned about, "that [Truman] supported partition."[67]

All the assurances Weizmann had been given came to a shocking halt the next day, March 19, when the U.S. ambassador to the United Nations, Warren Austin, made a speech to the Security Council as it was meeting to receive the Big Four's report on Palestine. The report gave "clear evidence," Austin began, that the Jews, Arabs, and Britain could not agree to implement the partition plan. If the British vacated Palestine as scheduled, Austin noted, it would lead to chaos, fighting, and loss of life. The Security Council could not permit that to happen and had the responsibility not to permit international peace to be threatened. It had to take necessary action to bring about the end of violence.[68] Zionists listening to the speech reacted to the words with "joyous surprise," the Zionist leader Julius Haber remembered. "The intimation seemed to be that at last, the United States was prepared to advocate a firm hand in Palestine."[69] Suddenly, without apparent reason, Austin asked for an intermission so that the Big Four representatives, plus Britain, could have a moment to concur.

A little less than an hour later, Austin went to the podium to resume his speech. What he said was a bombshell that reverberated through the country and the world. The U.S. government, he continued, "believes that a temporary trusteeship for Palestine should be established under the Trusteeship Council of the United Nations to maintain the peace and to afford the Jews and Arabs of Palestine . . . further opportunity to reach an agreement regarding the future government" of Palestine. Later that afternoon, Dean Rusk flew into New York from Washington and held a press conference announcing that the State Department was officially recommending a change in policy from partition to trusteeship.

Representing the Jewish Agency, Rabbi Silver immediately responded, denouncing Austin's statement in a speech to the Security Council as nothing but a "shocking reversal" of the U.S. position. "We are at an utter loss to understand the reason for this amazing reversal," he told the U.N. delegates, since it would lead to more violence and undermine the prestige and authority of the president and his administration. Not only would trusteeship not lead to peace, he pointed out, force would have to be used to implement it. It would mean the end of the sovereignty, territory, and immigration rights that were going to be established under the partition agreement. It could be imposed on the Yishuv only by force.[70]

Truman himself was stunned at the turn of events. He wrote in his diary:

> The State Dept. pulled the rug from under me today. I didn't expect that would happen. In Key West or enroute there from

> St. Croix I approved the speech and statement of policy by Senator Austin to UN meeting. This morning I find that the State Dept. has reversed my Palestine policy. The first I know about it is what I see in the papers! I am now in the position of a liar and a double-crosser. I've never felt so in my life.
>
> There are people on the third and fourth levels of the State Dept. who have always wanted to cut my throat. They've succeeded in doing it.[71]

When Clark Clifford saw Truman in his office, he found him "as disturbed as I have ever seen him. 'I don't understand this,' " Clifford remembered Truman as saying. " 'How could this have happened? I assured Chaim Weizmann that we were for partition and would stick to it. He must think I'm a plain liar.' "[72] Truman felt embarrassed and humiliated by his own State Department. In May he would deliver his own surprise.

Was Truman correct? Did the State Department, in fact, knowingly reverse U.S. policy on its own? It was not the first time that a speech by an administration official would appear to alter policy and Truman would deny that he had given the speaker approval.[73] The State Department personnel most concerned all swore that they had given Truman the speech and that he had known what Austin was going to say, only not when the speech would be delivered. Marshall had directed Austin to make his speech on March 16, Clifford remembered, and Austin and Rusk had been told not to delay it until after the Security Council took a final vote. Marshall and Lovett, Clifford claimed, "left no word that the President was to be informed when Austin was to speak," and that the "text of Austin's speech was not submitted to President for his approval."[74] Years later, Clifford provided further details in an article he wrote for *American Heritage,* the magazine of popular history. Truman, he wrote, had authorized Marshall and Austin to propose a U.N. trusteeship only *after* three qualifications were met: the Security Council had to exhaust all conciliatory measures; the Council would then recommend alternatives to partition; and finally, the Council would have voted to reject partition entirely. These conditions had not been met when Austin gave his speech.[75]

Clifford further explained it by using a legal analogy. There had been no meeting of the minds between Truman and the State Department's position on Palestine. Truman, according to Clifford, thought he was being presented with a contingency plan, a fallback position that he would take if he had to, but he did not want to change his policy of support for partition. The State Department, on the other hand, looked at it differently. It was convinced that partition would fail and, with their help, would be replaced with a trusteeship for which it now claimed they had Truman's consent. Clifford did not think State had served the president well, first by not making a fight for partition and then by not discussing its decision with Truman to change the direction of American policy. Before making such an important statement, he said, it should have gotten back in touch with Truman and told him that partition "is going very badly, and we think we ought to let you know this is going to end up on our having to agree to trusteeship."[76]

Although Truman did not want to believe it, Clifford thought Marshall and Lovett were not as innocent as they claimed to be. On five occasions he had told Truman that the State Department wanted to change his position. He had even taken a State Department memo to Truman as proof. But Truman had brushed off his warning and told him he was "unduly concerned." "I know how Marshall feels and he knows how I feel. They are not going to change our policy."[77]

The day after the speech, Truman convened a meeting with Clifford, Matthew Connelly, Charles Bohlen, Dean Rusk, and Charles Ross and instructed them to get the facts. The problem, Ross, Truman's childhood friend and press secretary, wrote in a memo to himself, was that "no final check had been made with the President before Austin spoke." Truman had thought the alternative of trusteeship would not be announced until after a final vote was taken and the Security Council judged partition impossible to achieve. "The whole business," Ross put it, "had been handled with singular maladroitness by State. No pronouncement of the momentous nature of Austin's should have been made without prior consultation with the President. . . . As it was, the reversal was without warning to the public and the President was placed in the most embarrassing position." Now, Ross wrote, Truman was forced to accept "the accomplished fact forced upon him by the precipitate State Department–Austin action." If the truth came out, Truman would appear "vacillating, or ignorant of something of the most vital importance . . . and the truth, moreover, could only have been accompanied with a wholesale repudiation of the State Department." At their meeting, Ross recorded, Truman told them, "They have made me out a liar and a double-crosser!–We are sunk."[78]

Truman called a meeting on March 24 with senior State Department officials and some White House staff. The meeting in the Cabinet Room included Secretary Marshall, Loy Henderson, Dean Rusk, Clark Clifford, Matt Connelly, Oscar Ewing, Howard McGrath, and David Niles. Truman told them he wanted to reconcile his support for partition with that of trusteeship. Clifford thought that State was acting as though they had the upper hand, especially when Henderson told the president, "That partition should be considered dead and buried." Truman rejected this suggestion right away. After a long discussion about their options, Truman told Clifford to prepare a statement that "would seek to adapt the trusteeship proposal to partition." Clifford worked on it all night with some of the State Department people.[79]

The next day, Truman formally issued a statement trying to undo the damage. Partition, he stated, "cannot be carried out at this time by peaceful means." The United States could not impose it on Palestine with the use of U.S. troops. When the British gave up its Mandate, there would immediately be fighting and bloodshed that could "infect the entire Middle East." Therefore, facing what Truman called "imminent" dangers, the United States proposed a temporary U.N. trusteeship that would help keep the peace after the British left Palestine. It was not, he emphasized, "a substitute for the partition plan" but only an attempt to fill the vacuum created by the British withdrawal. It would create the kind of order that would allow a final settlement. He ended by urging an immediate truce in fighting between the Jews and Arabs.[80]

Truman's statement hardly helped. It was, as the private meetings suggested, an attempt to square the circle and repudiate the State Department without openly disowning the Austin speech favoring trusteeship. Few were satisfied. Eleanor Roosevelt was so upset that she threatened to resign from her position on the U.S. delegation to the United Nations. Truman, realizing that her departure would be a political disaster, told her that the trusteeship proposal was "intended only as a temporary measure, not as a substitute for the partition plan."[81]

Trygve Lie, the secretary-general of the United Nations, felt betrayed by Austin. He had met with Austin before he gave his speech and told him that UNSCOP had already withdrawn trusteeship as a solution, realizing that all sides would be opposed. Lie did not understand why the United States had reversed its position. Perhaps, he thought, it was because the United States expected cooperation from the British and less opposition from the Arabs or that some in the State Department feared the effect on America's oil interests in the Middle East. In any case, Lie regarded the turnabout as "a blow to the United Nations" and one that showed "a profoundly disheartening disregard for its effectiveness and standing." The day after the speech, Lie went to see Austin at the Waldorf-Astoria and told him of "my sense of shock and of almost personal grievance. . . . This is an attack on the sincerity of your devotion to the United Nations cause." Lie then proposed that "as a measure of protest against your instructions, and as a means of arousing popular opinion to the realization of the danger in which the whole structure of the United Nations has been placed–I want to propose that we resign." Austin was not about to resign. "I didn't know you were so sensitive," he replied.[82]

Editorials in the most influential newspapers were almost universally critical of Austin and the White House. Even *The New York Times* editorialized that the fate of the Holy Land was now being "decided by expediency without a sign of the spiritual and ethical considerations." Palestine was often called the home of the prophets, the editorial noted, and now "it would take a prophet sitting on a rapidly spinning turntable to have foreseen the course which our Government has pursued during the last few months." The way that Palestine had been handled, the editorial continued, "has seldom been matched, in ineptness, in the handling of any international issue by an American Administration." The United States was bowing "to Arab threats," and proposing that the United Nations "retreat with us in the face of Arab scorn and fury." Such a surrender was "a blow to the authority of the United Nations" and a "shabby trick on the Jewish community in Palestine."[83]

The *New York Herald Tribune* argued that the Austin speech had been either a reversal of policy or an announcement that showed that the administration had "ignored facts which were patent to the most casual observer." Either, it declared, was not "to the credit of the Administration." It was, in addition, a heavy blow to the United Nations' prestige. With trusteeship, the United States would now "merely inherit Britain's 'squalid war' in the Holy Land."[84] The liberal pro-Zionist press was even more scathing. The left-wing New York paper *P.M.* called the day of Austin's speech "Black Friday" and said it was evidence of "American duplicity." The policy change, its editor wrote, revealed the "face of the deceiver," who had led Jews "to the precipice–and then dumped them."[85] And writing in the *New York Post,* Editor in Chief T. O. Thackrey called the speech a "dishonorable and hypocritical betrayal of Palestine" that had challenged the integrity of the president, and nothing less than a "betrayal."[86]

Eddie Jacobson was very upset. He had returned to Missouri secure in the knowledge that the president had guaranteed to back partition when talking to Chaim Weizmann. When Abe Granoff phoned him with the news of Austin's speech, Jacobson was "speechless" and "dazed as a man could be." In the immediate period following the speech, Jacobson could not "find one human being in Kansas City . . . who expressed faith and confidence in the word of the President." But on Monday, he received a phone call in his store from Weizmann, who told him:

> Mr. Jacobson, don't be disappointed and do not feel badly. I do not believe that President Truman knew what was going to happen in the United Nations on Friday when he talked to me the day before. I am seventy-two years old, and all my life I have had one disappointment after another. This is just another letdown for me. Don't forget for a single moment that Harry S. Truman is the most powerful single man in the world. You have a job to do; so keep the White House doors open.[87]

Truman felt dreadful about Weizmann. The president liked to think that people could count on his word, especially someone he had as high a regard for as Weizmann. He asked Sam Rosenman to go see him in order to explain what had happened. Rosenman reported back that when he had gone to his hotel room, he had found Weizmann calmly having coffee with Ben Cohen. Weizmann, Rosenman recounted, still had complete confidence in him. If Truman had changed his approach, Weizmann would simply overcome it. Being a politician, he was used to these sorts of things. Truman was relieved and greatly appreciated his attitude. In the future, Truman told Rosenman, he wanted to see only Weizmann if the Zionists needed to confer with him. Rosenman believed that Truman never forgot or forgave the State Department for its actions and from that point on watched its maneuvers closely and did not trust its judgment.[88]

Weizmann was not nearly as optimistic as he let on. He wrote a friend that the "unexpected and sudden let down by the American government will . . . have tragic effects." As for trusteeship, Weizmann thought it could never work. "It is a still-born project," he wrote, "produced on the spur of the moment by some fertile brain in the American State Department."[89] Now, from his room in the Waldorf-Astoria, Weizmann composed a lengthy letter to Truman explaining why he felt trusteeship was not the answer. First, he called attention not to the Austin speech but to Truman's statement that he had not abandoned partition as the ultimate settlement. "I welcome this assurance," Weizmann wrote, "because my long experience . . . has convinced me beyond doubt that no more realistic solution exists." Jews and Arabs, he argued, both wanted independence, and in the areas where they made up a majority, they were "virtually in control of their own lives and interests. The clock," he told the president, "cannot be put back" to the situation that existed before last November 29. Moreover, the psychological effect of "promising Jewish independence in November and attempting to cancel it in March" was too obvious to ignore.

It was two years since the report of the Anglo-American Committee, Weizmann reminded Truman. Yet the Jews of Europe were "still in those camps" and were subject to "dwindling resources of hope and morale," which had been lowered after they heard of the reversal of U.S. policy. "I cannot for a moment believe," he wrote Truman, ". . . that you would be a party to the further disappointment of pathetic hopes, which you yourself have raised so high." The reversal had also led to new Arab aggression, since the Arab leaders were more confident than ever because they believed partition had been revised. The only choice for the Jewish people, Weizmann ended, "is between Statehood and extermination. History and providence have placed this issue in your hands, and I am confident that you will yet decide it in the spirit of the moral law."[90]

The British would give up their Mandate in about a month. All were expecting an armed attack against the Yishuv by the Arab states, and U.S. policy was floundering. Weizmann was right: Harry Truman hadn't chosen it, but Providence had placed a heavy responsibility in his hands.

TWELVE

A NEW COUNTRY IS BORN: TRUMAN RECOGNIZES ISRAEL

No matter how it was spun, trusteeship, even if temporary, was the United States' new position on Palestine. It would be up to historians to decide whether Truman had approved Austin's speech but not the timing of it; had approved the speech but hadn't looked carefully enough at what he was signing or at its implications; or as Clark Clifford maintained, had believed he was signing on to a contingency plan that would probably never be activated. In any case, the government's anti-partition forces had succeeded, but at the price of earning Truman's enmity. Although Truman maintained that his long-term goal was partition, in the short run he did not want to see violent conflict or U.S. troops in Palestine. He decided to let State's trusteeship and then truce proposals wind their way through the U.N. Security Council. In any case, for the moment, he didn't have a choice.

The State Department had two months to achieve its trusteeship and truce plan before the British left, the Jews proclaimed their state, and anarchy reigned in Palestine. Henderson thought that for the scenario to work, Britain would have to be convinced to maintain its Mandate beyond May 15. On March 27, Henderson met with Sir John Balfour and Mr. Bromley of the British Embassy. Henderson stressed the importance of Anglo-American cooperation in regard to Palestine.

Just because the British were pulling out, Henderson told them, they could not ignore their responsibility to try to restore tranquillity to the Middle East. The U.S. government "hoped that the British Government would be willing to maintain British troops in Palestine beyond May 15." He then asked the diplomats for some "private assurances" that it would. Soon a top secret message arrived for him from the British Foreign Office. The U.S. proposals were giving them "great difficulty," it said. But although Britain sympathized with the efforts of United States to avert a civil war in Palestine, it was going to maintain its current line of "abstention" and would not be drawn into the fray again.[1]

On April 2, confident of its course of action, the State Department prepared a draft of a Trustee Agreement. It called for the appointment of a governor-general for Palestine with a three-year term, who would be appointed by the Trusteeship Council. There would also be a bicameral legislature (a House of Representatives and a Senate), and a prime minister, who would choose a cabinet and a judiciary.[2] When the U.S. delegation submitted these "tentative" proposals to the Security Council, both the Jews and the Arabs immediately raised strong objections. A spokesman for the Jewish Agency said the plan was "merely an attempt to perpetuate under UN auspices the White Paper policy of 1939" and would "meet with the most determined opposition of the Jewish people." Faris al-Khouri, the representative of Syria, said the plan was unacceptable to the Arabs, as did Jamal Husseini of the Arab Higher Committee.[3]

The United States faced greater embarrassment when the Soviet Union denounced trusteeship and adamantly defended partition. Speaking before the Security Council, Andrei Gromyko called partition a just solution that could be carried out peacefully. However, it was clear to him that the United States was now trying to bury it and to justify the new trusteeship proposal. If the United States wrecked partition, he bellowed, it would be because "of their own oil interests and military-strategic positions in the Middle East." The new U.S. trusteeship proposals, he told the Security Council, were meant to "convert Palestine into a military strategic base of the United States and England under the pretext of maintaining order in that country."[4]

Finding that trusteeship was gaining little traction, State decided to shift its focus to getting a truce and, once that was accomplished, to propose trusteeship. But this too was met with resistance. Speaking for the Jewish Agency, Moshe Shertok took the position that the quest for peace could not start with a truce. Rather, any plan had to take notice of the constant "Arab aggression from outside," sponsored by Arab states that were full U.N. members, which were trying to alter the U.N. resolution of November 1947 by force. What he preferred was that the United Nations curb the aggression or condemn it for the record. As for trusteeship, the Zionists were dead set against it. His people were "ripe for independence" and would refuse to accept its postponement after the Mandate came to an end.[5] An equally strong reason was given by Rabbi Silver, who was representing the Agency at the United Nations. Trusteeship, he pointed out, would need to be enforced by troops. The United States was not willing to use them to enforce partition; why would it now use them to enforce a trusteeship? The British, he argued, had had a trusteeship for twenty-five years and had not been able to produce a solution for Palestine.[6]

Undeterred, State brought the representatives of the Jewish Agency and the Arab Higher Command together at Lake Success to talk under U.N. auspices. But it took only one day for the negotiations to collapse. Shertok insisted that the negotiations had to be on the basis of the General Assembly's partition plan, while the Arab League said that the plan had to be "irrevocably" dropped by the United Nations.[7]

While success was eluding the State Department at the United Nations, the Yishuv was taking concrete steps to create a provisional government that would have all the trappings of an actual state. Dana Adams Schmidt, *The New York Times'* correspondent in Jerusalem, observed that "the problem of a Jewish state is no longer really one of being born but of getting a birth certificate." The Jewish state already exists, he wrote, "certified or uncertified." The Jews were going ahead with the organization of their state in accordance with the timetable set up by the November 29 U.N. resolution. They had agreed to a provisional council of government and had submitted it to the U.N. Palestine Commission well before the April 1 deadline. In addition, the Jewish Agency and the Vaad Leumi (Jewish National Council) had announced that on May 15 a provisional Jewish government would begin to function, consisting of a premier and twelve ministers, representing a coalition. They would then ask for international recognition and would prepare to hold elections. The Jews didn't lack for administrative staff. "Already they have a department for everything," Schmidt reported. But they would need international recognition mainly for financial matters. Schmidt asked, "Would the ordinary commercial banks lend the Jewish state money without the approval of the US Government?"[8]

In April, the Haganah began to turn the military situation around. It could not afford to continue to take a defensive stance, waiting it out until the British left. If it was doing so poorly now, what would happen when the expected Arab armies arrived on May 16? Ben-Gurion insisted they go on the offensive. The Haganah High Command put forth a new strategy. Its Plan D called for defending and capturing the areas lying within the proposed Jewish area. Especially important was seizing control of "Palestine's interior road network, as well as the country's important heights." The operation would require the capture of all Arab towns dominating vital arteries and communications, something new for the Haganah.[9]

The Haganah also began to receive the weapons it desperately needed. On April 1, the first transport plane containing arms from Czechoslovakia landed. Two days later a ship arrived with even more weapons, which were used to break the grip of the Arab forces that controlled the road to Jerusalem. The Haganah now had thousands of rifles and hundreds of machine guns. As Aubrey Eban later wrote, "We were reaping the fruits of Soviet support."[10] Between April 5 and May 14, 1948, four of the five cities of Palestine (Tiberias, Haifa, Jaffa, and Safed) had been conquered or had surrendered.[11]

The Jews launched another operation from April 5 to April 15, called Operation Nachshon, named after the first man to enter the Red Sea during the exodus from Egypt. The Haganah's mission was to clear the road to Jerusalem so that supplies could get through to the besieged Jews in Jerusalem's Old City. It was then that a major tragedy took place. The last Arab village along the western approach to the city was Deir Yasin, which had a nonaggression pact with the Haganah. Violating that agreement on April 9, the Irgun and Stern Gang decided to attack it.[12] During the fighting events took place that would shake the moral conscience of the world. Around one hundred ten people–half of them women and children–were killed by the Jewish forces. The Deir Yasin massacre, as it came to be called, would reverberate for decades. To the Arabs, it became the symbol of Jewish barbarism. The U.S. consul reported that the Arabs were indignant and resentful and determined to avenge the massacre, making any cease-fire or truce even more remote.[13] News of it spread and contributed to the flight of the Arabs from Palestine.

Ben-Gurion and the mainstream Zionist leadership strongly condemned the massacre, but the Arabs' revenge was swift in coming. Their forces ambushed an armored convoy heading to the Hadassah Hospital on Mount Scopus, wounding seventy-six people and killing thirty-four, most of them doctors, nurses, or other professionals. The Arab forces came with Sten and Bren guns and hand grenades, which hit the immobilized vehicles. The Jews fought for an hour, using machine guns and automatics, but eventually their armored cars and buses burst into flame and the occupants were shot down trying to escape. "The street," one reporter wrote, "was littered with dead bodies."[14]

Immediately, a group of eighty-six distinguished American scientists signed a letter of protest to the president. The convoy had been making a "routine journey with medical supplies, personnel and patients" to the Hadassah Hospital, considered the best medical facility in Palestine. "The whole Near East is bereft," they wrote. Moreover, the attack had been a gross violation of the Geneva Convention, which stated that all medical vehicles and personnel were immune to attack.[15] The next day, Truman agreed to meet with Dr. Louis Dublin of the Metropolitan Life Insurance Company, who presented him with the letter. "Those in whom we had placed all the hope of the future in building up a great medical center," he told Truman, "had been wiped out in spite of the fact that they were travelling towards the hospital on a mission of mercy, contrary to medical conventions and every instinct of decency."

Truman's response surprised Dr. Dublin. It appeared he hadn't known about the attack and "was horrified" when he heard about it. "What does it gain them," he said, "to do that sort of thing?" After making that comment and receiving the letter, Truman turned to other matters. Rather than address the issue that Dublin had come to talk about, "he immediately unburdened himself of his troubles," complaining of all he had gone through on the Palestine issue the previous months. "He wanted my sympathy," Dublin said, "as I wanted his." First he complained about the lack of unity among Jewish groups and argued that "they have engaged in very questionable practices themselves." He was upset that the Jews would not consider a truce. His main concern was what would happen on May 15 and thereafter. "All of his directions," Dublin thought, "were now bent in the direction of achieving peace in Palestine."

Dublin answered that he thought making the American public aware of what the Arabs were doing would help to mobilize public opinion. It seemed like minutes before Truman said anything. Finally Truman said "he was fearful that the publication of the letter . . . would embarrass him; . . . that it would be one more inflammatory incident; that there would be recriminations from the other side" and would work to hold up a truce. Truman pledged to read the letter again and take it up with Charlie Ross and see "if we can't issue a statement that would meet the situation."

Dublin knew that Ross and David Niles wanted the letter made public. Yet when Dublin opened the next day's newspaper, there was nothing in it about the letter. Realizing that Truman did not want their message to get out, Dublin advised Hadassah, the Zionist women's organization whose funds supported the hospital, to withdraw it from circulation and not publish it at that time. "If we take the matter into our own hands and issue the letter," he wrote, "it will have the effect of being one more insult." Truman felt that he had been "subjected to a series of insults" and that the Jewish groups' activities "nullified many of the things he hoped to accomplish." If they issued a letter, he concluded, it could result in having Truman "throwing his weight in the direction where we wouldn't want it." Rather than make it public immediately, Dublin suggested waiting a reasonable length of time and then making known what had happened. Then they could appeal to "the conscience of America."[16] Following his suggestion, Hadassah waited until April 26 to release the text of the letter, noting only that it had been sent to the White House at an earlier date.[17]

In mid-April, Warren Austin introduced a resolution at the U.N. special session for a temporary trusteeship that, he explained, was meant only "to ensure public order and the maintenance of public services." It would continue while there were negotiations on a final political settlement.[18] Austin's effort gained little support. No country would provide military support to enforce the plan, and hence no member states could support it. Secretary-General Trygve Lie observed that "skepticism about the practicality of trusteeship was everywhere." Many of the member states wanted the United Nations to implement partition, rather than go on talking. The representative from New Zealand, Sir Carl A. Berendsen, "called on the Assembly not to abandon partition in a capitulation to threats and violence" and declared "what the world needs today are not resolutions, it is resolution."[19] Nevertheless, the General Assembly ended its meeting by adopting three resolutions: it affirmed its support of the Security Council's efforts to bring about a truce in Palestine; it empowered a U.N. mediator to use his offices to promote a peaceful resolution of the situation and ensure protection of the holy places; and it relieved the Palestine Commission of further responsibilities. Most important, it did not rescind or amend the partition resolution of November 29, 1947.[20]

The State Department pressed for the British government to change its mind and continue on in Palestine. To accomplish this, Marshall and Lovett sent Lewis Douglas, the U.S. ambassador to Britain, on a number of missions to Ernest Bevin. On April 20, Douglas reported to Marshall on a long talk he had had with Bevin. Bevin was concerned, he informed the secretary, that the United States thought Britain was not cooperating, but he was at his "wit's end to know what to do." Churchill told Bevin that the Conservative Party would vigorously oppose any suggestion that British troops remain in Palestine. Most Labour members of Parliament would be equally opposed. The British public demanded that its troops be withdrawn from Palestine. It seemed that in Britain, as in the United States, public opinion had its influence on foreign policy.[21] Bevin explained to Lewis that trusteeship would require force to implement it and Britain would not use force against either the Jews or the Arabs to impose it.[22]

At the end of April, Robert McClintock, Dean Rusk's assistant in the State Department's U.N. Office, told Rusk that trusteeship was a bust and that they should focus on a French proposal to save Jerusalem. At least it would look as if they had accomplished something. He also wrote to Lovett that if there were security from Jerusalem to the sea it would bring about a de facto partition, with the Jewish state centered on Tel Aviv and extending north along the coast. Fatigued by the whole thing, he confided, "I should not care if Transjordan, Lebanon, Syria, and Egypt took over the rest of the country."[23]

By May 4, Lovett, admitting defeat, instructed the U.N. delegation to stop trying to "adjust" the November 29 partition vote. It should just concentrate on trying to get a truce agreement. A bigger problem was beginning to concern State: the real possibility of an Arab invasion of Palestine after May 15. The American consul in Cairo had sent his secretary to Damascus to ascertain Arab views about a new U.S. initiative: an informal truce agreement and ten-day cease-fire. When he met with Abdul Rahman Azzam, the head of the Arab League, he asked him if the Arabs had "considered the grave responsibilities which they were assuming before the world in invading Palestine when the matter was before the UN." Yes, Abdul Rahman Azzam replied, they had given very serious consideration to all of the consequences and had determined that they had no choice but to send armed forces into Palestine. The Arab armies were poised and ready to enter Palestine on May 15. And, he told the secretary, if they failed to invade Palestine, it would lead to "dissatisfaction and mutual recriminations" among Arabs; relatively moderate elements of the Arab League, including Abdul Rahman Azzamy himself, would be overthrown; and the unity of the Arab League would be endangered. He was also apprehensive that some Arab governments might be overthrown as a result of "rising passions among the Arab population."[24]

The Zionist response was equally negative. Warren Austin's deputy, John Ross, told him about a talk he had had with Silver and Shertok. They had discussed the idea for a truce for an hour, including the proposal for a ten-day extension of the mandate. Silver told him that the Jews "were not trying to shoot Arabs out of the Arab state but Arabs were trying to do this to the Jews." If the Arabs stopped their aggression and stopped shooting, the Jews would stop. He said they would not attempt to "take new positions nor attack Arab communities." Most important, Silver claimed, they would agree to postpone announcing a Jewish state if they could establish a provisional government "provided there was guarantee that at the end of the truce period they could go ahead and establish their state." Ross tried to argue him out of this position, but Silver, who by now was angry, told him he considered the U.S. government as a "hostile" government in which they had no confidence at all. Silver asked Ross, "How could we ask them to accept a truce and pretend to be friendly in doing so if we were not prepared to support the creation of a Jewish state and defend it against external aggression?"[25]

The State Department's plan to derail partition had proven to be an embarrassing failure. Austin acknowledged this when he told Marshall that the U.S. delegation would be in "a very weak and vulnerable position" in terms of American public opinion and at the United Nations if it didn't come up with any acceptable suggestions before the General Assembly session ended. If it did not, it would "make the United States vulnerable to accusations either of incredible naiveté or power-politics machinations."[26]

While the State Department's proposals were floundering at the United Nations, the president was going in another direction. On April 12, Eddie Jacobson visited Truman at the White House. Jacobson was relieved to hear "from my friend's own lips" what had really happened after Truman's meeting with Weizmann, when Truman had assured Weizmann of his support for partition, only to have it reversed by Austin at the United Nations the next day. Then, Jacobson said, Truman "reaffirmed, very strongly, the promises he had made to Dr. Weizmann and me; and he gave me permission to tell Dr. Weizmann so." Jacobson also discussed the matter of recognizing the new state with Truman, and *"to this he agreed with a whole heart"*[27] (emphasis in Jacobson's article).

Eleven days later, Truman sent another message to Weizmann, this time using Sam Rosenman as his emissary. Rosenman, who was laid up with a leg injury, called Weizmann to his hotel. He told him that he had been to see Truman and the president's first words to him had been "I have Dr. Weizmann on my conscience." Truman wanted Weizmann to know that "he would recognize the Jewish State as soon as it was proclaimed," but he asked Weizmann to keep it secret. Upon hearing that news, Weizmann for the first time felt certain that a Jewish state could be created. Rosenman added that Weizmann should write a letter to Truman requesting that he recognize the new state a few days before the end of the Mandate.[28]

Bartley Crum confirmed Truman's intentions. He had seen Truman on May 7 and told the Weizmanns that the president intended to recognize the Jewish state at six o'clock on May 14. Vera Weizmann wrote in her diary, "The President is anxious to do so before the Russians. What a joke." However, Crum advised them not to be too optimistic, because he said, "one never knows what influence will be brought to bear on the president by the British, the State Department, and anti-Zionists." "It was almost zero hour," she wrote, and she hoped Truman would carry out his plan on schedule because "the President will have no credit in the Jewish State's birth if his recognition comes too late."[29]

Weizmann confided Truman's intentions only to his closest associates. Eban and the Jewish Agency delegates at the United Nations got the message that they could not afford to lose the fight against the U.S. trusteeship proposal. Truman would be able to act on his pledge only if trusteeship were abandoned as an option. Eban didn't believe that Truman would "compete" with the United Nations in determining the "legal status of a previously mandated territory."[30]

Carrying forth with this agenda, Eban drafted a speech for Shertok to be given at the United Nations arguing why trusteeship was a bad idea. Shertok, in what Eban thought was a very generous gesture, asked Eban to give the speech he had written himself. It was par for the course to have a staff member write a speech for the head of a delegation. For the first time Eban gained visibility as a spokesman for the Zionist cause, and he would consider it one of the major turning points in his own life and career. Jewish statehood, Eban told the General Assembly, was a reality in everything but name. "More force would now be needed to prevent Jewish statehood than to let it take its course. . . . How absurd it would be," Eban said, "to ask a nation that had advanced to the threshold of independence to retreat back to tutelage! . . . The flight from partition would be a blatant acceptance of illicit force as the arbiter of international policy." Suddenly thrust into the limelight of international renown, Eban was invited that day to go to lunch with other Agency leaders and U.N. delegates at Trygve Lie's apartment. Among the other guests were Ernest Bevin and Andrei Gromyko. The Soviet representative came up to Eban and said, "Congratulations. You have killed American trusteeship." Bevin and his wife gave him icy looks.[31]

The Arabs were jockeying over who was to have power over Palestine once the British left. King Abdullah of Trans-Jordan pledged to take personal command of his own army, along with those of Syria and Lebanon, and move into Palestine when the Mandate ended. The Yishuv leadership had been hopeful that it could work out some kind of an agreement with Abdullah whereby he would take over the areas of Palestine designated for the Arab state but leave the Jewish areas alone. But now, referring to meetings he had had with Golda Meyerson in which they had agreed that the Mufti was their mutual enemy, Abdullah proclaimed, "I have advised the Jews before to content themselves and live as citizens in an Arab state, and my army is an Arab army. I shall do as I please."[32]

That same day, Jamal Husseini, the vice chairman of the Arab Higher Committee, announced that the Arabs would immediately set up their own state on May 15. Their goal was the creation of "a single democratic state" for all of Palestine. But despite their bravado, the Arabs lacked leadership and organization. When the United Nations searched for someone from the Arab Higher Committee in Palestine with whom it could discuss the truce proposal, it could not find anyone there to talk to.

In contrast, the Yishuv was a model of efficiency and organization. Its "blue and white flags were hoisted over public buildings in Tel Aviv and elsewhere, and the Jewish authorities began to issue their own stamps and levy their own taxes."[33] The U.S. consul at Jerusalem, reporting to Marshall, noted that "preparations for establishment of a Jewish State after termination of the Mandate are well advanced." The Jews had great confidence in their future, and the population's support for its leaders was simply "overwhelming." Once defensive operations by the Haganah had changed to offensive ones, its morale had increased due to its military victories and the flight of thousands of Arabs out of Palestine. True to the partition line, the consul reported, the Jews had not tried to seize any territories outside the boundaries proposed in the partition resolution.[34]

The consul's observations were reinforced in a lengthy memo on "The Future of Palestine" written by John E. Horner, an adviser to the U.S. delegation who was on loan from the State Department's Office of European Affairs. Trusteeship, he reported, had been totally abandoned by everyone at the United Nations. It was no longer an option. Most delegates were distrustful of what they felt was the oscillating policy of the United States and did not want to commit themselves to any U.S. program, fearing the United States would "suddenly commit an about-face, leaving them in an untenable position." One option might be considered, Horner thought: the annexation by Trans-Jordan of the area promised the Arabs in the partition resolution. Such a step would satisfy most of the parties and would end the influence of the Grand Mufti among Palestinian Arabs. Most important, Horner stressed, such a measure would force all sides to *"face up to the inescapable fact that a Zionist State already is in being in Palestine"*[35] (Horner's emphasis).

A few days later the State Department formally proposed a ten-day truce in the fighting between Jews and Arabs that would commence on May 5. Its purpose would be to fly Jewish Agency leaders on the president's plane *Sacred Cow* from New York to negotiate in Palestine with the Arab and Jewish authorities already there. It was hoped that once in Palestine, they would be able to negotiate a still longer truce and then begin negotiations for a "final political settlement," which obviously did not include creation or recognition of a Jewish state.[36]

The Jewish Agency was not buying into it. Shertok nevertheless took the proposal to his colleagues. "We do not consider," he answered Dean Rusk, "that the somewhat spectacular proceeding now suggested is warranted." The Jews of Palestine would accept a genuine cease-fire, he told Rusk, provided the Arabs did likewise. But Rusk's proposal for a plane flight to Jerusalem violated the action taken by the Security Council and would mean extension of the British Mandate.[37] Rusk understood, as he wrote Lovett, that Shertok's refusal "reveals the intention of the Jews to go steadily ahead with the Jewish separate state by force of arms . . . and rely on its armed strength to defend that state from Arab counterattack."[38]

Dean Rusk was becoming more realistic about the poor chances of the Arabs and the Jews reaching an agreement. The British couldn't do it, he reflected; perhaps it was hubris to assume that the Americans could do it in two months. A meeting with Prince Faisal of Saudi Arabia in New York confirmed this. Rusk told the prince, "I would not be frank . . . if I did not say that the President considers partition a fair and equitable solution for Palestine." The Arab states, Faisal answered, "could not ever accept a Jewish State," which would be nothing less than an "abscess to the political body of the Arabs."[39]

Truman and the White House staff were making their own preparations. David Niles had already written an announcement for the president on recognition of a Jewish state. He proposed that Truman say that he rejected trusteeship and that "Secretary Marshall and I have concluded that we should recognize the practical reality, since it conforms to the resolution of the U.N., to the security interests of the U.S., and to the announced and oft repeated objective of the U.S. Government." The president would then announce the intention of the United States "to accord formal recognition to the Jewish Government in Palestine when it is established."[40] Truman did not take Niles's advice and never gave any statement or speech making these points.

Even without a formal speech to that effect, the indications were that Truman was going to recognize the Jewish state. But how was he to explain yet another policy change to the American people? Other people were apparently pondering the same question. On May 5, Truman's military aide, General Harry Vaughan, received a letter from Dean Alfange, who headed the American Christian Palestine Committee in New York City. Vaughan passed the letter on to Truman. Alfange wrote from the perspective of politics and the potential fortunes of the Democratic Party in New York and throughout the nation. "Frankly," he said to Vaughan, "the President could not carry the State of New York in the present circumstances. The Jewish vote against him would be overwhelming." There was a dramatic step he could take that would "electrify the Jewish people": recognize the State of Israel when it was announced and nominate an American minister to it. Truman, he argued, would be on firm legal ground because the U.N. partition plan was "legal fact." As for the would-be temporary trusteeship, which the president had gone on record as supporting, Alfange noted that "recent events have knocked the props from under the Trusteeship proposal." It was "no longer tenable . . . because the Jewish military forces have since demonstrated by their decisive victories over the Arabs, that they can implement partition singlehanded."

This change in the situation, Alfange continued, gave Truman a justification for a policy change. He could tell the nation that "events and not he have reversed the Trusteeship plan and that the UN decision can be best carried out by recognizing the new Jewish State." Should Truman not want to do that, he could appoint a personal ambassador and send him to Palestine with the president's authority to negotiate a settlement on the basis of the U.N. partition plan. The Arabs, Alfange argued, "have been rudely awakened to the fact that they cannot dispose of the Jews as easily as they had thought." King

Abdullah of Jordan was a realist, he thought, and knew that his British-trained Arab Legion was no match for the Jewish forces, which now numbered 75,000 "disciplined and zealously devoted young men and women." Indeed, Alfange thought that the Haganah could beat the entire combined forces of the Arab League. Apart from Abdullah's British-trained legion, the other Arab forces were divided and hostile to one another, a fact that precluded any meaningful unity. Thus Truman could negotiate even with Abdullah, who Alfange thought was the "most practical and best situated of all the Arab leaders" and who desired "authority and influence over the Arab portion of Palestine" and thus would accept partition.[41]

Before Truman took any action, he had to deal with the State Department. At the beginning of May, Truman summoned Clark Clifford to his office and told him that he was scheduling an important meeting. "This temporary situation in the Middle East" would soon be coming to an end, Truman said, and it was very likely that there would be the announcement by the Jewish state declaring its independence. The meeting would be with Marshall, among others, on May 12, and he wanted Clifford to prepare an oral argument for recognition "just as though you were going to make an argument before the Supreme Court. Consider it carefully, Clark, organize it logically. I want you to be as persuasive as you possibly can be."

Clifford had come to the meeting already prepared with an outline of arguments for why the United States should recognize the new Jewish state when it was proclaimed:

PALESTINE

1. Recognition is consistent with U.S. policy from the beginning.
2. A separate Jewish state is inevitable. It will be set up shortly.
3. As far as Russia is concerned we would be better to indicate recognition.
4. We must recognize inevitably. Why not now.
5. State Dept. resolution doesn't stop partition.[42]

The forty-one-year-old Clifford was being asked to debate General Marshall. "Virtually every American," Clifford thought, "regarded General Marshall, then sixty-seven, with respect bordering on awe." Truman could not afford to lose Marshall. But General Marshall did not like Clifford and considered him a domestic political adviser who had no business interfering in foreign affairs. In this, Clifford thought he was mistaken. The position of national security advisor had not yet been created, but Clifford functioned in that role and had been very involved in the White House's Middle East policy. Nor did Clifford think very highly of the State Department, which, he believed, had "done everything in their power to prevent, thwart, or delay the President's Palestine policy. . . . Watching them find various ways to avoid carrying out White House instructions, I sometimes felt, almost bitterly, that they preferred to follow the views of the British Foreign Office rather than those of their President."[43] Knowing that he had Truman on his side gave Clifford the confidence he needed to carry out the daunting assignment.

Among the diplomats as well as the White House, a fear was developing that the Soviets might be the first to recognize a new Jewish state. Dean Rusk thought that if the Soviets recognized the Jewish state, as it was contemplating doing, it would allow them to come to the new state's defense if it were attacked by the Arabs. In light of this, Rusk thought that the United States had better be prepared to take a position once the new Jewish state was proclaimed.[44]

Before the May 12 meeting, Clifford requested a report from the State Department on its recent activities regarding Palestine. The department prepared a summary for him, going through all the measures it had proposed as an alternative to partition, including proposals for a truce and trusteeship as well as efforts to convince the British not to give up their Mandate authority. It mentioned its attempts to convince the Jewish Agency not to declare a Jewish state after May 15. Finally, they had imposed an arms embargo to the Middle East. That last step had been the only one implemented.[45] Clifford thought the only thing the State Department had managed to do was to embarrass the president.

May 15 was fast approaching, and Shertok requested a final audience with Marshall and Lovett before he left for Palestine, where plans were being made to announce the new Jewish state. Accompanying him was the Agency's U.N. representative, Eliahu Epstein. The Jewish Agency wanted clarification of the U.S. position. Rusk was also there. Sources had told him that there was a major debate taking place within the Jewish Agency on whether or not it should accept proposals for a truce or proceed with its announced intention to proclaim a Jewish state when the Mandate ended. Epstein and Shertok, Rusk thought, favored a truce. They were being opposed, however, by "more extreme elements such as Rabbi Silver and Ben-Gurion," who were "pressing for the immediate establishment of the Jewish State by force if necessary."[46]

The Jewish Agency had hoped to have the United States at its side when the new state was announced, but Marshall did not appear to favor it. Shertok told him that as much as the Agency wanted America's support, it would not "commit suicide to gain friendship." The main problem they had, Shertok candidly put it, was that to date the Jewish Agency did not know whether "the United States Government wanted or did not want a Jewish state to arise."[47]

State's U.N. truce proposal appeared to be based on the Jews of Palestine forgoing the announcement of a Jewish state when the Mandate ended. That "was tantamount to asking us to renounce our fundamental right," and if they did as State wanted, they might never have another chance. They were "on the threshold of the fulfillment of the hopes of centuries," and a new state "was within our physical grasp." Marshall was asking the Jews of Palestine to become a party to their own undoing. Now that the Jews were winning, the State Department wanted them to call a halt, which could be fatal. They had accomplished this on their own, without outside help. The U.S. truce proposal, in their eyes, would allow the Arabs to regroup and consolidate their forces. Shertok asked, "not having helped us, why should the United States Government now try to prevent us from attaining what was so imminently within our reach?"

Lovett and Rusk then joined in and cautioned that if the Jews fought the Arabs, the Arab Legion might wage a guerrilla campaign that would be costly to the Yishuv. A truce was preferable to any war. The Jews were taking great risks, Lovett argued. "The risk of forfeiting the chance of statehood," Shertok responded, "was graver." An hour and a half into the meeting, Secretary Marshall spoke up for the first time. It was 3:25 P.M., and Shertok's plane for Jerusalem was due to leave in ten minutes. Marshall phoned the airport to hold the departure and offered his own State car to take him to the airport.

Finally, Marshall expressed his opinion: The Zionists kept speaking out and engaging in "political pressure . . . blustering . . . misleading assurances." He had heard from the many Jews he knew that if partition occurred, peace and quiet would be a reality in Palestine. This had "turned out to be quite untrue." Second, Marshall complained about the continuing illegal immigration into Palestine, which was obviously sanctioned by the Jewish Agency and which caused him "no end of difficulty." Every time the U.S. tried to get Britain to act in a different fashion the Foreign Office would throw the illegal immigration in their face.

Then Marshall came to his main point, which was to address the "present dilemma." He used his experience as an intermediary between Chiang Kai-shek and Mao Tse-tung when he had tried to intercede in the Chinese Civil War. Shertok later summed up what Marshall had said:

> We [the Jews] had scored an initial military success; we should beware of relying on it. He could not help thinking about his experience in China when he went there as an intermediary. The analogy was striking. He had almost succeeded in arranging for a truce . . . during which time the Government was supposed to absorb the other people. But at that time the Government forces had just scored a success in the field and they were afraid that they would lose more than they would gain by the truce. The same thing happened on the other side and they too had the same fear. As a result the truce did not come off, and here, two years since . . . the outbreak of war in

> China . . . the Government in the meantime had lost Manchuria. He himself was a military man, but he wanted to warn us against relying on the advice of our military people. He indicated that flushed by victory, their counsel was liable to be misleading. If we succeed, well and good. He would be quite happy; he wished us well. But what if we failed? He did not want to put any pressure on us. It was our responsibility and it was for us to face it. We were completely free to take our decision, but he hoped we do so in full realization of the very grave risks involved.

Shertok was thrown by Marshall's presentation. The secretary had given military reasons to doubt the possibility of succeeding. He had not reiterated the arguments given by Loy Henderson or Robert Lovett. Indeed, he even seemed sympathetic to their cause. Shertok promised him that if the Jewish Agency decided to proceed with the announcement of a new state once he returned to Palestine, it would not be because they did not take his advice under consideration. Marshall's presentation so bothered him that Shertok was now questioning the Agency's plans.

Before Shertok was to leave for Palestine, Weizmann called to give him a message that he should also deliver to Ben-Gurion. "Moshe, don't let them weaken," he told Shertok, "don't let them swerve, don't let them spoil the victory–the Jewish State, nothing less." When he arrived at New York's International Airport, Shertok was surprised to be called to the phone. It was Weizmann again, with another message for him to deliver to Ben-Gurion: "Proclaim the Jewish State, now or never!"[48] Weizmann was putting his trust in Truman's word. Some sources have alleged that Shertok took Marshall's warning to heart and now supported putting off the creation of a Jewish state. Whether or not this is true, when the vote came before the National Council of Thirteen, the new ruling body in the Yishuv, Shertok took a strong stand for its creation and gave a forceful speech that turned the tide against the few moderates who wished for postponement. The vote was six to four to reject the truce. Shertok's vote in favor of proclaiming a state was the decisive one.[49]

Meanwhile, the important May 12 White House–State Department meeting was only a few days away, and Clifford turned to David Niles and especially Max Lowenthal for help with research and drafting his presentation. Lowenthal wrote at least six reports to help prepare Clifford for the meeting.[50] Later Truman would give Lowenthal credit for all the work he had done in bringing about the recognition of Israel, but, preferring to stay out of the limelight even decades later, Lowenthal claimed to have only heard about it secondhand from an anonymous source at the White House.[51] However, realizing the historic nature of the events taking place, he kept a detailed diary of the behind-the-scenes activities at the White House in the days running up to the announcement of the Jewish state.

In a May 9 memo, Lowenthal provided Clifford with his most powerful arguments: The Jews had on their own made partition a reality. Not only had they controlled the Jewish part of Palestine militarily, they had maintained and run what was in effect a government. Partition was a reality; the only question was whether it could be reversed. That would be impossible. The bottom line was that "it is unrealistic to believe that the Jews of Palestine could be persuaded to relinquish the State which they achieved largely through their own efforts."[52]

The best thing for Truman to do, Lowenthal argued, would be to make a statement announcing that he intended to recognize the Jewish state as soon as it was announced. That act alone "would retrieve the prestige which has been lost on this issue during the past few months by the President." It might also force the Arabs to accept what was inevitable, once they saw the U.S. support. In addition, most of the U.N. delegates had come to see that the State Department's proposals were "grandiose as well as futile." If Truman did not act, the Soviet Union certainly would, and any similar action by the United States would "seem begrudging–no matter how well-intentioned" and would amount to a "diplomatic defeat."

On the eleventh, a day before the critical meeting, Lowenthal sent

Clifford another long memo, writing on the front, "Clark: Please do not let anyone else read this dynamite." Lowenthal reviewed the alternatives confronting the president and posed some serious questions: What was the actual state of the Palestinian Jews' military preparedness? Was the United States "trying to advise the Palestinian Jews as their friends, or are we seeking to embark on a course of opposition . . . or of neutrality or of force?" How much further would the United States go? he asked? Would the United States issue economic sanctions against a Jewish state, would it send warships to blockade Jewish Palestine, and would it allow the Jewish immigrants to disembark in Palestine? The memo demonstrated that if the United States did not support a Jewish state, the alternatives would lead to a more difficult and eventually impossible situation in the Middle East.[53]

The day of the meeting, Clifford called Lowenthal and Niles into his office at 11:30 A.M. He wanted them to prepare two items for him before the 4 P.M. meeting. The first was a statement the president could make the next day announcing that he would grant recognition. The second was an analysis of the objections that the State Department would likely have to issuing such a statement.[54]

Clark Clifford called the May 12 meeting "Showdown in the Oval Office." On a "cloudless sweltering day," Clifford wrote in his memoir, the group met in the Oval Office. Truman "sat at his desk, his back to the bay window overlooking the lawn, his famous THE BUCK STOPS HERE plaque in front of him on his desk. In the seat to the president's left sat Marshall, austere and grim, and next to Marshall sat his deputy, Robert Lovett." Dean Rusk and Loy Henderson were supposed to be present. But Lovett, knowing that their very presence in the room would be inflammatory to Clifford and perhaps to Truman, had told them not to attend and to send their deputies Robert McClintock and Fraser Wilkins instead. To the right of Truman sat Clifford, Niles, and Truman's appointments secretary, Matthew Connelly.

Although the meeting began calmly enough, Clifford recalled, it soon developed into the most "confrontational and hostile" meeting he had ever attended during his five years in the Truman administration.[55] Truman did not raise the recognition issue when he opened the meeting. He wanted Clifford to do this, but only after Marshall and Lovett had their say. That way Truman would be able to see what Marshall really thought before Clifford revealed the White House's cards. The meeting started with a presentation from Robert Lovett. He reported that at his meeting on May 8 with Shertok and Epstein, the Jewish Agency had been optimistic that it could work out a deal with King Abdullah of Trans-Jordan, giving the king the Arab portions of Palestine and the Jews the Jewish areas already in their hands. As a result, the Jewish Agency was confident that the Jews would be able to establish their own state without having to agree to a truce.[56]

At that point, Secretary Marshall spoke up. He told the president and the others what he had told Shertok, "that it was extremely dangerous to base long-range policy on temporary military success" and that "they were taking a gamble." He also told Shertok that he was giving them notice that if the Jews got into trouble and came "running to us," there was no "warrant to expect help from the United States, which had warned them of the grave risk they were running."

As he spoke, an urgent message came from one of Marshall's aides, reporting on a press dispatch from Tel Aviv that Shertok had arrived in Tel Aviv, bearing Marshall's personal message to David Ben-Gurion. Marshall said he had not sent any message to Ben-Gurion and, furthermore, he had never heard of Ben-Gurion. This was a strange statement, since Ben-Gurion was the leader of the Jewish Agency and about to become the new state's prime minister. Marshall said he had no comment on the story and ended his presentation by saying he thought the United States should continue its efforts to support a U.N. trusteeship and not make any decisions on recognition at present.

Next, it was Clifford's turn to speak. Clifford's own account dovetails with that in Secretary Marshall's summary of the meeting. First, he announced his disagreement with State's position, which recommended efforts to gain a truce in Palestine. Rusk had said back on March 24 that he would be able to get one in two weeks; he still had not.

Second, Clifford noted that trusteeship, which the State Department favored, presupposed a single Palestine. "That," Clifford said, "is unrealistic. Partition into Jewish and Arab sectors has already happened. Jews and Arabs are already fighting each other from territory each side presently controls."

"Third, Mr. President," Clifford said, "I strongly urge you to give prompt recognition to the Jewish state immediately after the termination of the British Mandate on May 14. This would have the distinct value of restoring the President's firm position in support of the partition of Palestine. Such a move should be taken quickly, before the Soviet Union or any other nation recognizes the Jewish state." He should explain that since the Jewish Agency had complied with U.N. resolutions calling for a democratic government, recognition posed no problem. Furthermore, Truman should announce at his scheduled press conference the next day that he intended to recognize the new state when it was declared. He then handed his proposed press statement out to the group.

Finally, Clifford told the group that since the Balfour Declaration, the "Jewish people the world over have been waiting for thirty years for the promise of a homeland to be fulfilled. There is no reason to wait one day longer. Trusteeship will postpone that promise indefinitely. . . . The United States has a great moral obligation to oppose discrimination such as that inflicted on the Jewish people." Taking words directly from a speech Truman had given as a senator in 1943, he added, "There must be a safe haven for these people." A land of their own, he told them, would be one way of atoning for the atrocities committed by the Nazis. Addressing the issue of the United States' national interest, Clifford pointed to how important it was for the United States to have one friendly democracy in the Middle East "on which we can rely."[57]

As Marshall sat listening to Clifford's words, Clifford noticed that his face was turning red. When he finished, Marshall launched into what seemed to Clifford a furious, "emotional attack on the positions I had taken." "Mr. President," Marshall said, "I thought the meeting was called to consider an important and complicated problem in foreign policy. I don't even know why Clifford is here. He is a domestic policy adviser, and this is a foreign policy matter."

Truman answered simply, "Well, General, he's here because I asked him to be here." Marshall quickly retorted, "These considerations have nothing to do with the issue. . . . he [Clifford] is pressing a political consideration with regard to this issue. I don't think politics play any part in this."

At that point, Lovett added, it would be injurious to the United Nations to recognize the Jewish state before it actually existed "and while the General Assembly . . . was still considering the question of the future government of Palestine." It would also harm the president's prestige, being "a very transparent attempt to win the Jewish vote," which would actually "lose more votes than it would gain." The United States also did not know what kind of a Jewish state the Agency would establish. Lovett then pulled out intelligence reports claiming that the Soviet Union was sending Jewish Communist agents into Palestine. The reports had emanated from the British Foreign Office. Clifford found them ridiculous on their face and pointed out that no evidence had ever emerged to indicate that this was being done by the Soviet Union. As Weizmann and others had pointed out, Jewish Communists were a small minority in the Yishuv, and many Jews going to Palestine from Eastern Europe were in fact fleeing communism.[58]

When Lovett finished, Marshall again erupted, saying, *"If the President were to follow Mr. Clifford's advice and if in the elections I were to vote, I would vote against the President"*[59] (our emphasis). There was utter silence in the room. If Marshall's threat should become public, thought Clifford, it "could virtually seal the dissolution of the Truman Administration and send the Western Alliance, then in the process of creation, into disarray."[60] It would certainly ensure Truman's defeat in the upcoming presidential election. Truman ended the meeting quickly. Marshall was still agitated, and Truman turned to him and said, "I understand your position, General, and I'm inclined to side with you in this matter." When they had left, Clifford, who thought he had failed the president, picked up the papers he had brought with him for his presentation. Truman told him, "Well, that was rough as a cob. That was as tough as it gets. But you did your best." "But Boss," Clifford replied, "this isn't the first case I've lost. I never expected to win them all." Truman replied, "Well, let's don't proceed on the assumption, Clark, that you've lost it. Let the dust settle. I still want to do it. But be careful. I can't afford to lose General Marshall."

Max Lowenthal got a rundown of the meeting from Matt Connelly. "The last 24 hours have been something," Lowenthal wrote in his diary. Clifford wanted the administration to announce in advance that it intended to recognize a Jewish state when it was proclaimed; the State Department "wanted approval instead of a new resolution to be offered in the UN Assembly which I think it offered in part, to throw some taint of illegality on any declaration." State had "won the argument, saying that there was no precedent for this Government stating in advance of an application by a state for recognition, that recognition would be granted." Nevertheless, Connelly was hopeful because the State Department had not taken the "position that recognition should be refused after application made."[61]

Lowenthal also wrote in his diary that after he left the meeting, Clifford phoned Lovett to concede that State had won the decision on not announcing that the United States would recognize a Jewish state before it was proclaimed. But that did not mean that the United States should not recognize a state immediately after it was proclaimed. On Friday, Clifford phoned Lovett again and, at Lovett's invitation, joined him for lunch. Lovett showed him a draft, which said the president would consider recognition. "That won't do," Clifford told him. "Let's talk plainly . . . your staff placed the President in a very unfair position." The two men worked out a stronger draft and immediately showed it to Chip Bohlen, George Kennan, and Dean Rusk.[62]

Clifford's recollection years later of what happened was somewhat different from Lowenthal's. As Clifford recalled, he got a phone call from Lovett. "I have been deeply worried ever since the meeting," he told Clifford, "and I'm very concerned about it." It would be very unfortunate "for Truman and Marshall to have an open break." Clifford agreed and sought a way to calm things down. He went to Lovett's house. Despite his opposition to the policies Lovett espoused, Clifford considered him to be a reasonable person and a friend. Lovett made it clear that he and the others at the State Department were holding to their views. "Do you think," he asked Clifford, "that perhaps if you were to present these views to the President that the President might be persuaded to moderate his position and work out something with General Marshall?" Clifford said, "Bob, there is no chance whatsoever that the President will change his attitude . . . we've been in it for months, years maybe. I know how strongly he feels about it." Truman had asked him to speak at the meeting. "My views, Bob," Clifford stressed, "were the President's view . . . if anybody's going to have to give, it's going to have to be Marshall." Truman was "not going to give an inch."

The president, however, kept his word about not issuing a statement about his intention to recognize the Jewish state. At a press conference held on the thirteenth, Truman was asked whether or not the United States would recognize a Jewish state. "I will cross that bridge when I get to it," he replied.[63] Niles had a chance to speak with Truman afterward and later that day told Lowenthal about their conversation. Truman had told him, Niles said, that he thought that Marshall and Lovett meant well but they had "followed their subordinates." Moreover, he didn't buy Lovett's argument that the Soviet Union was sending Jewish Communists into Palestine.

That was good news, but Niles was still unsure about Truman's plans once the Jewish state was declared. He told Truman that he thought the United States should recognize the Jewish state before the Soviets or their satellites did. Truman, replied, "That is right, the western recognition should precede the Soviet bloc's recognition, so as to give it the right slant from the beginning." Niles also told Truman about all of the mass meetings of Jews that were going to be held to celebrate the Jewish state and of the great "opportunity of acclaim" for him if he recognized the Jewish state before the meetings. Truman said Ed Flynn had already called him to say that there were going to be at least three hundred mass meetings throughout the country. Niles then said that he and Lowenthal "were trying to prevent any adverse references to him at those meetings, that we were sure the President would work this whole affair out satisfactorily." Truman answered that someday he was going to show them "how much he appreciates what we have been doing."[64]

Clifford and Lovett knew a split between Truman and Marshall would be disastrous and had to be avoided. There were many important foreign policy issues that the White House and the State Department had to work closely on, chief among them the developing Cold War. The Berlin crisis was in full swing. To save Berlin from collapse when the Soviets closed down the city, the administration had airlifted food and supplies into the city on a round-the-clock basis. The negotiations continued. Lovett called Clifford and told him he had met with Secretary Marshall. Couldn't the president announce that he would recognize the new state but not announce it right away and hold off for an unspecified period? Again, Clifford rejected that offer. He had spoken with Truman, and the president was not going to budge.

Finally Clifford received the call he was waiting for. After meeting several times with Marshall, Lovett told him, "General Marshall says he cannot go along with Truman's policy . . . but he will not publicly oppose it." An open break between the president and the secretary of state had been avoided. Lovett even agreed to work with Clifford on the statement Truman would make announcing recognition. What had changed Marshall's mind? Years later, Lovett informed Richard Holbrooke that when he had gone to see Marshall on the thirteenth, "I told him it was the President's choice." Marshall, a good soldier, had accepted this. Some of Marshall's friends had urged him to resign, but he had replied, "No, gentlemen, you do not accept a post of this sort and then resign when the man who has the Constitutional authority to make a decision makes one. You may resign at any time for any other reason but not that one."

The administration did make one concession to the State Department. Truman agreed to give the new state de facto recognition and not de jure recognition. The difference meant little to the public but a great deal to State Department lawyers and opponents of partition. *New York Times* columnist Arthur Krock explained it to the paper's readers.[65] He based his article on a memo written to the administration by the Jewish Agency counsel in Washington, David Ginsburg, before recognition was declared.[66] By recognizing Israel de jure, Krock wrote, the United States would be finding "that the provisional government is permanent" and its boundaries recognized. It would be conditional on the new state holding elections.

Recognizing it de facto meant it was simply acknowledging that the Jewish state existed, that the new government could carry out its international obligations, and the country's inhabitants accepted its government. It was standard policy, Krock noted, going back to the 1790s. Not willing to give up, Lovett asked Clifford if the declaration could be delayed a day or two, because the United States might "lose the effects of many years of hard work in the Middle East with the Arabs and that it would jeopardize our position with the Arab leaders" as well as bring Americans in the Middle East into personal danger. Should we at least not warn Austin and the U.S. delegation to the United Nations in advance? he asked. Clifford answered that the president "could not afford to have any such action leak and that we should try to insure against it." Couldn't the U.S. announcement of recognition at least be delayed until the U.N. General Assembly was through meeting for the day? Again Clifford's answer was no. As it turned out, Lovett was given permission to phone Ambassador Austin with the news that recognition was forthcoming only at 5:45 P.M., a brief fifteen minutes before Truman was set to make his announcement. Lovett was upset. "My protests against the precipitate action and warning as to consequences with the Arab world," he wrote, ". . . have been outweighed by considerations unknown to me." He could only conclude "that the President's political advisers, having failed last Wednesday afternoon to make the President a father of the new state, have determined at least to make him the midwife."[67]

On May 14 at 4 P.M., the Jewish Agency leaders in Palestine, meeting at the Tel Aviv Art Museum, proclaimed the establishment of the State of Israel as of midnight, May 15. In order for the United States to grant recognition, someone representing the new Jewish state would need to formally apply for it. Clark Clifford had tried to reach David Niles but had been unable to. Failing to reach Lowenthal as well, he had then reached Ben Cohen, who in turn had called Eliahu Epstein. Epstein then called in David Ginsburg, and Clifford and Ginsburg worked together to draft the official request for recognition. When it was finished, Epstein submitted it on behalf of the provisional government of the new state and phoned Agency leaders in Palestine to ask their approval.[68]

The request was handed to the office of Secretary of State George C. Marshall. It informed State that the Act of Independence would be declared at one minute after 6 P.M. on the evening of May 14, Washington, D.C., time. Epstein was authorized by the provisional government to "express the hope that your government will recognize and will welcome Israel into the community of nations."[69] President Truman then announced the U.S. recognition at 6:11 P.M. on May 14, slightly after midnight in the time zone of the new State of Israel. Truman turned to one of his aides and said, speaking of Chaim Weizmann, "the old Doctor will believe me now."[70] He then called David Niles. Truman wanted him to know that he had just announced recognition. "You're the first person I called," he said, "because I knew how much this would mean to you."[71]

Truman's announcement, making the United States the first nation to recognize Israel, stunned everyone. Most surprised were Warren Austin and the U.S. delegation to the United Nations. At 5:45 P.M. on the fourteenth, Clark Clifford phoned Dean Rusk to tell him that the president would be recognizing Israel a short while later. Clifford asked Rusk to tell the U.S. delegation. Rusk was aghast: "But this cuts across what our delegation has been trying to accomplish in the General Assembly," referring to its efforts to attain a truce and then trusteeship. "Nevertheless," Clifford instructed him, "this is what the President wishes you to do." Rusk phoned Austin from Washington. Austin had to leave the floor of the General Assembly to take his call, and, after hearing the shocking news and without informing anyone in the U.S. delegation, he went home.

Then the acting ambassador of the United States, Francis Sayre, took the podium and told the delegates the news and that he had known nothing about Truman's recognition. The entire General Assembly was in a "state of pandemonium," as Rusk put it. Philip Jessup then came in to confirm that the news was accurate. One U.S. delegate physically restrained Cuba's delegate from going to the podium and withdrawing Cuba's membership in the United Nations. A few minutes after the president had made recognition public, Rusk received a phone call from Secretary Marshall. "Rusk," he said, "get up to New York and prevent the U.S. Delegation from resigning *en masse*." Rusk quickly got to the United Nations but by the time he arrived found that tempers had settled down.[72]

The next day, Truman heard from his friend Abe Granoff, who wrote on behalf of himself and Eddie Jacobson. "Of course," Granoff wrote, "you know that Eddie Jacobson's and my confidence in you on Palestine never wavered. . . . We felt that you were doing everything possible for the Jewish people abroad, consistent however with our own Country's best interests. We always recognized your anxiety to avoid bloodshed. . . . If Eddie and I and others shed a tear of gratitude, you above all can understand. So also will you understand if Chaim Weizmann's eyes were moist when he talked to Eddie a few minutes ago." He and other American Zionist leaders, Granoff wrote, had "never questioned [your] forthrightness. In short, Mr. President, if all American citizens of the Jewish faith throughout this land do not bless your name tonight in their houses of worship, then there is no gratitude in this world."[73]

Later, Truman recalled, after he had recognized Israel, Lovett said to him, alluding to his colleagues at the State Department, "They almost put it over on you." Noting that some at State had told him the announcement had come as a surprise, Truman wrote, "It should not have been if these men had faithfully supported my policy." Truman saw it all as part of his fight as president against those who sought to thwart the executive's right to make U.S. foreign policy. "No one in any department," he said, "can sabotage the President's policy."[74]

Celebrations of Israel's birth were numerous and dampened only by the apprehension of what was to come. As the next day dawned, Egyptian aircraft bombed Tel Aviv and the armies of five Arab countries began to mount an assault.

CONCLUSION

As the announcement of Israel's birth was taking place, the old order was coming to an end. That night, the last British official quietly boarded the cruiser *Euralysus* at Haifa harbor and slipped out of Palestine. As they had promised, Arab armies from Lebanon, Syria, Jordan, Iraq, and Egypt–equipped with tanks, armored cars, and abundant ammunition and using the air forces of Egypt, Iraq, and Syria–began an invasion. Their actions were denounced by Trygve Lie and many others for their "brazen defiance" of the United Nations. The Haganah, soon to be renamed the Israel Defense Forces, did not have weapons and equipment comparable to the Arabs' firepower. But they were highly motivated and mobilized, drawing on the resources of the Jewish community in Palestine and beyond. For them, as Weizmann said, the choice was between victory and annihilation.

On May 16, a Salute to Israel rally, organized by AZEC, took place in Madison Square Garden, with 19,000 jammed into the Garden and a throng estimated at 75,000 turned away. They listened as Senator Robert Taft, New York City Mayor William O'Dwyer, Henry Morgenthau, Jr., and Rabbis Wise and Silver celebrated the new state and called on the government to lift the arms embargo. Dr. Emanuel Neumann, AZEC's chairman, told the crowd, "The United States cannot recognize the new-born state on one day and then on the next abandon it to its fate."[1]

Max Lowenthal was also working on the issue and was at the White House early in the morning, typing "a memo on the need for action by our gov't in lifting embargo, and asking UN Security Council to act to stop invasion of Palestine." Britain's "helping Arabs," he added, "–while we maintain an embargo–puts us in an awkward position."[2] Israel's friends in the White House had some reason to believe that after he granted recognition, the president would help the new country. Truman had told Niles that the United States would have to go the "whole distance now with Israel." Niles wasn't sure exactly what that meant but thought that, among other things, Truman was contemplating an exchange of ambassadors.[3]

On May 17, Chaim Weizmann was elected president of Israel's Provisional Council of State, a symbolic post of honor. Since no official exchange of diplomats had as yet occurred, he immediately asked Jacobson if he would serve as his unofficial envoy to the White House. Jacobson was glad to fill in and raised the issue of the arms embargo and the granting of a loan with Truman.[4]

The State Department, still smarting over its failure to prevent Israel's creation and Truman's recognition surprise, took out its frustration in what seemed like petty ways. As president of the new country, Weizmann had been invited to Washington by Truman, who offered to put him up in the official government guest residence, Blair House. Lovett was angry. "Weizmann is only the head of a state recognized de facto, not de jure," he told Clifford. "If he stays at Blair House, that means de jure recognition." Charlie Ross simply said, "Nuts." The residence was government property to receive and lodge distinguished guests. He came up with an answer to satisfy Lovett and his needs of protocol. "If he stayed at the White House," he said, "that would be de jure recognition." Weizmann stayed at the guest house as planned.[5]

On May 24, Truman received the new president of Israel in the South Portico of the White House. Weizmann presented Truman with a Torah, a scroll containing in Hebrew the first five books of the Old Testament. Accepting it, Truman quipped, "Thanks, I always wanted one." Later Truman would say the Torah was one of the greatest things he owned and it was very special because Weizmann had had to issue an injunction authorizing a Baptist to handle it.[6] According to Vera Weizmann, the two men parted in a playful mood. "When Truman told Chaim that he was the President of so many millions of Americans," Chaim retorted, "But I am the President of a million presidents!" The joke was not lost on Truman.[7]

At the meeting, Weizmann told Truman that it was essential to Israel's safety to lift the embargo. Truman seemed sympathetic but gave him no assurances. Then Weizmann requested what he called "a medium sized loan," which Israel needed for military and reconstruction purposes. The first phase of reconstruction would be to bring in 15,000 DPs from Germany a month. They were destitute and would need housing, transportation, and food.[8] Truman was more positive about the loan and answered, "The Jews have a fine tradition for repaying their debts. I know it from my good friend Eddie."[9] Returning to Blair House before traveling back to Israel, Weizmann thanked the president for his first "official visit, coming soon after the recognition given to the new State of Israel," which he knew would be "a source of satisfaction and encouragement to my people." Weizmann then informed Truman that the new government had appointed Eliahu Epstein as its prospective first ambassador to the United States and expressed his hope that a minister to Israel would quickly be appointed.[10]

In June, when preparations were finally made to appoint a U.S. ambassador to Israel, State proposed one of its own, Charles Knox, who was well known to be an Arabist. Instead, Truman appointed James G. McDonald, a member of the Anglo-American Committee and supporter of the new state. Five weeks after the creation of Israel, "suddenly and as a complete surprise," McDonald wrote, the phone rang. Answering it, he heard Clark Clifford's voice. Clifford asked him to rush to the White House and to keep the visit confidential. "The President wants you to go to Israel as the Government's first representative," Clifford told him. "I am not canvassing a list; you are the one person I have been told to inquire about." Before McDonald even had time to think it over, Clifford phoned again an hour later. "I have just seen the President; he is delighted and wants to make the announcement immediately." McDonald was stunned. "I haven't yet accepted," he said. After promising him that his salary would be sufficient to support a family, a matter of concern for him, McDonald accepted. The president wanted McDonald to report directly to him. The appointment was made without the knowledge or approval of Secretary Marshall, who later told McDonald that he had opposed the appointment when he learned of it.[11]

Truman wanted to make some of his own personnel changes. Loy Henderson was number one on the White House's removal list. The president thought, among other things, that Henderson had deliberately lied to him by claiming that Israel had no possibility of winning a war with the Arabs, ignoring Jewish intelligence reports that indicated the opposite.[12] Using this as the reason, he asked Marshall to fire him. But as a career officer, Henderson could not be so easily disposed of. Learning that he could only be transferred from his Near East desk at State, Truman made him ambassador to India in 1948. Henderson went on to have a distinguished career as a diplomat but always remained bitter about the harsh and, in his opinion, false accusations made about him in regard to his role in the State Department's policy in Palestine from 1945 to 1948.

Truman would give Israel only some of what it desired. Israel would not officially receive American arms, although the FBI, with J. Edgar Hoover's permission, looked the other way and allowed arms to be smuggled into Israel through U.S. ports. After a series of truces (which gave the Jews a chance to receive new military equipment and arms) and despite the embargo, the Israelis outfought and outmaneuvered the Arab forces, which, in the end, suffered a humiliating defeat.

Fighting by themselves and with the new arms sold to them by Czechoslovakia, the Israelis would win and sign a truce on January 7, 1949.

The Israeli's victory, first in their fight with the Arab Palestinians and then with the Arab states, can be attributed to the Arabs' weaknesses as well as to the Israelis' strengths. Unlike the Yishuv, the Arabs had not prepared for war. The Palestinian Arabs were without any real government or national militia and their villages and towns were dispersed and under local control. And although the Arab states had an overwhelming demographic advantage over Israel, they only recently had achieved independence and, except for Jordan, had inexperienced commanders and armies. Aside from aiming to destroy or prevent Israel from being created, they also had conflicting goals. A pragmatist, King Abdullah of Jordan aimed to seize part of what was supposed to be the Arab Palestinian State, mainly the West Bank. Egypt had its own territorial designs on Palestine, as did Syria.

In May 1949, one year after recognition, Israel gained full membership in the United Nations. The Truman administration gave Israel a $100 million loan. De jure recognition would finally be announced on January 31, 1949, after the Israeli elections. At the signing were Eddie Jacobson; Frank Goldman, the national president of B'nai B'rith; and Maurice Bisgyer, its secretary. When Eliahu Epstein became Israel's first ambassador to the United States, he changed his last name to the Hebrew Elath, in honor of the area around the Gulf of Aqaba.[13]

After Israel was created, the historical and religious meaning of what he had done became more important to Truman, especially his role in the return of the Jews to Palestine. Truman shared his thoughts with Clifford about biblical prophecies concerning the Jews' return to Zion in the Old Testament. Clifford, who considered himself an amateur Bible student, recalled exchanging passages with the president that dealt with the subject. One of the most striking quotes Clifford found was from Deuteronomy: "Moses went up from the plains of Moab unto the Mount of Nebo, the top of Pisgah that is before Jericho, and the Lord showed him all the land from Gilead unto Dan . . . all the land of Judah unto the western sea and the south and the plain." As Clifford remarked in 1988, "You can take an old biblical map and you could do quite a lot with that with the present boundaries of Israel." That, he added, was what the Old Testament had promised to the Jewish leaders of the day.[14] One of the president's favorites, which he often quoted, was also from Deuteronomy: "Behold, I have given up the land before you. Go in and take possession of the land to which the Lord has sworn unto your fathers, to Abraham, to Isaac, and to Jacob." Others were from Genesis, which referred to "an everlasting possession."

Similar remembrances came from Alfred Lilienthal, an Arabist and opponent of partition and then recognition. Lilienthal served during the Truman administration in the State Department and was a consultant to the U.S. delegation to the United Nations during its founding conference. Truman, he wrote, "was a Biblical fundamentalist who constantly pointed to these words of the Old Testament," citing the same passage from Deuteronomy 1:8 that Clifford alluded to.[15] These prophecies, for Truman, lent a stamp of approval from a higher authority to the decision to recognize Israel.

In the spring of 1949, Eliahu Elath accompanied Israel's Chief Rabbi, Isaac Halevi Herzog, to a meeting with Truman. Truman asked him if he knew what he had done for the refugees and to establish Israel. Herzog "reflected for a moment and replied that when the President was still in his mother's womb . . . the Lord had bestowed upon him the mission of helping his Chosen People at a time of despair and aiding in the fulfillment of His promise of Return to the Holy Land." In ancient times, Rabbi Herzog continued, "a similar mission had once been imposed on the head of another great country, King Cyrus of Persia, who had also been given the task of helping to redeem the Jews from their dispersion and restoring them to the land of their forefathers." At that point, Elath recalled, the Rabbi read aloud the words of Cyrus: "The Lord God of heaven hath . . . charged me to build him a House at Jerusalem, which is in Judah." As Truman heard the Rabbi's quote, "he rose from his chair and with great emotion, tears glistening in his eyes, . . . turned to the Chief Rabbi and asked him if his actions for the sake of the Jewish people were indeed to be interpreted thus and the hand of the Almighty was in the matters." The Rabbi told Truman, "he had been given the task once fulfilled by the mighty king of Persia, and that he too, like Cyrus, would occupy a place of honor in the annals of the Jewish people."[16] Truman took the Rabbi's words to heart. When Eddie Jacobson told an audience at the Jewish Theological Seminary in New York that he was introducing the man who had helped create Israel, Truman quickly responded, "Helped create Israel? I am Cyrus I am Cyrus."

Eliahu Elath, speaking years later as the president of Hebrew University, told his audience that the Bible was Truman's "main source of knowledge of the history of Palestine in ancient times." Quoting Truman's farewell address as president, given on January 15, 1953, Elath singled out a passage that revealed his hopes for the Holy Land. Truman had said, "The Tigris and the Euphrates Valley can be made to bloom as it did in the time of Babylon and Nineveh. Israel can be made into the country of milk and honey as it was in the time of Joshua." This passage, Elath thought, was not inserted "by mere chance."[17]

When David Ben-Gurion, now the first prime minister of Israel, met Truman in New York City in 1961, he was surprised by Truman's intense emotions. "I told [Truman]," Ben-Gurion wrote, "that as a foreigner I could not judge what would be his place in American history; but his helpfulness to us, his constant sympathy with our aims in Israel, his courageous decision to recognize our new State so quickly and his steadfast support since then had given him an immortal place in Jewish history." After he spoke these words, Truman's eyes were suddenly filled with tears. "I had rarely seen anyone so moved," Ben-Gurion wrote. He held Truman back for a few moments until he regained his composure, so that the press waiting outside the hotel suite where they were meeting would not notice.[18]

In the next few years, Truman would lose many of the people he credited with helping to bring about the creation of Israel. In May 1951, Niles, who claimed fatigue after fifteen years of government service, submitted his resignation. Truman was reluctant to accept it and wrote to him, "You have been a tower of strength to me during the past six years and I can't tell you how very much I appreciate it." Niles wanted to travel, especially to Israel, but his health declined and he passed away from cancer on September 28, 1952, before making the trip.[19]

After a year of being seriously ill, Chaim Weizmann died of a heart attack on November 9, 1952. Truman had come to think of Weizmann as a friend and someone he could identify with.

When Weizmann wrote Truman a congratulatory letter after he won the presidential election in 1948, Truman wrote back to him on November 29, the anniversary of the U.N.'s partition decision, that he was struck by the similarities in their experiences. "We had both been abandoned by the so-called realistic experts to our supposedly forlorn lost cause. Yet we both kept pressing for what we were sure was right—and we were both proven to be right. My feeling of elation on the morning of November 3rd must have approximated your own feelings one year ago today, and on May 14th and on several occasions since then." Truman ended with "I want to tell you how happy and impressed I have been at the remarkable progress made by the new State of Israel. What you have received at the hands of the world has been far less than was your due. But you have more than made the most of what you have received, and I admire you for it."[20]

The hardest loss Truman suffered was that of Eddie Jacobson. The founding of Israel had brought the old friends closer together. In the difficult period before recognition, Eddie Jacobson had become an important intermediary between the Zionist movement and the president. He had then acted as an unofficial ambassador to Israel, until it got its officials up and running. Writing to Jacobson in 1952, Vera Weizmann told him, "Only the most intimate friends knew the extraordinary role that was played by you in swinging the scale in our favor when the future looked so precarious and ambiguous."[21] Perhaps the most important of his contributions had been convincing Truman to receive her husband at the critical moment. Jacobson and his wife, Bluma, did go to Israel and stayed at the home of the McDonalds.

After Truman returned home from the White House, he and Jacobson were often seen lunching together in Kansas City. In 1955, they began planning a trip of a lifetime.[22] Truman was excited about the itinerary and wrote to Jacobson that they would leave by ship from New York for England, where Truman was to receive an honorary degree from Oxford. Then they would visit Winston Churchill and the

Queen. Next on the itinerary was Holland, where they would meet the royal family, and Paris, where they would meet political leaders. Then would be Rome, where Truman and Jacobson would have an audience with the Pope. Finally, Truman said, they would travel to Israel by ship, arriving in the port of Haifa around mid-October.[23] Jacobson died of a heart attack in October of that year. Truman called off the trip and never went to Israel.

After Jacobson's death, Truman said, "I don't think I have ever known a man that I thought more of outside my own family than I did of Eddie Jacobson. He was an honorable man. . . . he was one of the finest men that ever walked on this earth. . . . Eddie was one of those men that you read about in the Torah . . . if you read the articles in Genesis concerning two just men. . . . [Enoch and Noah] you'll find those descriptions will fit Eddie Jacobson to the dot."[24] After the funeral, he went to the family's home to offer condolences. As tears clouded his eyes, he told Jacobson's daughters, Gloria and Elinor, that their father was the "closest thing he had had to kinfolk."[25] "Eddie was one of the best friends I had in this world," Truman told the press. "He was absolutely trustworthy. I don't know how I'm going to get along without him."[26]

The British and the State Department viewed Truman's actions regarding the creation of Israel as the result of political expediency, i.e., a need for the Jewish vote or Jewish campaign contributions. Truman himself had encouraged this belief by complaining to Bevin and others about the pressures of the Jewish vote, especially in New York. It is true that Truman did not support a Jewish state in the way the Zionist movement did and would have preferred a democratic pluralist country like the United States to take root in Palestine. So we must ask, was gaining the Jewish vote the most important factor in motivating the president to recognize Israel and in taking the actions he did leading up to it? Would he have done it unless he thought it was in America's national interests to do so?

Harry Truman was a politician. As president, he was the leader of the Democratic Party, and he could not ignore how his actions on Palestine would impact not only his own political fortunes, but those of his fellow Democrats, whether through votes or contributions. In a democracy, public opinion is bound to affect governmental policy. As much as they tried to separate it, in the case of the Jews and Palestine, domestic and foreign policy issues were tied together by very strong sentiments. With the horrors of the Holocaust and the continuing deprivations of the DPs fresh in the minds of Americans, there was widespread support for a just solution to the Jewish condition both within the American Jewish community and beyond. Because public opinion was behind it, the Republicans and the Wallace campaign in 1948 sought to outdo Truman on the issue and kept up a steady stream of criticism at how he was handling it. Truman often said that what he had done to help the Jews had backfired on him. It never seemed to be enough, and he deeply resented the pressure put on him by competing politicians and especially by the American Zionist Emergency Council. He said he was going to do what he thought was right despite the actions of those he called the "extreme Zionists."

Richard Crossman, the British member of the Anglo-American Committee, was offended by the cynical arguments about Truman's motivations, given that British democracy also had to respond to public opinion. Speaking in the House of Commons in January 1949, Crossman said it was easy to "make jibes about the votes in New York and to insult the President. . . . But if we had a million Jews in this country, our Cabinet might have been slightly more careful to keep their election pledges. Do not let us attack American politicians for what we ourselves would have done . . . anyone reckless enough to take the word of a State Department official who thought that he would wangle the White House into letting down the Jews cannot have had much experience and should not believe it."[27]

Truman was angered and upset by the charge that he was motivated by political concerns. His actions, he maintained, were guided by humanitarian and moral considerations, and by his understanding that America's national interests stood above all else. While he always seriously considered the State Department's arguments and was often conflicted himself, in the end he concluded that recognizing Israel was in the national interest of the United States.

Truman had come full circle on the issue. He became president after serving in the Senate for ten years, from whose perch he had witnessed Hitler's rise to power and the Nazis' exterminationist policies against the Jews. In 1939, he denounced the British White Paper as making "a scrap of paper out of Lord Balfour's promise to the Jews" for a national home in Palestine. He joined the Christian Zionist group, the American Palestine Committee, and even briefly was a member of Peter Bergson's Committee for a Jewish Army. After taking office as president, one of his first meetings was with Rabbi Stephen Wise. He told Wise that he was confident that he both could look out for the long-term interests of the United States and still help the persecuted Jews of Europe find a home. In the end, Truman believed that he had achieved this goal.

The president's counselor, Clark Clifford, was also outraged by this characterization of Truman. "To portray President Truman as risking the welfare of his country for cheap political advantage," Clifford wrote, "is bitterly resented by all of us who admired and respected him."[28]

To refute these charges, Clifford, considered to be the architect of Truman's 1948 campaign strategy, pointed to a forty-three-page memorandum he presented to Truman in November 1947. Drafted by James Rowe, a former administrative assistant to FDR, and perfected by Clifford, it covered every aspect of the campaign. In the memo, Clifford argued that Truman should concentrate on winning the South and the states west of the Mississippi, which would allow him to discount the electoral votes of New York, as well as New Jersey, Illinois, and Ohio, states with a substantial Jewish population. Moreover, Clifford argued that a Jewish vote, if it existed, was an issue only in New York City. Truman would be more likely to win, Clifford advised, if he made policy based on an issue's "intrinsic merit."[29]

Clifford turned out to be prophetic. If the Jewish vote was a key to victory in the 1948 presidential election, making policy to gain that vote proved to be a dismal failure. The largest Jewish vote was in three East Coast and midwestern states: New York, Pennsylvania, and Illinois. New York City was home to the largest Jewish population in the world and the center of pro-Zionist sentiment. Yet Truman lost New York, whose Electoral College votes went to the Republican candidate, Thomas E. Dewey. Truman's loss of the state had more to do with the third-party candidacy of Henry A. Wallace, who siphoned off labor, liberal, and Jewish votes, thereby enabling Dewey to win the state. Of the three states with a significant Jewish population, Truman won only Illinois.

Moreover, there is evidence that when Truman had a chance to make a blatant appeal for Jewish votes before the election, he turned down the opportunity. Chester Bowles, the Democratic candidate for governor of Connecticut, suggested during the 1948 presidential campaign that Truman grant de jure recognition to Israel on the eve of Yom Kippur, just as he had called for partition before the holiest night in 1946. Dewey was rumored to be preparing such a statement. If Truman lost the opportunity and delayed granting full recognition until later in October, Bowles wrote, he risked losing Connecticut. Action along those lines "is vital," Bowles wrote. "I know how important [the issue] is in Connecticut; and if we are up against it here, it must be infinitely tougher in New York."[30]

On September 9, Marshall announced that de jure recognition would not be granted to Israel until after it held an election, which was customary. The elections were scheduled for mid-October but, because of the Arab-Israeli war, did not take place until January 1949, when the war ended. Clifford suspected that Israel might not be able to hold its elections until after the American elections for president in November and argued that it wasn't necessary to wait until Israel's elections for the president to grant it.[31] At that critical moment, when Truman knew that not granting de jure recognition could harm his presidential bid, he refused to do it.

It might be helpful in evaluating Truman's efforts to define a U.S. policy for Palestine and his role in the creation of Israel to compare his path to that taken by Abraham Lincoln at the time of the American Civil War. In 1860, it was not clear that slavery would be abolished. Lincoln campaigned on the platform only of preventing slavery's expansion into the new western territories, not on abolishing it where it already existed. For this view, Lincoln was castigated by the abolitionists and the radical Republicans, who saw the abolition of slavery as the only real issue. Lincoln was seen by them as a prevaricator, much as the Zionists saw Truman.[32] When the Civil War erupted, Lincoln refused to free the slaves or enroll free blacks in the Union Army. Although he was criticized for his halfway measures, his pragmatism allowed him to move at the critical moment and both enroll blacks in the Union Army after the issuance of the Emancipation Proclamation and to wage the Civil War as a vehicle for abolishing slavery instead of just limiting its expansion. Yet he continued to decline, even after victory, to take further steps demanded by the radicals and former abolitionists, such as insisting on the right to vote for the former slaves. In the same manner, Truman recognized Israel yet refused to lift the embargo, supply arms, or immediately grant full recognition. Yet, just as the radicals such as Frederick Douglass embraced Lincoln as a hero despite his limitations, so did most Zionists and the founders of Israel embrace Harry Truman.

Truman, if not Cyrus, was probably essential to Israel's birth. Despite all of the reversals of his administration along the way, Eliahu Elath recognized the concrete steps Truman had taken that made Israel possible. They were, he wrote, "of truly historic significance" in the Zionist struggle for statehood: "His appeal to the British Government to grant a hundred thousand permits to Jewish survivors of the Holocaust to enter Palestine; his support for the Palestine Partition Plan with the inclusion of the Southern Negev and Elath within the Jewish State; United States recognition of the State of Israel upon its establishment; and the granting of the first international loan to be received by the Government of Israel."[33]

Elath, like Weizmann and Ben-Gurion, would have differences with Truman in the period after recognition. They were not satisfied with the limits of his support. Truman did not move to end the arms embargo, even though the new state was attacked by five Arab armies. Truman was upset about reports of the plight of new Palestinian refugees and wanted more of them to be able to return to their homes. The State Department had been pressing him to force Israel to take back Arab Palestinian refugees, but the Israelis were reluctant to allow more than a small number to return. In their eyes, the only reason there were Arab refugees was that many Palestinian Arabs as well as the neighboring Arab countries had gone to war against their new state. Those who remained were welcome to stay, but the Israeli government worried that to allow all those who had fled to return would be to create a potential fifth column. This was of special concern because some of the Arab leaders were threatening another round of hostilities.

As a result of the creation of Israel and the war, an almost equal number of Jews had been forced out of the Arab countries as there had been Arab Palestinian refugees who fled Israel. Most of these new Jewish immigrants from the Arab lands arrived in Israel without money or resources, which they had been forced to leave behind. Israel had absorbed them and granted them citizenship. The Israeli leadership argued that the Arab countries should do likewise by admitting and integrating the Palestinians into their own societies. For the most part, the Arab states failed to do this, leaving many Palestinians living in refugee camps on the outskirts of their cities or in the West Bank and Gaza, sustained by international charity.

Ben-Gurion took the position that Israel would discuss repatriation and compensation for Arab refugees only within the context of peace treaty negotiations. But the Arab states insisted that Israel meet its repatriation and compensation demands before any negotiations took place and refused to engage in talks with Israel. Truman was disappointed that no provisions had been made for the refugees when the conflict officially ended with only the signing of armistice agreements.

Despite their differences with the Truman administration on the refugees and other issues, as time passed, most Israelis and their supporters understood the importance of what Harry S. Truman *had* done. As Weizmann observed about the Balfour Declaration, "Even if all the governments of the world gave us a country it would be a gift of words, but if the Jewish people will go and build Palestine, the Jewish state will become a reality and a fact." And so they did. The new state, however, needed legitimacy and recognition for it to survive. It is fair to conclude as, David Niles did, that if FDR had lived and Truman not been president, there probably would not have been an Israel. Certainly, if Franklin D. Roosevelt had been in office, support at critical moments would most likely not have been offered. Without Truman, the new State of Israel might not have survived its first difficult years, and succeeded thereafter. For this, Truman will continue to be viewed as a hero in Israel and continue to have a place of honor in the history of the Jewish people.

ACKNOWLEDGMENTS

When writing a book on any historical subject, authors are indebted to those who have gone before them, and on whose work they build. We have learned much from Truman's various biographers and to those scholars who have written previously on both the Truman era and on Truman and Israel. Some of them may differ wih our interpretation and analysis, but our book is richer because of their pioneering work.

We want to thank the archivists at the various archives we visited for sharing their expertise and pointing out valuable sources. At the Harry S. Truman Library, we thank Michael Devine, director of the library, and his staff. We are especially indebted to that master of the archive, Randy Sowell, who knew which hidden files might contain gems we could have overlooked. He also helped us by promptly replying to e-mail requests, and in answering any questions we had as we went along in our research and writing. Others who helped us at the library were Ray Geselbracht, special assistant to the director, Amy Williams, supervisory archivist, and audiovisual archivist Pauline Testerman.

The resources of the Center for Jewish History in New York City, which houses the archival collections of the American Jewish Historical Society and the YIVO Institute, among others, was important to our study. We are especially indebted to archivist Susan Woodland, director of the Hadassah Archives, who helped us with our queries and with finding relevant material.

At our hometown public library in Martinsburg, West Virginia, librarian Bernadette Whalen got us rare and otherwise unavailable books from libraries far and wide. We could not have written this book without her help in obtaining access to these materials.

At the Library of Congress, our friend and colleague John Earl Haynes, archivist of Recent American History, obtained articles for us from relevant journals. He also led us to manuscript collections that he thought would be useful. We also received help at the LOC from Jennifer Brathovde and Bruce Kirby.

At the David S. Wyman Institute for Holocaust Studies, Executive Director Rafael Medoff discussed issues with us and made available photos for our use from their photo archive.

At the University of Minnesota, University Archivist Elizabeth Kaplan and Assistant archivist Karen Klinkenberg helped us with access to the Max Lowenthal papers. We would also like to thank Sean Martin, associate curator for Jewish history at Western Reserve Historical Society.

In Israel, our research was facilitated by the following people: Merav Segal, director of the Weizmann archive at Rehovot, not only brought relevant files to our attention, but arranged a personal tour of the late Zionist leader's home. At the Central Zionist Archives, we were guided to relevant files by Rochelle Rubinstein, deputy director of archival matters. At the Harry S. Truman Institute of World Peace at Hebrew University in Jerusalem, our host was Jill Twersky, coordinator of events and publications, who arranged our participation at a panel on Truman and Israel presented at the University's 60th Anniversary and the Institute's 40th. Finally, we would like to thank Josh Block for arranging a trip to Israel for us under the auspices of the American-Israel Educational Foundation. The incredible tour gave us a good sense of life in Israel past and present.

The former archivist of the United States, Allen Weinstein, not only offered his enthusiastic support, but alerted others to our project. He gave us his own writings on the subject and shared other material with us. Jeffrey Herf, professor of history at the University of Maryland and scholar of German history, shared his insights on radical Islam with us. Tuvia Friling, former National Archivist of Israel, was living in Washington, D.C., when we were writing the book, when he was visiting scholar at the National Holocaust Museum. He not only discussed material with us, but was good enough to translate material we needed from Hebrew to English. Sol Stern, an expert on Israel, discussed at length with us the role played by Peter Bergson, whom he knew well. Eric Fettman at *The New York Post* provided us with material on Abe Feinberg. Peter Collier read parts of the manuscript and gave us his expert advice.

At HarperCollins, we thank our editor, Claire Wachtel, who showed great faith in our project from the beginning, made incisive comments on the manuscript, and gave us helpful recommendations for its improvement. Her able assistant editor, Julia Novitch, handled myriad problems and was always available for dealing with issues as they came up. We are indebted to copy editor Lynn Anderson, who carefully went through our manuscript, improved it, and prevented us from committing errors. We take responsibility for any that remain.

And finally, good friends and relatives provided us with useful information and shared our enthusiasm. These include Shari Nadell, Bob Cohen, Louis Menashe, and Danny Kalb. Our friends Nancy Lieber and Robert J. Lieber of Georgetown University, whose understanding of the Middle East is second to none, helped in many invaluable ways. Special thanks and gratitude goes to our children and their spouses: Michael and Jennifer Radosh, Daniel Radosh and Gina Duclayan, Laura Radosh and Silke Radosh Hinder, and our grandchildren, Milo, Margalit, and Malka.

NOTES

Note: Full references for all books can be found in the bibliography. See bibliography for a list of abbreviations.

INTRODUCTION

1. Alonzo L. Hamby, *Man of the People: A Life of Harry S. Truman*, p. 278.
2. Wilson D. Miscamble, *From Roosevelt to Truman: Potsdam, Hiroshima, and the Cold War*, p. 308.
3. Arthur M. Schlesinger, "Our Presidents: A Rating by 75 Historians," *The New York Times Magazine*, July 29, 1962, p. 12.
4. Margaret Truman, *Harry S. Truman*, pp. 416–19.
5. Harry Truman, *Memoirs; Vol. 2, Years of Trial and Hope*, paper edition, p. 185.

CHAPTER 1: FDR'S LEGACY

1. "Term IV," *The New York Times*, Jan. 21, 1945, p. 63.
2. Bertram D. Hulen, "Shivering Thousands Stamp in the Snow at Inauguration," *The New York Times*, Jan. 21, 1945, p. 1.
3. C. P. Trussell, "Truman Hastens to Call Mother," *The New York Times*, Jan. 21, 1945, p. 1.
4. Tony Judt, *Postwar: A History of Europe Since 1945*, pp. 17–18.
5. Meyer Weisgal, *Meyer Weisgal . . . So Far*, p. 218.
6. "Schwartz Says Only 1,500,000 Jews are Left in Europe as a Result of German Murders," *The New York Times*, Feb. 17, 1945.
7. Meyer Weisgal, *Meyer Weisgal . . .* , p. 217.
8. Martin Gilbert, *Churchill and the Jews: A Lifelong Friendship*, pp. 27–28.
9. Howard M. Sacher, *The History of Israel from the Rise of Zionism to Our Time*, pp. 96–99.
10. Herbert G. Feis, *The Birth of Israel: The Tousled Diplomatic Bed*, p. 13. Theories abound as to what prompted the British government to authorize the Balfour

Declaration. Weizmann believed that important British leaders were deeply religious and wanted the biblical injunction of returning the Jews to Palestine to become a reality. Others argued that at the start of the First World War, Britain wanted American intervention, and its leaders hoped that American Jews would help influence public opinion on behalf of intervention. Support of Jewish aspirations in Palestine after the dismantling of the Ottoman Empire at the war's end, they hoped, would lead the American Jews to offer Britain their support. See James Malcolm, "Origins of the Balfour Declaration: Dr. Weizmann's Contributions," National Archives of Great Britain, Kew Gardens, London; FO 371 45383/E8500.

11. Chaim Weizmann, *Trial and Error: The Autobiography of Chaim Weizmann*, p. 30.
12. Ibid., pp. 171–72.
13. Nahum Goldmann, *The Autobiography of Nahum Goldmann: Sixty Years of Jewish Life*, p. 109.
14. Ibid., p. 111.
15. Harlan B. Phillips, ed., *Felix Frankfurter Reminisces*, pp. 180–85.
16. Samuel Halperin, *The Political World of American Zionism*, p. 9.
17. Quoted in Abba Eban, *My People: The Story of the Jews*, 2nd ed., p. 366.
18. Chaim Weizmann, *Trial and Error*, p. 68.
19. Howard M. Sachar, *The History of Israel*, p. 134.
20. Herbert G. Feis, *The Birth of Israel: The Tousled Diplomatic Bed*, pp. 13, 23.
21. Walter Laqueur, *A History of Zionism*, p. 561; also see Abba Eban, *My People*, p. 366.
22. Howard Sachar, *The History of Israel: From the Rise of Zionism to Our Time*, p. 224.
23. Emanuel Celler, *You Never Leave Brooklyn: The Autobiography of Emanuel Celler*, p. 117.
24. FDR told this to both Senator Robert A. Taft and Senator Robert F. Wagner. The text of his letters was published in the Zionist magazine *New Palestine*, Oct. 27, 1944. Also see Stephen Wise's draft of the letter he asked FDR to publish, CZA Z5/388, Oct. 12, 1944.
25. See, e.g., Arthur D. Morse, *While Six Million Died: A Chronicle of American Apathy*, and David S. Wyman, *The Abandonment of the Jews: America and the Holocaust, 1938–1945*. For the opposing view, by writers who claimed FDR did all possible for the Jews, see Robert N. Rosen, *Saving the Jews: Franklin D. Roosevelt and the Holocaust*, and Robert L. Beir with Brian Josephser, *Roosevelt and the Holocaust*.
26. Meyer Weisgal, *Meyer Weisgal*, p. 185.
27. Eliahu Elath, *Israel and Elath: The Political Struggle for the Inclusion of Elath in the Jewish State*, pp. 183, 310.
28. Emanuel Neumann, *In the Arena: An Autobiographical Memoir*, p. 206.
29. Quoted in Marc Lee Raphael, *Abba Hillel Silver: A Profile in American Judaism*, p. 79.
30. Ibid., p. 128. Also see: Peter Grose, *Israel in the Mind of America*, pp. 166–67.
31. Marc Lee Raphael, *Abba Hillel Silver*, p. 94. Also see Peter Grose, *Israel in the Mind of America*, pp. 172–76.
32. Marc Lee Raphael, *Abba Hillel Silver*, p. 79.
33. Proskauer, quoted in Menahem Kaufman, *An Ambiguous Partnership: Nonzionists and Zionists in America, 1938–1948*, pp. 148, 160.

34. Zvi Ganin, *American Jewry and Israel: 1945–1948*, p. 13.
35. Stimson to Chairman of the Senate Committee on Foreign Relations (Connally), Feb. 7, 1944; *FRUS*, vol. 5 (1944), p. 563. Also see the discussion in Michael T. Benson, *Harry S. Truman and the Founding of Israel*, pp. 56–58.
36. Diary entry, Nov. 15, 1944, in *The Diaries of Edward R. Stettinius, Jr., 1943–1946*, ed. Thomas M. Campbell and George C. Herring (New York: New Viewpoints/Franklin Watts, 1974), p. 174.
37. Rosenman, quoted in Eliahu Elath, "Samuel Irving Rosenman and his Contribution Before the Establishment of the State of Israel" [Hebrew], *Molad*, 7, no. 37–38 (Spring 1976): 448–54.
38. "Minutes of Conversation with Judge SR. Washington, DC, April 27, 1944," CZA, no. 1380.
39. Thomas M. Campbell and George C. Herring, *The Diaries of Edward R. Stettinius Jr., 1943–1946*, pp. 187–88, diary entry of Dec. 2, 1944. Also see Marc Lee Raphael, *Abba Hillel Silver: A Profile in American Judaism*, pp. 117–29.
40. Abba Hillel Silver, speech of March 21, 1945, in Abba Hillel Silver, *Vision and Victory: A Collection of Addresses by Dr. Abba Hillel Silver, 1942–1988*, p. 83.
41. "Minutes of Conversation with Judge SR," April 27, 1944, CZA, no. 1388.
42. Weisgal to Weizmann, Sept. 23, 1944, WA, no. 2521.
43. Neumann to Weizmann, Sept. 28, 1944, WA, no. 2522.
44. Emanuel Neumann, *In the Arena: An Autobiographical Memoir*, p. 198.
45. David S. Wyman and Rafael Medoff, *A Race Against Death: Peter Bergson, America and the Holocaust*, pp. 13–14.
46. Material on Bergson is culled from David S. Wyman, *Abandonment of the Jews*, pp. 90–92; David S. Wyman and Rafael Medoff, *A Race Against Death*, pp. 13–28; Marc Lee Raphael, *Abba Hillel Silver*, pp. 93–95; Isaac Zaar, *Rescue and Liberation: America's Part in the Birth of Israel*, pp. 19–37; Francisco Gil-White, "The Problem of Jewish Self-Defense," at www.hirhome.com/israel/leaders1.htm; Peter Grose, *Israel in the Mind of America*, pp. 161–62.
47. Quoted in Grose, *Israel in the Mind of America*, pp. 175–76.
48. David S. Wyman and Rafael Medoff, *A Race Against Death*, pp. 39–42.
49. Undated letter on American Jewish Conference stationery, quoted in Isaac Zaar, *Rescue and Liberation: America's Part in the Birth of Israel*, pp. 73–74.
50. Diary entry of Oct. 6, 1943, in William D. Hassett, *Off the Record with F.D.R., 1942–1945*, p. 209.
51. Minutes of Silver interviews with Rosenman and Bergson, October 12, 1943, WA, no. 2462.
52. Our discussion on the Bergson group is based on David Wyman and Rafael Medoff, *A Race Against Death: Peter Bergson, America and the Holocaust*, pp. 50–52.
53. See Zvi Ganin, *American Jewry and Israel, 1945–1948*, p. 6.
54. "Minutes of Conversation," Judge Samuel Rosenman and Dr. Nahum Goldman, April 27, 1944, Washington, D.C.; Central Zionist Archive, Jerusalem, Z5/388.
55. Evan M. Wilson, "The Palestine Papers: 1943–1947," *Journal of Palestine Studies* 2, no. 4 (Summer 1973): pp. 34–37.
56. Secretary of State Cordell Hull to President Roosevelt, May 7, 1943, enclosing "Summary of Lieutenant Colonel Harold B. Hoskins' Report on the Near

East," in *Foreign Relations of the United States* (hereafter *FRUS*), vol. 4, 1943, pp. 781–85.

57. Patrick Hurley to FDR, May 5, 1943, *FRUS*, vol. 4, pp. 776–80.
58. FDR to Robert Wagner, Dec. 3, 1944; in Elliott Roosevelt, ed., *FDR: His Personal Letters*, vol. 2, *1928–1945*, pp. 1559–60.
59. Evan M. Wilson, "The Palestine Papers, 1943–1947," *Journal of Palestine Studies* 2, no. 4 (Summer 1973): pp. 37–38.
60. Ibid., p. 39.
61. Nancy MacLennan, "Roosevelt Backs Palestinian Plan as Homeland for Refugee Jews," *The New York Times*, March 10, 1944, p. 1.
62. Cordell Hull, *Memoirs*, vol. 2, pp. 1535–36.
63. Thomas M. Campbell and George C. Herring, eds., *The Diaries of Edward R. Stettinius, Jr., 1943–1946*, p. 170.
64. Herbert L. Feis, *The Birth of Israel: The Tousled Diplomatic Bed*, p. 16.
65. Diary entry of Oct. 6, 1943, in William D. Hassett, *Off the Record with F.D.R., 1942–1945*, p. 209.
66. Campbell and Herring, *Stettinius Diaries*, p. 170.
67. Ibid., p. 211.
68. Sumner Welles, *We Need Not Fail*, pp. 29–30.
69. W. C. Lowdermilk, " 'TVA' Reclamation Project for the Jordan Valley and a Post-war Solution for the Jewish Refugee Problem," Aug. 10, 1942, Ben Cohen Papers, Box 12, Folder 1, Library of Congress, Washington, D.C.
70. Ibid.
71. Ibid.
72. Ibid.
73. "Memorandum by the Secretary of State to President Roosevelt," Jan. 4, 1945, in *FRUS*, vol. 8, *The Near East and Africa*, p. 678.
74. Joseph P. Lash, *Eleanor: The Years Alone*, p. 103.
75. Wise to FDR, Jan. 24, 1945, Central Zionist Archives, Jerusalem, Israel; also see Samuel Halperin and Irvin Oder, "The United States in Search of a Policy: Franklin D. Roosevelt and Palestine," *The Review of Politics* 24, no. 3 (July 1962): 336–37.
76. Landis to FDR, Jan. 20, 1945, in *FRUS*, vol. 8, pp. 680–83.
77. William Eddy to State Department, Feb. 1, 1945, in *FRUS*, vol. 8, p. 687.
78. Stettinius to FDR, Jan. 9, 1945, in *FRUS*, vol. 8, p. 679.
79. Memo from State Department, Jan. 30, 1945, in *FRUS*, vol. 8, pp. 684–87.
80. Charles E. Bohlen, *Witness to History: 1929–1969*, p. 212.
81. C. L. Sulzberger, "American Interest in Arabia Growing," *The New York Times*, Feb. 12, 1945, p. 13.
82. Frank Freidel, *Rendezvous with Destiny: A Biography of Franklin D. Roosevelt*, pp. 593–94.
83. "U.S. Warship Becomes Arab Court in Miniature for Ibn Saud's Voyage," *The New York Times*, Feb. 21, 1945, p. 1.
84. "White House Announcement of New Talks," *The New York Times*, Feb. 21, 1945, p. 8.
85. Bohlen, *Witness to History: 1929–1969*, pp. 211–12.
86. William A. Eddy, *FDR Meets Ibn Saud*, pp. 31–32.

87. Ibid., p. 34.
88. Bohlen, *Witness to History*, pp. 211–12; also see p. 33.
89. Eddy, *FDR Meets Ibn Saud*, p. 34.
90. Bohlen, *Witness to History*, p. 213.
91. Quoted in Michael B. Oren, *Power, Faith and Fantasy: America in the Middle East, 1776 to the Present*, p. 472.
92. Martin Gilbert, *Churchill and the Jews: A Lifelong Friendship*, p. 65.
93. Ibid., p. 202.
94. "Summary of Reports Sent by Dr. Weizmann," Jan. 17, 1944; sent by Weizmann to Judge Samuel Rosenman; David Niles manuscripts, Harry S. Truman Library.
95. William A. Eddy to Edward Stettinius, Feb. 22, 1945, in *FRUS*, vol. 8, pp. 689–90. Ibn Saud had given a private audience to Eddy to keep him informed of what had transpired at the king's meeting with Churchill.
96. Rabbi Stephen Wise to Chaim Weizmann, March 21, 1945, Chaim Weizmann Archives, Rehovoth, Israel, no. 2575.
97. "Report of President Roosevelt in Person to the Congress on the Crimea Conference," *The New York Times*, March 2, 1945, p. 12.
98. Harry S. Truman, *Memoirs*, vol. 1, *Years of Decision*, p. 3.
99. "Report of President Roosevelt in Person to the Congress on the Crimea Conference," *The New York Times*, March 2, 1945, p. 12.
100. Emanuel Celler, *You Never Leave Brooklyn*, p. 118.
101. Quoted in Melvin R. Urofsky, *We Are One!*, p. 62.
102. Sumner Welles, *We Need Not Fail*, p. 30.
103. Rabbi Stephen Wise to Chaim Weizmann, March 21, 1945, Chaim Weizmann Archives, Rehovoth, Israel, no. 2575.
104. Philip J. Baram, *The Department of State in the Middle East, 1919–1945*, p. 322; also see Michael J. Cohen, *Truman and Israel*, pp. 87–88.
105. Memorandum by the Director of the Office of Near Eastern and African Affairs (Murray) to the Acting Secretary of State, March 20, 1945, in *FRUS*, vol. 8, pp. 694–95.
106. Lieutenant Colonel Harold B. Hoskins to the Deputy Director of the Office of Near Eastern and African Affairs (Alling), March 5, 1945, in *FRUS*, vol. 8, pp. 690–91.
107. Evan M. Wilson, "The Palestine Papers: 1943–1947," *Journal of Palestine Studies* 2, no. 4 (Summer 1973): p. 35.
108. Joseph M. Proskauer, *A Segment of My Times*, pp. 69–70.
109. Memorandum by Joseph Proskauer to the leadership of the American Jewish Committee, April 1945, American Jewish Committee files, YIVO Institute for Jewish Research; American Jewish Historical Society, New York City.

CHAPTER 2: TRUMAN INHERITS A PROBLEM

1. Diary entry of April 12, 1945, in William D. Hassett, *Off the Record with F.D.R., 1942–1945*, p. 333.
2. Ibid., p. 333.

3. Arthur Krock, "End Comes Suddenly at Warm Springs," *The New York Times*, April 13, 1945, pp. 1, 3.
4. "Oral History Interview with Judge Samuel I. Rosenman," Oct. 15, 1968, and April 23, 1969, conducted by Jerry N. Hess, Harry S. Truman Library.
5. "Oral History Interview with Edward D. McKim," Feb. 19, 1964, conducted by James R. Fuchs, Harry S. Truman Library. Also see David McCullough, *Truman*, p. 327.
6. Margaret Truman, *Harry S. Truman*, p. 203.
7. Harry S. Truman, *Memoirs*, vol. 1, *Years of Decision*, p. 2.
8. Quoted in Robert J. Donovan, *Conflict and Crisis: The Presidency of Harry S Truman, 1945–1948*, p. 4. Donovan quotes from his oral history interview with Lewis Deschler, conducted on Feb. 21, 1972.
9. C. P. Trussell, "Truman Is Sworn in the White House," *The New York Times*, April 13, 1945, p. 1.
10. Arthur Krock, "End Comes . . . ," *The New York Times*, April 13, 1945, p. 3; Harry S. Truman, *Memoirs*, vol. 1, *Years of Decision*, pp. 6–8.
11. Quoted in Margaret Truman, *Harry S. Truman*, p. 232.
12. Trussell, "Truman Is Sworn in the White House."
13. George McKee Elsey, *An Unplanned Life: A Memoir*, p. 82; also see Robert J. Donovan, *Conflict and Crisis*, p. 4.
14. Roy Roberts, "Truman to Shelve Personal Rule; Senate Will Be His Adviser," *The New York Times*, April 15, 1945, p. 6.
15. Harry S. Truman, *Memoirs*, p. 12.
16. Ibid., pp. 14–17, 19.
17. Quoted in Jonathan Daniels, *The Man of Independence*, p. 263.
18. See the discussion in David McCullough, *Truman*, pp. 369–72, and Elizabeth Edwards Spalding, *The First Cold Warrior*, pp. 24–26.
19. Oral history interview with Abraham Feinberg, Aug. 23, 1972, HSTL; Jonathan Daniels, *The Man of Independence*, p. 263.
20. William Hassett, *Off the Record with F.D.R.*, p. 36.
21. Obituary, *The New York Times*, June 25, 1973, p. 1.
22. Alonzo Hamby, *Man of the People: A Life of Harry S. Truman*, p. 271.
23. Quoted in Zvi Ganin, *American Jewry and Israel, 1945–1948*, p. 24.
24. Oral history interview with Matthew J. Connelly, conducted by Jerry N. Hess, Aug. 21, 1968, HSTL.
25. Eliahu Elath.
26. Our discussion is based on Peter Grose, *Israel in the Mind of America*, p. 219; Robert Donovan, *Conflict and Crisis*, p. 316; and Benjamin V. Cohen, "David Niles Memorial Lecture," April 27, 1945.
27. Peter Grose, *Israel in the Mind of America*, p. 220.
28. Eliahu Elath, "Samuel Irving Rosenman and His Contribution Before the Establishment of Israel," *Molad* 7, no. 37–38 (Spring 1976): 448–454. Translated from the Hebrew by Tuvia Friling.
29. Peter Grose, *Israel in the Mind of America*, p. 221.
30. Oral history interview with Judge Samuel I. Rosenman, HSTL, Oct. 15, 1968.
31. Alben W. Barkley, *That Reminds Me: The Autobiography of the Veep* (Garden City,

N.Y.: Doubleday, 1954), p. 197; Jonathan Daniels, *The Man of Independence*, p. 298. Barkley is cited in Robert J. Donovan, *Conflict and Crisis*, p. 123.

32. Jonathan Daniels, *The Man of Independence*, pp. 295–96; Alonzo Hamby, *Man of the People*, pp. 362–64.
33. Wilson D. Miscamble, *From Roosevelt to Truman: Potsdam, Hiroshima, and the Cold War*, pp. 307–32. Miscamble's deep analysis of Truman's new foreign policy is the best historical account of the importance of Truman's postwar leadership.
34. Alonzo Hamby, *Man of the People*, pp. 12–14; David McCullough, *Truman*, p. 35.
35. Harry S. Truman, *Mr. Citizen*, p. 102.
36. Ibid., pp. 103–4.
37. Ibid., p. 14.
38. Eliahu Elath, "Harry S. Truman–The Man and Statesman," First Annual Harry S. Truman Lecture, May 18, 1977, published by the Harry S. Truman Institute at the Hebrew University, Jerusalem, pp. 47–50.
39. Michael Benson, *Harry S. Truman and the Founding of Israel*, p. 54.
40. Robert J. Donovan, *Conflict and Crisis*, p. 386.
41. John Adams to Abigail Adams, Aug. 14, 1776, quoted in Peter Grose, *Israel in the Mind of America*, p. 6.
42. "Statements of the Presidents of the United States," in *America and Palestine*, ed. Reuben Fink, pp. 87–88.
43. Quoted in Michael J. Cohen, *Truman and Israel*, pp. 44–45. Truman's comments appear in *Congressional Record*, 76th Congress, 1st session, 1939, vol. 84, pt. 13, appendix, pp. 2231–32. Also see Michael T. Benson, *Harry S. Truman and the Founding of Israel*, p. 60, fn. 17.
44. Wise to Truman, Feb. 25, 1941, HSTL, Senate files.
45. "Barkley Praises Palestine Leaders," *The New York Times*, May 1, 1941, p. 9; Emanuel Neumann, *In the Arena*, pp. 152–54.
46. Neumann, *In the Arena*, p. 154.
47. "68 Senators Back Palestine Refuge," *The New York Times*, April 20, 1941, p. 28.
48. "To the Political Dept. Note on the New President of the United States," April 13, 1945. Weizmann Archives, Rehovoth, Israel.
49. James H. Becker, Chairman, Joint Emergency Committee, to Truman, April 16, 1943, HST Senate File, Box 108, HSTL.
50. Harry S. Truman, "Speech to Be Delivered by Senator Harry S. Truman Before the United Rally to Demand Rescue of Doomed Jews at Chicago Stadium, Chicago Illinois," April 14, 1943; HST Senate File, Box 108, HSTL.
51. Truman to Andrew L. Somers, Jan. 28, 1942, HSTL, HST Senate File, Box 108, HSTL.
52. David S. Wyman and Rafael Medoff, *A Race Against Death*, p. 85.
53. *The New York Times*, Dec. 7, 1942, pp. 14–15.
54. David S. Wyman, *The Abandonment of the Jews*, p. 342. Wyman concludes elsewhere that the Bermuda Conference was a "pretense" and "diplomatic hoax" meant only to defuse pressure for rescue: see David S. Wyman, "The American Jewish Leadership and the Holocaust," in *Jewish Leadership During the Nazi Era*, ed. R. L. Braham, p. 14.
55. David S. Wyman and Rafael Medoff, *A Race Against Death*, p. 11.

56. Ibid., pp. 37–38; *The New York Times*, May 4, 1943, p. 17.
57. Quoted in David S. Wyman and Rafael Medoff, *A Race Against Death*, p. 37. The authors had access to the FBI's file on Bergson, "Memorandum for Mr. Ladd, May 13, 1943," FBI File on Bergson.
58. Ibid., p. 85; Bergson oral history, April 13–15, 1973, interview by David Wyman. Also see Peter Grose, *Israel in the Mind of America*, p. 189. Grose argues that Truman resigned because he took a "personal slight" to his friend seriously. As a result, he argues, "Truman had become more cautious about associating himself with causes that he was not sure he understood." He attributes Truman's decision not to endorse the proposed 1944 resolution of Congress to this new caution.
59. Harry S. Truman to Peter Bergson, May 6, 1943, HSTL, Senate and Vice-Presidential Files. We are indebted to Michael Cohen's discussion in his book *Truman and Israel*, pp. 38–43. But we disagree with him on his claim that the "key point was that Truman did not share Bergson's jaundiced view of the Bermuda Conference" (p. 41), since all advocates of greater immigration rights for the Jews were united in their conclusion that the conference was a farce.
60. Rabbi Stephen Wise to Harry S. Truman, May 20, 1943, HSTL, Senate and Vice-Presidential Files.
61. Harry S. Truman to Rabbi Stephen Wise, June 1, 1943, HSTL, Senate and Vice-Presidential Files. Again, we disagree with Michael Cohen, who argues that Truman's letter proved that he too favored inaction and was using the war as an excuse for any "lack of meaningful Allied action to save Europe's doomed Jewish communities" (p. 43).
62. See, e.g., Michael J. Cohen, *Truman and Israel*, pp. 46–47.
63. "To the Political Dept. Note on the New President of the United States." The Jewish Agency memo took the statement from a book, *America and Palestine*, published by the American Zionist Emergency Council in 1944. Truman used the same statement in letters to various correspondents. The italics in the text are our emphasis.
64. Truman to Wagner, June 1, 1943, HSTL, Senatorial and Vice-Presidential Files.
65. Speech on the floor of the Senate by Senator Bennett Champ Clark, March 28, 1944, in Reuben Fink, *America and Palestine*, pp. 101–20, quote on pp. 116–17.
66. "Roosevelt Tribute by Free Synagogue," *The New York Times*, April 16, 1945, p. 8.
67. "Change Advocated in Palestine Rule," *The New York Times*, April 17, 1945, p. 17.
68. Emanuel Neumann, *In the Arena*, pp. 208–9.
69. Eliahu Elath, *Zionism at the UN: A Diary of the First Days*, p. 46.
70. Ibid., p. 288.
71. Ibid., p. 300.
72. Secretary of State Edward Stettinius to the President, April 18, 1945, HSTL, Subject File, Box 160.
73. Memo for the AZEC reporting on the meeting, April 23, 1945, CZA Z5/1206.
74. Merle Miller, *Plain Speaking*, p. 215.

75. Harry S. Truman, *Memoirs*, vol. 1, p. 69.
76. Merle Miller, *Plain Speaking*, pp. 216–17.
77. Eliahu Elath, *Zionism at the UN*, entry of Tuesday, June 19, p. 292.
78. Grew to Truman, May 1, 1945, in *FRUS*, vol. 8, pp. 705–6.
79. Grew to Truman, May 14, 1945, in *FRUS*, vol. 8, p. 706.
80. Truman to King Abdullah of Trans-Jordan, May 17, 1945, in *FRUS*, vol. 8, p. 707; Truman to President Nokrashy, Egyptian Council of Ministers, in *FRUS*, vol. 8, pp. 708–9.
81. "Comments with Regard to Some of the Questions Raised by Dr. Lilienthal in his letter of Feb. 16, 1977," Loy Henderson Papers, Box 11, HSTL.
82. "Memorandum of a Conversation Between the Secretary of State and the British Ambassador (Halifax)," Oct. 19, 1945, in *FRUS*, vol. 8, pp. 777–79.

CHAPTER 3: FROM SAN FRANCISCO TO POTSDAM

1. Eliahu Elath, *Zionism at the U.N.*, p. 39. Epstein changed his name to Elath after he became the first Israeli ambassador to the United States. To avoid confusion, we will refer to him by the name Elath.
2. Zvi Ganin, *Truman, American Jewry, and Israel*, p. 25.
3. Eliahu Elath, *Zionism at the U.N.*, p. 40.
4. Chaim Weizmann, *The Letters and Papers of Chaim Weizmann*, vol. 11, p. 551.
5. Quoted in Melvin I. Urofsky, *We Are One!*, p. 97. See also David Ben-Gurion, *New Palestine* 35 (Israel), May 18, 1945.
6. Ibid. Urofsky, May 7, 1945, p. 72.
7. Ibid., May 2, 1945, p. 36.
8. "Zionists Condemn 'Liberation' Group," *The New York Times*, May 18, 1945, p. 13.
9. Melvin I. Urofsky, *We Are One!*, pp. 97–99.
10. "60,000 at Rally Back Zionist Plea," *The New York Times*, April 30, 1945, p. 22.
11. Eliahu Elath, *Zionism at the U.N.*, pp. 82–83.
12. Ibid., pp. 84–85.
13. Ibid., pp. 89–90.
14. Ibid., pp. 131–32.
15. Ibid., pp. 250–51.
16. Ibid., p. 81.
17. Ibid., p. 258.
18. Ibid., pp. 59–60.
19. Ibid., pp. 141–42.
20. Ibid., pp. 7–10.
21. Ibid., pp. 165–69.
22. J. C. Hurewitz, *The Struggle for Palestine*, pp. 227–29.
23. C. L. Sulzberger, "Fires That Flame Behind the Arab Crisis," *The New York Times Magazine*, June 17, 1945, pp. 9, 45–46.
24. Hurewitz, *The Struggle for Palestine*, p. 211.
25. David Niles papers, April 4, 1945, Israel file, Box 29, HSTL.
26. Report of Lord Halifax to the Foreign Office, July 1, 1945, British National Archives, Kew Gardens, London, England, CAB 121/643, vol. 1.

27. Gene Currivan, "Nazi Death Factory Shocks Germans on a Forced Tour," *The New York Times*, April 18, 1945, p. 1.
28. "Dachau," *Time*, May 7, 1945.
29. Harold Denny, "The World Must Not Forget," *The New York Times Magazine*, May 6, 1945, pp. 8, 42.
30. "Congressmen Plan to See More Camps," *The New York Times*, April 27, 1945, p. 3.
31. William S. White, "Congress, Press, to View Horrors," *The New York Times*, April 22, 1945, p. 13.
32. "Buchenwald Tour Shocking to M.P.'s," *The New York Times*, April 23, 1945, p. 5.
33. "The Conflicts of Harry S. Truman: At War with the Experts," CBS TV, 1964, HSTL, Film Collection.
34. "Shift of 1,000,000 to Palestine Set," *The New York Times*, April 13, 1945, p. 30.
35. Eliahu Elath, *Zionism at the UN: A Diary of the First Days*, p. 294.
36. Leonard Dinnerstein, *America and the Survivors of the Holocaust*, pp. 24–36; quote is on p. 33.
37. Melvin I. Urofsky, *We Are One!*, p. 105.
38. Quoted in Dinnerstein, *America and the Survivors of the Holocaust*, pp. 16–17.
39. Kurt Grossman, "The Jewish DP Problem," in Harry N. Rosenfeld Papers, box 16, HSTL.
40. This discussion is based on Leonard Dinnerstein, *America and the Survivors of the Holocaust*, pp. 34–36; and Zvi Ganin, *Truman, American Jewry and Israel*, pp. 28–30.
41. Eliahu Elath, *Zionism at the UN*, pp. 303–4.
42. "Memorandum by the Director of the Office of Near Eastern and African Affairs (Henderson)," June 22, 1945; Elath, *Zionism at the UN*, pp. 712–13.
43. Bartley Crum, *Behind the Silken Curtain: A Personal Account of Anglo-American Diplomacy in Palestine and the Middle East*, p. 173.
44. "Memorandum of Conversation, by Mr. Evan M. Wilson," June 27, 1945, in *FRUS*, vol. 8, pp. 713–15.
45. American Zionist Emergency Council, Minutes of Meeting held July 12, 1945, "Near Eastern Division of State Department," confidential, no. 26, CZA, 25/1205.
46. Ibid., "Conference with Acting Secretary of State."
47. Allen H. Podet, "Anti-Zionism in a Key U.S. Diplomat: Loy Henderson and Zionism at the End of World War II," Loy Henderson Papers, Israel-Palestinian Folder, box 11, HSTL. Podet's article was published in *American Jewish Archives* 30, no. 2 (November 1978).
48. Robert D. Kaplan, *The Arabists: The Romance of an American Elite*, p. 90.
49. Oral history interviews with Loy W. Henderson, June 14, 1973, and July 5, 1973, HSTL.
50. Our discussion of this historic meeting is based on the indispensable book by Leonard Slater, *The Pledge*, pp. 21–28. We are indebted to Slater for his thorough and essential account of this event.
51. Quoted in Michael Makovsky, *Churchill's Promised Land: Zionism and Statecraft*, p. 144.

52. Ibid., p. 25.
53. Ibid., p. 27.
54. Quoted in Michael Bar-Zohar, *Ben-Gurion: A Biography*, p. 127.
55. Marc Lee Raphael, *Abba Hillel Silver: A Profile in American Judaism*, p. 119.
56. Eliahu Elath, *Zionism at the UN*, pp. 287–88.
57. Robert F. Wagner, "Palestine–A World Responsibility," *The Nation*, Sept. 15, 1945, pp. 247–49.
58. Douglas to Senator Robert A. Taft, May 18, 1945, Robert A. Taft Papers, Box 735, in Folder "Palestine and the Jews," Library of Congress, Washington, D.C.
59. Memorandum from Arthur Lourie, Political Secretary, AZEC, May 21, 1945, CZA 25.
60. AZEC, minutes of meeting held July 12, 1945, CZA, 25/1205.
61. Wagner to Truman, July 3, 1945, Official Files, Box 771, HSTL. Truman's response is scribbled on the first page of the letter.
62. Ibid.
63. Stephen Wise and Herman Shulman, "Memorandum on Palestine," July 3, 1945, Official Files, Box 771, HSTL.
64. Quoted in Zvi Ganin, *Truman, American Jewry and Israel*, p. 31; Proskauer to Truman, July 6, 1945, papers of Jacob Blaustein. (Ganin gained access to this depository from family members in Baltimore, Maryland.)
65. "Palestine: Jewish Immigration," June 22, 1945, in *FRUS*, vol. 1, *The Conference in Berlin (The Potsdam Conference)*, pp. 972–74.
66. "Memorandum by President Truman to the British Prime Minister," July 24, 1945, in *FRUS*, vol. 8, pp. 716–17.
67. Chaim Weizmann, *Trial and Error*, p. 436.
68. Emanuel Neumann, *In the Arena*, p. 210.
69. Attlee to Truman, July 31, 1945, in *FRUS*, vol. 8, p. 719.
70. David Horowitz, *State in the Making*, p. 4.
71. On Potsdam and Truman's policy, see: Wilson D. Miscamble, *From Roosevelt to Truman: Potsdam, Hiroshima, and the Cold War*, pp. 172–217.
72. Ibid., pp. 221–23, 235–40.
73. Truman to Momma and Mary, Aug. 17, 1945, in Robert M. Ferrell, ed., *Off the Record: The Private Papers of Harry S. Truman*, p. 62.
74. "Truman Discloses Plea on Palestine," *The New York Times*, Aug. 17, 1945, p. 8.
75. AZEC, Minutes of meeting held August 28, 1945, CZA, Z5/1205.
76. "Memorandum of Conversation by the Director of the Office of Near Eastern and African Affairs (Henderson)," Aug. 17, 1945, in *FRUS*, vol. 8, p. 721.
77. "Memorandum of Conversation by the Deputy Director of the Office of Near Eastern and African Affairs (Allen)," Aug. 18, 1945, in *FRUS*, vol. 8, pp. 722–23.
78. Loy W. Henderson to Secretary of State James F. Byrnes, Aug. 20, 1945, State Department Papers, NLT-691, National Archives of the United States, College Park, Maryland.
79. "Roosevelt Pledge to Arabs Alleged," *The New York Times*, Aug. 25, 1945, p. 12.
80. "Memorandum Regarding Conversation Between President Roosevelt and the

King of Saudi Arabia," n.d., attached to Rosenman to Truman, Sept. 7, 1945, Samuel I. Rosenman papers, box 4, HSTL.

81. Rosenman to Truman, Sept. 7, 1945, Samuel I. Rosenman papers, box 4, HSTL.

CHAPTER 4: THE PLIGHT OF THE JEWISH DPs

1. Eben A. Ayers, diary entry, Aug. 25, 1944, in Diary 1941–1953, Eben A. Ayers Papers, box 19, HSTL.
2. Abraham J. Klausner, *A Letter to My Children from the Edge of the Holocaust*, pp. 67–73.
3. Harrison to Truman, n.d., 1945, in David Niles Papers, Box 29, HSTL. All further citations from the Harrison Report are from this document.
4. Oral history interview with Phileo Nash, conducted by Jerry N. Hess on Oct. 13, 1966, HSTL.
5. Eliahu Elath, *Harry S. Truman: The Man and Statesman*, First Annual Harry S. Truman Lecture, May 18, 1977, Hebrew University of Jerusalem, pp. 25–26. Truman told Elath this in an interview he conducted at his home in Independence, Missouri, in 1961.
6. Truman to Attlee, Aug. 31, 1945, in *FRUS*, vol. 8, pp. 737–39.
7. Niles is quoted in Elath, *Harry S. Truman*, p. 27.
8. Dean Acheson, *Present at the Creation*, p. 169.
9. "Immigration into Palestine Previous to a Final Decision with Regard to the Future Status of Palestine," Aug. 29, 1945, Memo from Division of Near Eastern Affairs, Subject File, Box 160, HSTL. A précis of this memo was also issued by State: Gordon P. Merriam to Loy Henderson, Aug. 31, 1945, Subject File, Box 160, HSTL.
10. "Gillette Heads Palestine League," *The New York Times*, Aug. 2, 1945, p. 11.
11. Lord Halifax to British Cabinet, Sept. 15, 1945, CAB 121/643, Vol. 1, Public Office Records, Kew Gardens, London, England. Halifax quotes a Jewish Telegraphic Agency Bulletin dated September 13.
12. Harry Shapiro to Chairmen of Local Emergency Committees, Sept. 18, 1945, CZA Z5/1204.
13. William Roger Louis, *The British Empire in the Middle East: 1945–1951*, pp. 388–89. Louis quotes Yehuda Bauer, who wrote that the Jewish Agency "soon came to regard its proposal of the 100,000 as rather less than wise and was very much afraid that the British might accept it."
14. "Report of Meeting with James F. Byrnes," Sept. 22, 1945, in *The Letters and Papers of Chaim Weizmann*, Series A, vol. 22, May 1945–July 1947, Letter no. 66, pp. 56–57; also see Minutes of Office Meeting, Sept. 24, 1945, CZA A172/3/6.
15. Weizmann to Byrnes, October 3, 1945, in *The Letters and Papers of Chaim Weizmann*, Series A, vol. 22, May 1945–July 1947, Letter no. 75, p. 65.
16. Alan Bullock, *Ernest Bevin: Foreign Secretary, 1945–1951*, pp. 49–51; Zvi Ganin, *Truman, American Jewry and Israel*, pp. 49–50.
17. David Horowitz, *State in the Making*, p. 12.

18. Ibid., pp. 55–56.
19. William Roger Louis, *The British Empire in the Middle East, 1945–1951*, p. 386.
20. Alan Bullock, *Ernest Bevin*, pp. 167–68.
21. Leonard Dinnerstein, *America and the Survivors of the Holocaust*, pp. 78–79.
22. Report of the Palestine Committee, September 8, 1945, Cabinet Papers, 1946; PRO, Kew Gardens, London; also see Alan Bullock, *Ernest Bevin*, p. 170; also see "Truman Said to Aid Jews on Palestine," *The New York Times*, Sept. 23, 1945, p. 17. The *Times* story reports on the Sept. 8 Labour Palestine Committee's proposals.
23. Attlee to Truman, Sept. 14, 1945, in *FRUS*, vol. 8, p. 739.
24. Attlee to Truman, Sept. 16, 1945, in *FRUS*, vol. 8, pp. 730–41.
25. Truman to Attlee, Sept. 17, 1945, in *FRUS*, vol. 8, p. 741.
26. Quoted in Eliahu Elath, *Harry S. Truman*, pp. 29–30; quote is on p. 30.
27. Ibid., pp. 30–31. Elath relates what Niles told him years later.
28. "UNO Decision on Palestine Will Be Sought by Britain," *The New York Times*, Sept. 24, 1945, p. 1.
29. "Weizmann Rejects Offer on Palestine," *The New York Times*, Sept. 25, 1945, p. 13.
30. Quoted in Zvi Ganin, *Truman, American Jewry and Israel*, p. 40. Ganin cites Silver's notes of Sept. 29, 1945, from the Abba Hillel Silver Papers, the Temple, Cleveland, Ohio.
31. Quoted in ibid., p. 197, citing the notes of Blaustein, Sept. 29, 1945, from the Blaustein Papers, Baltimore, Maryland.
32. "Truman Asked to Aid Jewish Immigration," *The New York Times*, Sept. 30, 1945, p. 39.
33. "Dewey Backs Plea for Jewish State at Big Rally Here," *The New York Times*, Oct. 1, 1945, p. 1.
34. Lord Halifax, Report, Cabinet Distribution, Oct. 4, 1945, PRO, CAB /21/643/ vol. 1.
35. Halifax Report, Amended Distribution, Cabinet Distribution, Oct. 4, 1945, PRO, FO 371, 45/280.
36. "Jews' Treatment Hit by Barkley," *The New York Times*, Oct. 16, 1945, p. 12.
37. Chaim Weizmann, *Trial and Error: The Autobiography of Chaim Weizmann*, p. 440.
38. Introduction, *The Letters and Papers of Chaim Weizmann, Series A: Letters*, vol. 22, p. xi.
39. Christopher Sykes, *Cross Roads to Israel*, pp. 336–37.
40. Introduction, *The Letters and Papers of Chaim Weizmann*, vol. 22, Series A, Letters, May 1945–July 1947, p. xi.
41. Weizmann to Ben-Gurion, Oct. 10, 1945, in ibid., letter no. 78, pp. 66–67.
42. Ibid., Introduction, p. xii.
43. "Palestine 'Pledge' Denied by Truman," *The New York Times*, Sept. 27, 1945, p. 14.
44. Henderson to Byrnes, Oct. 9, 1945, Department of State, Political Files, Oct. 1, 1945–Dec. 12, 1945, Lot 56-D359, NA.
45. Henderson to Byrnes, Oct. 10, 1945, Department of State, Political Files, Oct. 1, 1945–Dec. 12, 1945, Lot 56-D359, NA.

46. Samuel I. Rosenman, "Memorandum for the President," Oct. 17, 1945, Harry S. Truman Subject File, Box 160, HSTL; also in Samuel I. Rosenman Papers, Box 4, HSTL.
47. Samuel I. Rosenman. "Memorandum," Oct. 18, 1945, Samuel I. Rosenman Papers, Box 4, HSTL.
48. Samuel I. Rosenman, "Memorandum for the President," Oct. 23, 1945, Samuel I. Rosenman Papers, Box 4, HSTL.
49. The letters were published verbatim in *The New York Times*, Oct. 19, 1945, p. 4.
50. "U.S. Bars Decision on Palestine Without Consulting Jews, Arabs," *The New York Times*, Oct. 19, 1945, p. 1.
51. AZEC Emergency Council, "Minutes of Meeting Held October 20, 1945," 25/1205, CZA.
52. "Text of Memorandum submitted by the American Zionist Emergency Council to the State Department . . ." Oct. 23, 1945, attached to Charles G. Ross to Leo R. Sack, Oct. 31, 1945, in Official File, 204 Miscellaneous (Oct. 1945–1951), Box 771, HSTL.
53. Letter from "Washington correspondent," Oct. 20, 1945, L35/102, CZA and Chaim Weizmann Archives, Rehovoth, Israel, copy in Weizmann File, HSTL.
54. Elath, memo, Oct. 24, 1945, L35/102, CZA.
55. Lord Halifax to Cabinet, July 1, 1945, CAB 121/643, vol. 1, PRO.

CHAPTER 5: THE SEARCH FOR CONSENSUS

1. Richard Crossman, *Palestine Mission*, p. 65.
2. Halifax to Byrnes, Oct. 19, 1945, in *FRUS*, vol. 8, pp. 771–75.
3. Alan Bullock, *Ernest Bevin: Foreign Secretary, 1945–1951*, p. 179.
4. British Embassy to the Department of State, Informal Record of Conversation, Oct. 19, 1945, in *FRUS*, vol. 8, pp. 775–76.
5. "Memorandum of Conversation Between the Secretary of State and the British Ambassador (Halifax)," Oct. 22, 1945, in *FRUS*, vol. 8, pp. 779–83.
6. Rosenman to Truman, "Memorandum for the President," Oct. 23, 1945, Samuel I. Rosenman Papers, box 4, HSTL.
7. "Memorandum for the President" (Rosenman to Truman), Nov. 1, 1945, Rosenman Papers, box 4, HSTL.
8. Byrnes to Halifax, Oct. 23, 1945, in *FRUS*, vol. 8, pp. 785–86.
9. John H. Crider, "Truman Discloses U.S. Palestine Role," *The New York Times*, Nov. 14, 1945, p. 13.
10. Paul W. Ward, "Truman Drops Palestine Idea," *The New York Sun*, Nov. 14, 1945, p. 1.
11. Crider, "Truman Discloses U.S. Palestine Role."
12. Albert J. Gordon, "Goldstein Insists on Free Palestine," *The New York Times*, Nov. 18, 1945, p. 12.
13. Alvin Rosenfeld, "Survival of Jewry Hinges on Palestine, Weizmann Says," *New York Post*, Nov. 21, 1945.
14. Text of address delivered by Chaim Weizmann, Nov. 19, 1945, at the 48th Convention of the Zionist Organization of America, American Jewish Conference

papers, American Jewish Historical Society, New York City. Unless otherwise noted, all quotes from Weizmann's speech are from this source.

15. Quoted in Norman Rose, *Chaim Weizmann*, p. 407.
16. Excerpts from address delivered by Dr. Abba Hillel Silver, Nov. 18, 1945, at the 48th Convention of the Zionist Organization of America.
17. Eleanor Roosevelt to Truman, Nov. 20, 1945, in *Eleanor and Harry: The Correspondence of Eleanor Roosevelt and Harry S. Truman*, ed. Steve Neal, pp. 47–48.
18. Truman to Eleanor Roosevelt, Nov. 25, 1945, in ibid., pp. 47–48.
19. Truman to Ball, Nov. 24, 1945, President's Secretary File, Box 184, HSTL. Truman wrote on the letter, "Do not send." He might have thought twice before letting anyone know his response in such a candid fashion.
20. "Truman Held Firm on Open Palestine," *The New York Times*, Dec. 5, 1945, p. 14.
21. Ibid.; c.f., Halifax to FO, Dec. 6, 1945, FO 371, 45403/E9543, PRO, Kew Gardens, London.
22. Halifax to Cabinet, Dec. 4, 1945, FO 371, 45403/E9437, PRO, Kew Gardens, London.
23. Weizmann to Truman, Dec. 12, 1945, Official File, Box 772, HSTL. The paragraphs in the text following this citation are all from this letter.
24. Ibid.
25. Our discussion is indebted to Zvi Ganin, *Truman, American Jewry and Israel*, p. 44.
26. Wagner and Taft to Truman, Dec. 6, 1945, Official File, box 771, HSTL; L5/3462, CZA; and Robert A. Taft Papers, LOC. The letter was made public by the senators and published as a pamphlet by Jewish groups in America.
27. Dean Acheson, *Present at the Creation*, p. 172.
28. White House Press Release, "Statement by the President," Dec. 10, 1945, David Niles Papers, box 29, folder 1940–1945, HSTL.
29. Richard Crossman, *Palestine Mission*, 1947, p. 25.
30. James G. McDonald, diary entry, April 8, 1933, insert by McDonald in his diary, in *Advocate for the Doomed: The Diaries and Papers of James G. McDonald, 1932–1935*, ed. Richard Breitman et al., pp. 47–48.
31. Enclosure in letter from Loy W. Henderson to Allen H. Podet, March 29, 1976, Loy W. Henderson Papers, box 11, folder Israel-Palestine, Lectures and Notes, HSTL.
32. Michael J. Cohen, *Truman and Israel*, p. 126.
33. Bartley Crum, *Behind the Silken Curtain*, pp. 4–5.
34. Oral history interview with Loy W. Henderson, June 14, 1973, conducted by McKinzie, HSTL; enclosure in letter from Henderson to Podet, March 29, 1976.
35. Leo Rabinowitz to Eliahu Epstein (Elath), Dec. 27, 1945, L35/142, CZA.
36. James G. McDonald, diary entry, Friday, January 18–Wednesday, January 23, James G. McDonald Papers, Columbia University, New York City.
37. Ibid.
38. Eliahu Epstein (Elath), Dec. 15, 1945, in Confidential Report to members of the Executive Committee of the Jewish Agency for Palestine, Dec. 28, 1945, L35/79, CZA.

39. Richard Crossman, *Palestine Mission*, pp. 23–28. Crossman quotes extensively from his own diaries of his work with the committee.
40. Bartley Crum, *Behind the Silken Curtain*, p. 7.
41. Richard Crossman, *Palestine Mission*, p. 32.
42. Ibid., pp. 43–46.
43. Ibid., pp. 42, 48–49.
44. Meyer Weisgal, *So Far: An Autobiography*, p. 222.
45. Bartley Crum, *Behind the Silken Curtain*, pp. 23–24.
46. Richard Crossman, diary entry, Jan. 3, 1946, in Richard Crossman, *Palestine Mission*, p. 47.
47. James G. McDonald, diary entry, Jan. 11, 1946, James G. McDonald Papers, Columbia University, New York City.
48. Bartley Crum, *Behind the Silken Curtain*, pp. 24–28.
49. Richard Crossman, diary entry, Jan. 13, 1946, p. 47.
50. Bartley Crum, *Behind the Silken Curtain*, p. 31.
51. Ibid., p. 33.
52. Ibid., p. 37.
53. Richard Crossman, *Palestine Mission*, p. 66.
54. Richard Sykes, *Cross Roads to Israel*, p. 343.
55. Richard Crossman, *Palestine Mission*, p. 70.
56. James G. McDonald, diary entry, Feb. 1, 1946, James G. McDonald Papers, Columbia University, New York.
57. James G. McDonald, diary entry, Jan. 30, 1946, James G. McDonald Papers, Columbia University, New York.
58. Richard Crossman, *Palestine Mission*, pp. 84–85.
59. Bartley Crum, *Behind the Silken Curtain*, p. 79.
60. Ibid., p. 85.
61. Alan Bullock, *Ernest Bevin*, p. 168.
62. Leonard Dinnerstein, *America and the Survivors of the Holocaust*, pp. 110–12.
63. Bartley Crum, *Behind the Silken Curtain*, p. 130.
64. Report of Frank Buxton, quoted in ibid., p. 127.
65. James G. McDonald, diary entry, Feb. 17, 1946, James G. McDonald Papers, Columbia University, New York.
66. Crum, *Behind the Silken Curtain*, pp. 110–11.
67. Ibid., pp. 114, 124, 130.
68. Ibid., pp. 101–4.
69. Ibid., p. 132.
70. David Bernard Sacher, *David K. Niles and United States Policy*, pp. 21–22. On p. 22, Sacher quotes Niles to Crum, Feb. 20, 1945.
71. Richard Crossman, *Palestine Mission*, p. 119.
72. James G. McDonald, diary entry, March 2, 1946, James G. McDonald Papers, Columbia University, New York.
73. James G. McDonald, diary entry, March 8, 1946, James G. McDonald Papers, Columbia University, New York.
74. Richard Crossman, diary entry, March 8, 1946, in Crossman, *Palestine Mission*, p. 133.

75. Richard Crossman, diary entry, March 10, 1946, in Crossman, *Palestine Mission*, pp. 124–37.
76. Richard Crossman, diary entry, March 24, 1946, in Crossman, *Palestine Mission*, pp. 168–69.
77. Bartley Crum, *Behind the Silken Curtain*, p. 213.
78. Richard Crossman, diary entry, March 17, 1946; diary entry, March 24, 1946; in Crossman, *Palestine Mission*, pp. 149, 157.
79. Richard Crossman, diary entry, March 7, 1946, in Crossman, *Palestine Mission*, p. 132.
80. Richard Crossman, diary entry, March 20, 1946, in Crossman, *Palestine Mission*, pp. 164–65.
81. James G. McDonald diary entry, March 11, 1946, James G. McDonald Papers, Columbia University, New York.
82. Richard Crossman, diary entry, March 26, 1946, in Crossman, *Palestine Mission*, pp. 171–72.
83. Bartley Crum, *Behind the Silken Curtain*, pp. 175–82.
84. James G. McDonald, diary entry, March 12, 1946, James G. McDonald Papers, Columbia University, New York.
85. *The Problem of Palestine*, Evidence presented by the Arab Office in Jerusalem to the Anglo-American Committee of Inquiry, March 1946, Robert A. Taft Papers, March 1946 file, Library of Congress.
86. James G. McDonald, diary entry, March 13, 1946, James G. McDonald Papers, Columbia University, New York.
87. Richard Crossman, diary entry, March 14, 1946, in Richard Crossman, *Palestine Mission*, p. 142.
88. "Memorandum of the Jewish Resistance Movement to the Anglo-American Committee on Palestine," March 25, 1946, David Niles Papers, Box 29, HSTL; and John W. Snyder Papers, box 22, HSTL.
89. Bartley Crum, *Behind the Silken Curtain*, pp. 267–68.
90. Emanuel Celler to Truman, March 20, 1946, Emanuel Celler Papers, Israel-Palestine File, Library of Congress.
91. Quoted in Leonard Dinnerstein, *America and the Survivors of the Holocaust*, p. 87.
92. James G. McDonald, diary, Monday, April 1.
93. Bartley Crum, *Behind the Silken Curtain*, pp. 270–73.
94. Richard Crossman, *Palestine Mission*, pp. 180–84.
95. Ibid., pp. 187–89. Crossman reprints the memorandum he wrote at the time.
96. Bartley Crum, *Behind the Silken Curtain*, pp. 265–66.
97. The gist of Hutcheson's comments are taken from the summary and portions of a transcript in Bartley Crum, *Behind the Silken Curtain*, pp. 267–69; and James G. McDonald, diary, Monday, April 1.
98. Report of the Anglo-American Committee, Chapter 1: Recommendations and Comments; The European Problem, available on the Internet at Jewish Virtual Library, at www.jewishvirtuallibrary.org/jsource/History/anglotoc.html. Also see: "The Acting Secretary of State (Acheson) to Certain American Diplomatic and Consular Officers," April 25, 1946, *FRUS*, vol. 8, pp. 585–86.
99. Richard Crossman, *Palestine Mission*, pp. 192–93.

CHAPTER 6: IMPASSE

1. David Horowitz, *State in the Making*, p. 94–95.
2. James G. McDonald to David Ben-Gurion, April 22, 1946, James G. McDonald Papers, Columbia University, New York.
3. Quoted in Michael Cohen, ibid., p. 128. From Ben-Gurion diary, April 22–23, 1946; Ben-Gurion Papers.
4. Quoted in Michael Cohen, *Truman and Israel*, pp. 127–28; from Ben-Gurion to AZEC, April 22, 1946, David Ben-Gurion Papers, Sde Boker, Israel.
5. Ibid., p. 143. The longhand draft by Silver is in the Silver Papers and his confidential notes of April 28, 1946. Also see Zvi Ganin, *Truman, American Jewry and Israel*, p. 63.
6. Acheson to Byrnes, April 30, 1946, *FRUS*, vol. 8, pp. 588–89. Acheson sent the text of Truman's statement to Secretary Byrnes. Attached was an addendum saying it should urgently be sent to Ambassador Harriman in London.
7. James Reston, "Truman Comment Arouses Concern," *The New York Times*, May 1, 1946, p. 13.
8. Halifax to the Cabinet, May 7, 1946, FO 271 52521/E4175.
9. Francis Williams, *Ernest Bevin: Portrait of a Great Englishman*, p. 260.
10. Memorandum of Conversation, by the Director of the Office of European Affairs (H. Freeman Matthews), April 27, 1946, *FRUS*, vol. 7, pp. 587–88.
11. Minutes of the Meeting of the Prime Ministers, London, April 30, 1946, FO371 52520/E4016.
12. Richard Crossman, "War in Palestine?," *The New Statesman and Nation* (Great Britain), May 11, 1949.
13. Richard Crossman, *Palestine Mission*, p. 198.
14. Lawrence Resner, "Truman Said to Plan Start of Jewish Entry 'Forthwith,' " *The New York Times*, May 1, 1946, p. 1.
15. Halifax to Foreign Office, May 3, 1946, FO 371 52520/E40551. Halifax encloses Resner's article, which was obviously a longer unedited version than the one published in *The New York Times*.
16. James Reston, "Britain Demands We Share Responsibility in Palestine as Prelude to Immigration," *The New York Times*, May 1, 1946, p. 1.
17. Abba Hillel Silver, Nahum Goldmann, Eliahu Epstein (Elath), Stephen S. Wise, Louis Lipsky, and Meyer W. Weisgal to Truman, May 2, 1946, HSTL.
18. Copy of telegram sent by Abba Hillel Silver and Stephen Wise to Chairman of Local AZEC chapters, May 9, 1946, CZA L35/142.
19. Memorandum of conversation of Loy Henderson with Arab Foreign Ministers, May 10, 1946, *FRUS*, vol. 7, pp. 604–5.
20. The Consul General at Jerusalem (Pinkerton) to the Secretary of State, May 2, 1946, *FRUS*, vol. 7, pp. 590–91.
21. The Nation Associates, "The Record of Collaboration of King Farouk of Egypt with the Nazis and their Ally, the Mufti," Memorandum submitted to the United Nations, June 1948, in Clark Clifford Papers, Box 13, HSTL.
22. A summary of the mufti's life with links to various sources may be found in the Wikipedia entry: www.en.wikipedia.org/wiki/Amin_Al-Hysanyi.
23. J. H. Hilldring to Dean Acheson, May 3, 1946, *FRUS*, vol. 7, pp. 591–92.

24. Dean Acheson to Truman, May 6, 1946, *FRUS*, vol. 7, pp. 595–96.
25. Truman to Clement Attlee, May 9, 1946, ibid., pp. 596–97.
26. Byrnes to Truman, May 9, 1946, Telegram 2260, *FRUS*, vol. 7, p. 603; Acheson to Byrnes, May 10, 1946, Telegram 2266, p. 603.
27. Byrnes to Truman, May 9, 1946, *FRUS*, vol. 7, pp. 601–3.
28. Attlee to Truman, May 13, 1946, *FRUS*, vol. 7, p. 606.
29. Attlee to Truman, undated, *FRUS*, vol. 7, pp. 612–15. Sent to Secretary Byrnes by British Minister Balfour on May 27 and forwarded by Byrnes to the president on the twenty-eighth.
30. Henderson to Acheson, May 27, 1946, in *FRUS*, vol. 7, citing 867N.01/52746 in the State Department files.
31. Truman to Attlee, June 5, 1946, *FRUS*, vol. 7, pp. 617–18.
32. Attlee to Truman, June 10, 1946, *FRUS*, vol. 7, pp. 623–24.
33. Inverchapel to Foreign Office, June 13, 1946, FO 37152529/E 5226.
34. Text of speech by Ernest Bevin at the Labour Party Conference, Bournemouth, June 12, 1946, General Public Statements, FO 371, 52529/E5546. This is a separate document with the same file number as that cited in footnote 55.
35. "Bevin Assailed at Garden Rally," *New York Herald Tribune*, June 13, 1946.
36. Niles to Charles Ross, June 13, 1946, quoted in David Bernard Sacher, *David K. Niles and United States Policy*, p. 33.
37. Quoted in ibid., p. 33, Wise to Niles, June 18, 1946.
38. Inverchapel to Foreign Office, June 13, 1946, FO 371 52529/E 5444, PRO.
39. Nahum Goldmann, Louis Lipsky, Abba Hillel Silver, and Stephen S. Wise to Truman, CZA, L365/121 PRO.
40. Truman to Niles, June 19, 1946, OF, 204 Misc. HSTL.
41. "MESSAGE FROM THE PRESIDENT TO PRIME MINISTER ATTLEE, DATED JUNE 14," June 14, 1946, State Department Palestine File, Jan. 5, 1946–June 29, 1946, National Archives, College Park, Maryland. Also see: Byrnes to Harriman, June 10, 1946; *FRUS*, vol. 7, pp. 624–26. Byrnes informs Harriman about a group of experts put together to aid him in the discussions, to be made up of Evan Wilson, the assistant chief of the Office of Near Eastern and African Affairs; W. Cramer from Assistant Secretary Hilldring's office; and three Army officers.
42. Byrnes telegram, June 23, 1946, cited in a footnote to Harriman to Byrnes, June 27, 1946, *FRUS*, vol. 7, pp. 638–39.
43. Attlee to Truman, n.d., 1946, *FRUS*, vol. 7, pp. 639–40. Attlee was actually acting on a threat previously made, when he told the president that since tension was mounting, he feared that "precipitate action on the immigration question would provoke widespread violence." Attlee to Truman, June 14, 1946, *FRUS*, vol. 7, pp. 626–27.
44. "Memorandum by the Joint Chiefs of Staff to the State-War-Navy Coordinating Committee," A. J. McFarland, U.S. Army Secretary, June 21, 1946, *FRUS*, vol. 7, pp. 631–33.
45. Matthew Connelly to Niles, June 19, 1946, OF, Box 772, HSTL.
46. Celler to Connelly, June 25, 1946, OF, Box 772, HSTL.
47. Emanuel Celler, *You Never Leave Brooklyn*, p. 118.
48. Daniel J. Fellman, *An American Friendship*, p. 42.

49. Alonzo Hamby, *Man of the People*, p. 61.
50. Robert H. Ferrell, ed., *Dear Bess: The Letters of Harry Truman to Bess Truman*, Truman to Bess Wallace, Oct. 28, 1917, p. 233.
51. David McCullough, *Truman*, p. 107.
52. Margaret Truman, *Bess W. Truman*, pp. 21–26.
53. Alonzo Hamby, *Man of the People*, p. 85.
54. Maurice Bisgyer, *Challenge and Encounter*, p. 189.
55. Harry S. Truman, *Years of Trial and Hope*, p. 133.
56. David McCullough, *Truman*, pp. 145–48.
57. Frank Adler, *Roots in a Moving Stream*, pp. 198–207.
58. Samuel Halperin, *The Political World of American Zionism*, p. 145.
59. Joseph Lelyveld, *Omaha Blues: A Memory Loop*, pp. 61, 75.
60. Frank Adler, *Roots in a Moving Stream*, p. 204.
61. Memorial Tribute to A. J. Granoff by Daniel Brenner, January 1970, TL, A. J. Granoff Papers.
62. Oral history of A. J. Granoff, April 9 and August 27, 1969, HSTL.
63. Maurice Bisgyer, *Challenge and Encounter*, pp. 188–89.
64. *St. Louis Post-Dispatch*, June 14, 1946; also see Frank J. Adler, *Roots in a Moving Stream*, p. 207.
65. The Lelyveld report is printed in Zvi Ganin, *Truman, American Jewry, and Israel*, pp. 74–75. Ganin cites Lelyveld to Silver, July 1, 1946, AZEC Papers, which were then in New York City.
66. George McKee Elsey, *An Unplanned Life*, p. 165.
67. David Horowitz, *State in the Making*, pp. 95, 102.
68. Unpublished dispatch sent to "a prominent newspaper in New York" from a correspondent in Palestine, London, Cairo, and Ankara, in Eliahu Epstein (Elath) to Executive of the Jewish Agency, June 22, 1946, L35/73, CZA.
69. David Horowitz, *State in the Making*, p. 103.
70. Ibid., pp. 101–2.
71. Attlee to Truman, June 28, 1946, FRUS, vol. 7, pp. 639–40.
72. Freda Kirchwey, "Palestine and Bevin," *The Nation*, June 22, 1946.
73. David Horowitz, *State in the Making*, p. 103.
74. Attlee to the House of Commons, CAB 104/264, July 1, 1946, pp. 1806–7, PRO, Kew Gardens, London.
75. Quoted in Address by Dr. Nahum Goldmann, to a meeting of the National Board of Hadassah at the home of Mrs. Siegried Kremarsky, Oct. 28, 1946, Jewish Agency Files, 1924–1928, Box 6, G4, American Jewish Historical Society, New York.
76. "U.S. Zionists Score Raids," *The New York Times*, June 30, 1946, p. 20.
77. Address by Dr. Nahum Goldmann to meeting of the National Board of Hadassah at the home of Mrs. Siegried Kremarsky, Oct. 28, 1946, Jewish Agency Files, 1924–1948, Box 6, G4, American Jewish Historical Society, New York.
78. Truman quoted in Eric Pace, "Abraham Feinberg, Philanthropist for Israel," the obituary column on Feinberg, *The Jerusalem Post*, p. 10, Dec. 30, 1994. According to Elinor Borenstein, Eddie Jacobson's daughter, her father told her that Silver had more than once pounded his fist on Truman's desk and shouted at him. Quoted in Frank J. Adler, *Roots in a Moving Stream*, p. 209, fn. Abe Granoff's son,

Loeb Granoff, reported that he had heard that Silver pounded on Truman's desk and that Truman had left strict orders "not to let that son-of-a-bitch into the White House again." Quoted in Daniel J. Fellman, *An American Friendship*, interview with Loeb Granoff, July 27, 2004.

79. "Zionists' Release is Truman's Hope," *The New York Times*, July 3, 1946, p. 1; also see the official press release from the White House, July 2, 1946, *FRUS*, vol. 7, pp. 642–43.
80. Truman to Attlee, July 2, 1946, *FRUS*, vol. 7, p. 642.
81. Neumann to Professor Paul Hanna, June 26, 1946, quoted in Zvi Ganin, *Truman, American Jewry and Israel*, pp. 121–22. The letter is to be found in the Silver Papers.

CHAPTER 7: CONFLICT BETWEEN ALLIES

1. Herbert L. Matthews, "British Stiffening in Palestine Talks," *The New York Times*, July 14, 1946, p. 17.
2. Herbert L. Matthews, "Grady Plane Delayed," *The New York Times*, July 12, 1946, p. 8.
3. Matthew J. Connelly to the Jewish Agency, July 18, 1946, OF, Misc., HSTL. The letter, which bore press secretary Connelly's signature, was actually written by David Niles. See Truman to Niles, July 10, 1946, OF, Misc., HSTL.
4. Dean Acheson, *Present at the Creation*, p. 174.
5. Harriman to Byrnes, July 19, 1946, *FRUS*, vol. 7, pp. 646–47.
6. Byrnes to Harriman, July 22, 1946, *FRUS*, vol. 7, pp. 648–49.
7. Grady to Harriman, in Harriman to Byrnes, July 24, 1946, *FRUS*, vol. 7, pp. 651–52.
8. Dean Acheson, *Present at the Creation*, pp. 174–75. The complete terms of the report may be found in Grady to Byrnes, in Harriman to Byrnes, July 24, 1946, *FRUS*, vol. 7, pp. 652–67. Also see Summary of the plan prepared by the Near East Division of State, July 1946, Department of State Files, Palestine File, Box 1, NA.
9. Grady to Harriman, in Harriman to Byrnes, July 24, 1946, *FRUS*, vol. 7, pp. 651–52.
10. Michael L. Hoffman, "Divided Palestine Is Urged by Anglo-U.S. Cabinet Body, Delaying Entry of 100,000," *The New York Times*, July 26, 1946, p. 1.
11. Grady to Henderson, July 24, 1946, *FRUS*, vol. 7, p. 652.
12. Michael L. Hoffman, "Divided Palestine Is Urged by Anglo-U.S. Cabinet Body," *The New York Times*, July 26, 1946.
13. Diary of Henry A. Wallace, "Telephone Conversation Between the President and Wallace, July 26, 1946," in *The Price of Vision: The Diary of Henry A. Wallace*, ed. John Morton Blum, pp. 603–4.
14. "Memorandum for President Truman from James G. McDonald," n.d., James McDonald Papers, Columbia University, New York.
15. "J.M.D.'s Report on his, Wagner's and Mead's Conversation with the President on July 27, 1946," L35/121 CZA. The rest of the discussion is from this document.
16. Truman to McDonald, July 31, 1946, L35/121 CZA.

17. John D. Morris, "Truman 'Rebuffs' Palestine Plea," *The New York Times*, July 31, 1946, p. 5.
18. Fitzpatrick to Truman, Aug. 2, 1946, Misc. File, HSTL.
19. Edward J. Flynn to Truman, July 30, 1946; enclosing Bernard A. Rosenblatt to Flynn, n.d., 1946, OF, HSTL.
20. Truman to Flynn, n.d., 1946, OF, HSTL.
21. Niles, quoted in Michael Cohen, *Truman and Israel*, p. 135.
22. Matthew Connelly to Truman, July 30, 1946, OF, Box 775, HSTL.
23. Byrnes to Truman, July 29, 1946, *FRUS*, vol. 7, pp. 671–72.
24. Wallace Diary, July 30, 1946, in John Morton Blum, ed., *The Price of Vision*, pp. 606–7.
25. T. E. Bromley to Burrows, relating the story of Miall, the BBC correspondent in London, British Embassy, June 10, 1948, FO/371 68650/E 8350. If such a telegram was sent by Byrnes to Truman, it is not to be found in any of the archives. Most likely, Acheson recalled the events through the prism of his own point of view, which was favorable to Morrison-Grady.
26. Dean Acheson, Memorandum, July 30, 1946, *FRUS*, vol. 7, pp. 673–74.
27. Inverchapel to Foreign Office, July 31, 1946, FO 371 52548/ E7325, PRO.
28. James Reston, "U.S. Politics Has Big Role in Decision on Palestine," *The New York Times*, Aug. 1, 1946, p. 10.
29. Walter H. Waggoner, "Truman Recalls Palestine Group," *The New York Times*, Aug. 1, 1946, p. 1.
30. Inverchapel to Foreign Office, Aug. 1, 1946, FO 371 52548/E7422.
31. Chargé in Egypt (Lyons) to Byrnes, Aug. 2, 1946, *FRUS*, vol. 7, pp. 676–77. The message conveys the views of Azzam Bey.
32. Nahum Goldmann, Address to Hadassah National Board, Oct. 28, 1946, Jewish Agency Files, 1924–1945, box 6, AJHS, New York.
33. Nahum Goldmann, *The Autobiography of Nahum Goldmann*, pp. 131–32. Also see Resolution adopted at meeting of the Executive of the Jewish Agency in Paris, Aug. 5, 1946, CZA L35/121.
34. Nahum Goldmann, Address to Hadassah National Board, Oct. 28, 1946, Jewish Agency Files, 1924–1945, box 6, AJHS, New York.
35. Nahum Goldmann to Assistant Secretary of the Treasury Ed Foley, Aug. 8, 1946, John W. Snyder Papers, Box 22, HSTL.
36. Nahum Goldmann, *The Autobiography of Nahum Goldmann*, p. 235.
37. Truman to Attlee, Aug. 12, 1946, *FRUS*, vol. 7, p. 682.
38. Ernest Bevin to Clement Attlee, Aug. 14, 1946, FO 371 52552/E7941 #500 E8005. Bevin reported to Attlee about a meeting he had held with Nahum Goldmann. The next set of paragraphs in the text are taken from this report.
39. Acheson to Harriman, Aug. 20, 1946, *FRUS*, vol. 7, pp. 687–88.
40. Attlee to Truman, Aug. 19, 1946, *FRUS*, vol. 7, p. 687.
41. Gellman, Charge in the UK, to Secretary of State James Byrnes, Sept. 17, 1946, *FRUS*, vol. 7, pp. 692, 696.
42. Gallman to Byrnes, Sept. 23, 1946, *FRUS*, vol. 7, pp. 698–99.
43. Clifton Daniel, "Arabs Ask Parley by World on Jews," *The New York Times*, Oct. 4, 1946, p. 8.

CHAPTER 8: TRUMAN'S OCTOBER SURPRISE

1. Truman to Bess, Sept. 15, 1946, in *Harry S. Truman, Dear Bess: The Letters from Harry to Bess Truman, 1910–1959*, ed. Robert H. Ferrell, p. 537.
2. Elath to Goldman, Oct. 9, 1946, Chaim Weizmann Archives; FO 371 CZA, L35/92, PRO.
3. Margaret Truman, *Bess W. Truman*, p. 223.
4. Hannegan to Truman, Oct. 1, 1946, Crum to Hannegan, n.d., 1946, Subject File, Box 160, HSTL.
5. Michael J. Cohen, *Truman and Israel*, p. 189.
6. Felix Frankfurter, *Felix Frankfurter: Reminiscences*, pp. 136–37.
7. Kirchwey, obituary of Lowenthal, *The Nation*, June 14, 1971, p. 741; cf. obituary of Lowenthal, *The New York Times*, May 20, 1971, p. 50.
8. "Talk with President Truman, Oct. 3, 1946, 12:20 PM, for about fifteen minutes," Max Lowenthal Papers, University of Minnesota, Minneapolis. Lowenthal's memo reflects some confusion on either his or Truman's part about whether Truman was talking about the Anglo-American Committee or the Morrison-Grady Commission as a basis for solving the problem.
9. Sue Fishkoff, "He Ushered Israel to the White House," *The Jerusalem Post*, Dec. 30, 1994, p. 10.
10. Oral interview with Abraham Feinberg, Aug. 31, 1973, by Richard D. McKinzie, HSTL.
11. Inverchapel to Foreign Office, Oct. 3, 1946, FO 371 52560/E9938, PRO.
12. Truman to Attlee, Oct. 3, 1946, *FRUS*, vol. 7, pp. 701–3.
13. Statement by the President, Oct. 4, 1946, David Niles Papers, Box 33, HSTL. The statement has been published in many other sources. It may, for example, be found in *The New York Times*, Oct. 5, 1946, p. 2; Truman to Attlee, Oct. 3, 1946, *FRUS*, vol. 7, pp. 701–3.
14. Attlee to Truman, received Oct. 4, 1946, *FRUS*, vol. 7, p. 704.
15. Truman to Attlee, Oct. 4, 1946, *FRUS*, vol. 7, p. 704.
16. "Most Unfortunate," *Time*, Oct. 14, 1946.
17. Attlee to Truman, Oct. 4, 1946, *FRUS*, vol. 7, pp. 704–5.
18. Inverchapel to Attlee, FO 371 52 5 60/E9987.
19. Perhaps the first and most dogmatic presentation of this may be found in John Snetsinger, *Truman, the Jewish Vote, and the Creation of Israel*. Truman, for the first time in his presidency, Snetsinger writes, "was on record as supporting the basic Zionist goal." American Zionists delightedly watched as Thomas E. Dewey and Harry Truman "tried to outbid each other in an attempt to win favor among Jewish voters." He concludes his book by saying that Truman's statement was given "to counter apparent Republican inroads among Jewish voters by endorsing the most crucial part of the Zionist program." He held no firm convictions on the issue, Snetsinger argues. When his policies "were in accord with the Zionists' program," he writes, "he was motivated primarily . . . by political exigencies"; pp. 42, 140.
20. Truman to Attlee, Oct. 10, 1946, *FRUS*, vol. 7, pp. 706–8.
21. Dean Acheson, *Present at the Creation*, p. 176.
22. George to Truman, Oct. 5, 1946; Truman to George, Oct. 6, 1946, Subject File,

Box 160, HSTL; and Presidential Secretary's File, Box 61, HSTL. Truman had been making efforts to bring DPs to the U.S. In Dec. 1945, he issued a directive mandating preferential treatment for all DPs, especially orphans, within the existing U.S. immigration laws. This resulted in about 27,000 Jewish DPs being resettled in the U.S. He was frustrated in other attempts to get Congress to pass new pro-immigrant legislation, because anti-immigrant sentiment was strong after the war. In the end, between 1945 and 1952, 137,450 Jews were admitted to the U.S. (For more information, see: Leonard Dinnerstein, *America and the Survivors of the Holocaust*, pp. 285–87.)

23. Zvi Ganin, *Truman, American Jewry and Israel*, pp. 107–8.
24. Ibn Saud to Truman, transmitted to Acting Secretary of State on Oct. 15, 1946; *FRUS*, vol. 7, pp. 708–9; OF, Box 771, HSTL.
25. Truman to Ibn Saud, Oct. 25, 1946, *FRUS*, vol. 7, pp. 714–17.
26. Memorandum on the Jewish People of Europe, enclosed with Edwin W. Pauley to Truman, Oct. 9, 1946, HST Subject File, "Palestine-Jewish Immigration Folder," Box 161, HSTL.
27. Truman to Pauley, Oct. 22, 1946, HST Subject File, "Palestine-Jewish Immigration Folder," Box 161, HSTL.
28. Quoted in William Roger Louis, *The British Empire in the Middle East, 1945–1951*, p. 443.
29. Inverchapel to Bevin, "Jewish Affairs in the United States," Nov. 22, 1946; FO 371 52571/E 11651, PRO.
30. Alan Bullock, *Ernest Bevin: Foreign Secretary*, p. 333. Also see Bevin's own report on the New York trip to the cabinet, Nov. 26, 1946, FO800/48/PA/46/131, PRO. Strangely, *The New York Times*, which had the reputation as "the paper of record," did not tell its readers anything about the negative reception Bevin received in New York City.
31. *New York Post*, Nov. 4, 1946.
32. Minutes of meeting between Truman and Bevin at the White House, Dec. 8, 1946, FO 371 61762/E221, PRO.
33. Dean Acheson, *Present at the Creation*, p. 177.
34. Memorandum of conversation between Truman and Amir Faisal, Dec. 13, 1946, *FRUS*, vol. 7, pp. 729–33.
35. Dean Acheson, *Present at the Creation*, p. 177.
36. Quoted in Meyer Weisgal, *Meyer Weisgal*, p. 241.
37. For discussion of events at the congress, see Chaim Weizmann, *Trial and Error*, pp. 442–43, and Joseph B. Schechtman, *The United States and the Jewish State Movement*, pp. 183–84. Perhaps the best summary of what took place is in Zvi Ganin, *Truman, American Jewry and Israel*, pp. 111–18.
38. General Counsel Sholes at Basel to Byrnes, Dec. 30, 1946, *FRUS*, vol. 7, Addendum, dispatch 517 of Jan. 16, 1947.
39. Truman to Byrnes, Jan. 7, 1947, in *Public Papers of the Presidents of the United States: Harry S. Truman*, 1947, pp. 12–13.
40. Jan. 3, 1947; Harry S. Truman, 1947 diary, HSTL and www.trumanlibrary.org/diary/index.html.
41. Jan. 8, 1947; Harry S. Truman, 1947 diary, HSTL and www.trumanlibrary.org/diary/index.html

42. Emanuel Neumann, speech to American Zionist Council, "REPORT FROM LONDON," Statler Hotel, Washington, D.C., Feb. 17, 1947, Zionist Organization of America papers, Box 1, Folder 10, AJHS, New York.
43. Gellman, Chargé in the UK, to Marshall, Jan. 28, 1947, *FRUS*, vol. 7, pp. 1015–17.
44. David Horowitz, *State in the Making*, pp. 128–29.
45. Ibid., pp. 133–46.
46. Bevin to Marshall, n.d., February 1947, *FRUS*, vol. 7, pp. 1035–37.
47. Pinkerton to Marshall, Jan. 31, 1947, *FRUS*, vol. 7, pp. 1023–24.
48. Bevin, speech to the House of Commons, Feb. 25, 1947, cited in Memorandum by Mr. William J. McWilliams of the Executive Secretary to Marshall, Feb. 25, 1947; *FRUS*, vol. 7, pp. 1056–57.
49. Harry S. Truman, *Years of Trial and Hope*, pp. 182–83.
50. Clifton Daniel, "British Seem Trapped in Fortress to Visitor Returning to Palestine," *The New York Times*, March 1, 1947, p. 4.
51. "Martial-Law Rule Defined by British," *The New York Times*, March 2, 1947, p. 2.
52. David Horowitz, *State in the Making*, pp. 142–43.

CHAPTER 9: UNSCOP

1. British Representative at the UN, Cadogan, to Assistant Secretary General of the UN, Hoo, April 2, 1947, *FRUS*, vol. 5, pp. 1067–68.
2. Marshall to Truman, April 17, 1947, *FRUS*, vol. 5, pp. 1070–72.
3. Memorandum of conversation by Acheson with Moshe Shertok and Loy Henderson, April 23, 1947, *FRUS*, vol. 5, pp. 1073–77.
4. Thirty Members of Congress to Marshall and Austin, April 22, 1947, State Department File, National Archives, College Park, Maryland.
5. Marshall to Congressmen, May 5, 1947, in *Congressional Record* 23, 80th Congress, 1st Session, pp. A2201–3.
6. Bunche's work on UNSCOP is chronicled in Brian Urquhart, *Ralph Bunche: An American Odyssey*, pp. 139–51. Our comments are based on his book.
7. Clayton Knowles, "U.S. Would Allow Zionists U.S. Voice," *The New York Times*, May 2, 1947, p. 1.
8. Chaim Weizmann to Selig Brodetsky, Feb. 19, 1947, *The Letters and Papers of Chaim Weizmann*, pp. 245–49.
9. Sara Alpern, *Freda Kirchwey: A Woman of* The Nation, pp. 197–200; quote is on p. 198.
10. "The Mufti's Henchman," *The Nation*, May 17, 1947, pp. 561–62.
11. "The Grand Mufti in World War II," *The Nation*, May 17, 1947, pp. 597–99.
12. "3 Arab Delegates Called Axis Aides," *The New York Times*, May 12, 1947, p. 3.
13. *The Nation* issue of May 17, 1947, contains a supplement, "The Palestine Problem: Report at the United Nations–April–May 1947," an abridged version of the longer memorandum the U.N. delegates received.
14. David K. Niles, "MEMORANDUM FOR THE PRESIDENT," May 12, 1947; Subject File, Palestine–1945–1947, Box 160, HSTL.
15. Ibid.; also see the typewritten transcript of Truman's note to Niles, sent as

"Memorandum for David K. Niles," President's Secretary File, Box 246, HSTL. The manuscripts also have a typed version of this note, prepared by the president's secretary. In transferring it to type, the secretary, while correcting the president's incorrectly spelled words, added one word that changes the meaning of a critical sentence. Truman had written, in handwriting, "Terror and Silver are the contributing causes of *some*, not all of our troubles." The typed version says, Terror and Silver are the contributing causes of *some*, if not all, of our troubles." The word "if" drastically changes the meaning of the sentence.

16. Yaacov Ro'i, *From Encroachment to Involvement*, p. 37.
17. Andrei Gromyko, speech, May 14, 1947, in *General Assembly, 1st Special Session, 77th Plenary Meeting*, vol. 1, pp. 127–35. Also see reprint of speech in vol. 1, 127–35. Also see reprint of speech in Yaacov Ro'i, *From Encroachment to Involvement*, pp. 37–41.
18. Dubrow to Marshall, May 10, 1947, *FRUS*, vol. 5, pp. 1081–82.
19. Smith's memo is summarized in Howard M. Sachar, *A History of Israel*, p. 286.
20. Dean Rusk to Acheson, May 27, 1947, *FRUS*, vol. 5, pp. 1088–89.
21. Laurent Rucker, "Moscow's Surprise: The Soviet-Israeli Alliance of 1947–1949," Working Paper 48, Woodrow Wilson International Center for Scholars, p. 20.
22. David Horowitz, *State in the Making*, p. 157.
23. Memorandum of conversation of Acheson with Moshe Shertok, Eliahu Epstein, and Loy Henderson, May 29, 1947, *FRUS*, vol. 5, pp. 1094–96.
24. James Reston, "Britain Prods U.S. to Stop Funds to Defy Palestine Law," *The New York Times*, May 20, 1947, p. 1.
25. Gene Currivan, "Zionists Tending to Middle Course," *The New York Times*, May 23, 1947, p. 5.
26. Gene Currivan, "Ben-Gurion Favors Palestine Division," *The New York Times*, May 23, 1947, p. 12.
27. "Irgun Leader Skeptical of Solution for Palestine Through U.N. Inquiry," *The New York Times*, May 21, 1947, p. 12. The article contains the text of an interview with Begin. See also "Zionist Dissidents Reject Partition," *The New York Times*, July 13, 1947, p. 15.
28. Statement by the President on June 5, 1947, Clark Clifford Papers, HSTL.
29. Marshall to Certain Diplomatic and Consular Offices, June 13, 1947, FRUS, vol. 5, p. 1103.
30. Marshall to Austin, June 13, 1947, *FRUS*, vol. 5, pp. 1103–5.
31. Epstein to Silver, June 27, 1947, CZA L35/83.
32. "Marshall Reaffirms Policy on Palestine," *The New York Times*, July 2, 1947, p. 3.
33. "2 Senators Say U.S. Lags on Palestine," *The New York Times*, July 7, 1947, p. 3.
34. Quoted in Joseph B. Shechtman, *The United States and the Jewish State Movement*, p. 212. Shechtman quotes from the *JTA Bulletin*, July 17, Aug. 11, 1947.
35. Clifton Daniel, "Zionist Sees U.S. Opposing Terror," *The New York Times*, June 12, 1947, p. 10.
36. "2 Senators Say U.S. Lags on Palestine," *The New York Times*, July 7, 1947, p. 3.
37. Leonard Slater, *The Pledge*, pp. 94–96.
38. Gene Currivan, "Haganah Foils Irgun Plot to Bomb British Army Center in Tel Aviv," *The New York Times*, June 19, 1947, p. 1.

39. Macatee, Consul General in Jerusalem, to Marshall, June 11, 1947, *FRUS*, vol. 5, p. 1102.
40. "New War on Jews Forecast by Arab," *The New York Times*, June 19, 1947, p. 16.
41. Robert St. John, *Eban*, pp. 17–25. St. John's biography is the basic study of Eban's life, and our discussions of Eban are based on this book.
42. Ibid., p. 163.
43. Ibid., p. 167.
44. Clifton Daniel, "U.N. Body Worried on Jews' Hanging," *The New York Times*, June 22, 1947, p. 1. See also Macatee to Marshal, June 23, 1947, *FRUS*, vol. 6, pp. 1107–12.
45. Jorge García Granados, *The Birth of Israel*, pp. 51–55.
46. Ibid., pp. 119–25; quote is on p. 123.
47. Clifton Daniel, "Zionist Rules Out Unity in Palestine," *The New York Times*, June 18, 1947, p. 13.
48. Evan M. Wilson, *Decision on Palestine: How the U.S. Came to Recognize Israel*, pp. 110–11.
49. David Horowitz, *State in the Making*, pp. 166–69.
50. Jorge García Granados, *The Birth of Israel*, pp. 81–84.
51. Ibid., pp. 31–39.
52. David Horowitz, *State in the Making*, p. 171.
53. Jorge García Granados, *The Birth of Israel*, pp. 90–91.
54. Macatee to Marshall, June 30, 1946, *FRUS*, vol. 5, pp. 1113–16.
55. David Horowitz, *State in the Making*, p. 171.
56. Ibid., pp. 166–69.
57. Ibid., pp. 169–70.
58. Ben-Gurion, speech excerpted in Macatee to Marshall, July 7, 1947, *FRUS*, vol. 5, p. 1118. Macatee's report of the committee's third week is on pp. 1117–20.
59. Quoted in Jorge García Granados, *The Birth of Israel*, pp. 129–30.
60. Ibid., p. 134.
61. Macatee to Marshall, July 14, 1947, *FRUS*, vol. 5, pp. 1123–28. Weizmann's speech is summarized on pp. 1125–26. The entire report of week four is on pp. 1123–28.
62. Macatee to Marshall, July 14, 1947, *FRUS*, p. 1125.
63. David Horowitz, *State in the Making*, pp. 173–74.
64. Norman Rose, *Chaim Weizmann*, pp. 299–300. The authors visited the Weizmann home in 2007, but the accurate description is taken from Rose's biography.
65. David Horowitz, *State in the Making*, p. 177.
66. Macatee to Marshall, July 21, 1947, *FRUS*, vol. 5, pp. 1128–31.
67. Bevin to Marshall, June 30, 1947, *FRUS*, vol. 5, pp. 1112–13.
68. I. F. Stone, *Underground to Palestine*, pp. 15–16; pp. 221–24.
69. Ruth Gruber, *Inside of Time: My Journey from Alaska to Israel*, pp. 272–89. The following paragraphs in the text are based on Gruber's book.
70. Quoted in Robert St. John, *Abba Eban*, p. 168.
71. Ruth Gruber, *Inside of Time*, pp. 272–89.
72. Jose García Granados, *The Birth of Israel*, pp. 173–82.
73. Ibid., pp. 183–88.

74. Harry S. Truman, diary, July 21, 1947, President's Secretary's File, Box 232, HSTL.
75. "My Day," syndicated column, Aug. 23, 1947, quoted in Joseph Lash, *Eleanor: The Years Alone*, pp. 113–14.
76. Truman to Eleanor Roosevelt, Aug. 23, 1947, President's Personal File, HSTL.
77. Memorandum by Lovett, Notes on Cabinet Meeting, Aug. 22, 1947, *FRUS*, vol. 5, pp. 1138–39.
78. Lovett to Embassy, Aug. 22, 1947, *FRUS*, vol. 5, pp. 1139–40; Lovett to Embassy, Aug. 22, 1947, pp. 1140–41.
79. Douglas to Marshall, Aug. 26, 1947, *FRUS*, vol. 5, pp. 1141–42.
80. Jorge García Granados, *The Birth of Israel*, pp. 198–207.
81. Ibid., p. 200. The following paragraphs in the text are based on Garcías's account. See pp. 216–33.
82. Ibid., pp. 208–15.
83. David Horowitz, *State in the Making*, p. 184.
84. Ibid., pp. 202–9.
85. Evan M. Wilson, *Decision on Palestine*, p. 112.
86. Quoted in Brian Urquhart, *Ralph Bunche*, pp. 149–50.
87. David Horowitz, *State in the Making*, pp. 222–23; Robert St. John, *Eban*, p. 174.
88. "Palestine Compromise," *The Christian Science Monitor*, Sept. 3, 1947.
89. "A Plan for Palestine," *The Atlanta Constitution*, Sept. 2, 1947.
90. "Time to Act in Palestine," *Chicago Sun*, Sept. 3, 1947.
91. "Our Palestinian Policy," *St. Louis Post-Dispatch*, Sept. 3, 1947.
92. "Make Palestine Justice Real," *The New York Post*, Sept. 3, 1947.
93. "The Palestine Report," *The New York Times*, Sept. 1, 1947.
94. Arthur Krock, "For a Jewish Homeland: *The New York Times*, Sept. 3, 1947.
95. I. F. Stone, "The UNSCOP Proposals on Palestine," *P.M.*, Sept. 4, 1947.

CHAPTER 10: THE FIGHT OVER PARTITION

1. Clifton Daniel, "Arabs Threaten Force if Holy Land Is Split," *The New York Times*, Sept. 7, 1947, p. E4.
2. J. C. Hurewitz, *The Struggle for Palestine*, p. 308.
3. "Palestine Frees 3 in Exodus Crew," *The New York Times*, Sept. 9, 1947, p. 2.
4. Samir Rifai to Bevin, Sept. 22, 1947, FO 371 61880/E9354, PRO.
5. Macatee to Marshall, Sept. 2, 1947, *FRUS*, vol. 5, pp. 1143–44.
6. Clifton Daniel, "Zionists Ask U.N. to Pass New Plan," *The New York Times*, Sept. 2, 1947, p. 1.
7. "Irgun Denounces Partition Support," *The New York Times*, Sept. 7, 1947, p. 7.
8. Hawkins to Marshall, Sept. 11, 1947, *FRUS*, vol. 5, pp. 1145–46.
9. David Horowitz, *State in the Making*, pp. 253–54.
10. Chaim Weizmann, *Trial and Error*, p. 457.
11. Lourie to Celler, Sept. 8, 1947, enclosing three different drafts of proposed letter to Truman; Emanuel Celler Papers, LOC.
12. Edmund I. Kauffman and Paul Tishman to recipients (Miss Leibel), Sept. 29, 1947, AJHS Papers, Box 42, Folder 10, New York. The role of Freda Kirchwey and the Nation Associates in working for the partition resolution

is discussed in Ronald Radosh and Allis Radosh, "Righteous Among the Editors: When the Left Loved Israel," *World Affairs* 171, no. 1 (Summer 2008): 65–75.

13. Eleanor Roosevelt to Lessing J. Rosenwald, Oct. 1, 1947; quoted in Joseph P. Lash, *Eleanor: The Years Alone*, p. 116.
14. George Barrett, "Arab War Threat Discounted to U.N.," *The New York Times*, Oct. 6, 1947, p. 4.
15. "Role of the Nation Associates in Bringing About First American Statement Favoring UNSCOP Proposal," October 1947, enclosed in "The Work of the Nation Associates on Palestine," March 22, 1948, CZA47 L35, 137.
16. Evan M. Wilson, *Decision on Palestine*, p. 116. Herbert A. Fierst, "A Personal Memoir on the Establishment of Israel," *The Record*, vol. 23 (1998–99) pp. 25–33; Evan M. Wilson, *Decision on Palestine*, p. 116. The story behind General John Hilldring's appointment is not well known. Herbert A. Fierst, who worked at the Pentagon and then the State Department, alerted David Niles that State was planning to appoint Arabist George Wadsworth to the United States' U.N. delegation. Fierst, who had worked with General Hilldring in Europe on the Displaced Persons issue, especially the Jewish DPs, highly recommended Hilldring for the post. Truman thwarted States plans, convincing Hilldring, who was set to retire to accept the U.N. Delegate position. He requested that Hilldring report directly to him at the White House.
17. George E. Jones, "All U.N. Members in Palestine Group," *The New York Times*, Sept. 24, 1947, p. 5.
18. Joseph P. Lash, *Eleanor: The Years Alone*, pp. 114–15.
19. Thomas F. Powers, Jr., Excerpts from the Minutes of the Sixth Meeting of the U.S. Delegation to the 2nd Session of the General Assembly, New York, Sept. 15, 1947; *FRUS*, vol. 5, pp. 1147–51; also see Forrest C. Pogue, *George Marshall*, pp. 345–47.
20. Statement by Marshall to the General Assembly on 9/17/47, *FRUS*, vol. 5, pp. 1151–52.
21. Oral history interviews with Loy W. Henderson, conducted June 13 and July 5, 1973, HSTL. The following paragraphs in the text are from this interview.
22. Henderson to Marshall, Sept. 22, 1947, "Certain Considerations Against Advocacy by the U.S. of the Majority Plan," *FRUS*, vol. 5, pp. 1153–59.
23. Interview with Clark Clifford by Richard Holbrooke and Brian VanDenMark, May 4, 1988, HSTL.
24. Oral history interview with Clark Clifford, HSTL.
25. George McKee Elsey, *An Unplanned Life*, pp. 133, 141.
26. Oral history interviews with Loy W. Henderson, conducted June 14 and July 5, 1973, HSTL.
27. Memorandum by Hilldring to the Acting U.S. Rep. of the U.N., Herschel Johnson, Sept. 24, 1947, *FRUS*, vol. 5, pp. 1162–63.
28. Memo prepared by the Department of State, "United States Position with Respect to the Question of Palestine," *FRUS*, vol. 5, Sept. 30, 1947, pp. 1166–70.
29. Daniel Mandel, *H. V. Evatt and the Establishment of Israel*, p. 132.
30. Thomas J. Hamilton, "U.N. to Hear Arabs; 'Three Noes' on Palestine Inquiry Report Today," *The New York Times*, Sept. 29, 1947, p. 1; "Palestinian Arabs Re-

ject U.N. Plans; Warn of a Battle," *The New York Times*, Jan. 30, 1947, p. 1; and "Text of the statement by Jamel el-Husseini before the U.N.," p. 13.

31. "Text of Statement before the U.N. by Abba Hillel Silver," *The New York Times*, Oct. 3, 1947, p. 20; and "Palestine Division Is Accepted in U.N. by Jewish Agency," *The New York Times*, Oct. 3, 1947, p. 1.
32. St. John, *Eban: A Biography*, pp. 178–85.
33. Memorandum by Mr. Gordon Knox to Herschel Johnson, Oct. 3, 1947, *FRUS*, vol. 5, pp. 1173–74.
34. Eddie Jacobson to Truman, Oct. 3, 1947, Eddie Jacobson Papers, HSTL.
35. Truman to Jacobson, Oct. 8, 1947, Eddie Jacobson Papers, HSTL.
36. Arthur Vanderberg to Robert A. Taft, Oct. 8, 1947, Robert A. Taft Papers, LOC.
37. "State Democrats Ask Palestine Aid," *The New York Times*, Oct. 7, 1947, p. 13.
38. Clifton Daniel, "Arab States to Send Troops to the Borders of Palestine," *The New York Times*, Oct. 10, 1947, p. 1.
39. Harry S. Truman, *Years of Trial and Hope*, p. 183. A slightly different claim was made by Eliahu Elath, who wrote that Truman met at the White House with Lovett, Niles, and Clifford where he confirmed the State Department speech, which made U.S. support for partition public. Quoted in Michael J. Cohen, *Truman and Israel*, p. 157.
40. Thomas J. Hamilton, "Arabs Are Warned," *The New York Times*, Oct. 12, 1947, p. 1.
41. Marshall to Lovett, Oct. 23, 1947, *FRUS*, vol. 5, pp. 1200–1201.
42. Marshall to Austin, Nov. 12, 1947, *FRUS*, vol. 5, pp. 1255–56.
43. Johnson and Hilldring to Lovett, in Warren Austin to Marshall, Nov. 18, 1947, *FRUS*, vol. 5, pp. 1266–68.
44. Memorandum of conversation of Lovett, Oct. 15, 1947, *FRUS*, vol. 5, pp. 1181–82.
45. Eliahu Elath, *Israel and Elath*, pp. 4–6. This is the text of the Lucien Wolf Memorial Lecture to the Jewish Historical Society, London, June, 1966.
46. Dean Acheson, *Present at the Creation*, p. 178.
47. Eliahu Elath, *Israel and Elath*, pp. 4–6.
48. Ibid., pp. 18–20. The memorandum is on pp. 19–20.
49. Eliahu Elath, *Israel and Elath*, pp. 20–21.
50. Chaim Weizmann, *Trial and Error*, pp. 458–59. Also see Robert Donovan, *Conflict and Crisis*, p. 327.
51. Robert McClintock, memorandum to the file, Nov. 19, 1947, *FRUS*, vol. 5, pp. 1271–72.
52. Bohlen to Lovett, Nov. 19, 1947, *FRUS*, vol. 5, p. 1271. Bohlen's memo is attached to the McClintock memo in the file and appears as a footnote on the *FRUS* page.
53. Chaim Weizmann, *Trial and Error*, pp. 458–59.
54. Herschel Johnson, text of address, *The New York Times*, Nov. 23, 1947, p. 58.
55. Frank S. Adams, "Britain Mystifies U.N. on Palestine," *The New York Times*, Nov. 14, 1947, p. 1.
56. Henderson to Lovett, Nov. 24, 1947, *FRUS*, vol. 5, pp. 1281–82.
57. Ibid., p. 1282. This appears as a marginal notation to the Henderson letter to Lovett.

58. Memorandum of telephone conversation by Acting Secretary of State Lovett with Johnson and Hilldring, Nov. 24, 1947, *FRUS*, vol. 5, pp. 1283–84.
59. Gabriel Sheffer, *Moshe Sharett*, p. 265.
60. Emanuel Neumann, *In the Arena*, p. 251.
61. David Horowitz, *State in the Making*, pp. 298–99.
62. Nahum Goldmann, *The Autobiography of Nahum Goldmann*, pp. 244–45.
63. David Horowitz, *State in the Making*, p. 299.
64. Abba Eban, *Personal Witness*, p. 121.
65. Emanuel Neumann, *In the Arena*, p. 251.
66. Julius Haber, *The Odyssey of an American Zionist*, pp. 314–18.
67. David Horowitz, *State in the Making*, p. 299.
68. Reports from the British delegation to the United Nations, cited in Michael J. Cohen, *Palestine and the Great Powers*, p. 294.
69. Telegram of U.S. Ambassador to the Philippines O'Neal to Lovett, in Memorandum of Lovett to Truman, Dec. 10, 1947, *FRUS*, vol. 5, p. 1306.
70. Jorge García Granados, *The Birth of Israel*, pp. 263–65; Sumner Welles, *We Need Not Fail*, p. 63.
71. Oral history interview with Loy Henderson, June 14, 1973, HSTL.
72. Quoted in Zvi Ganin, *Truman, American Jewry and Israel*, p. 145.
73. David Horowitz, *State in the Making*, p. 300.
74. Quoted in Melvyn Urofsky, *We Are One!*, p. 145.
75. Dan Kurzman, *Genesis 1948*, p. 40.
76. Eban A. Ayers, diary, Nov. 29, 1947, HSTL.
77. Memorandum of Acting Secretary of State to Truman, Dec. 10, 1947, *FRUS*, vol. 5, pp. 1305–7; Emanuel Neumann, *In the Arena*, pp. 252–53.
78. Emanuel Neumann, *In the Arena*, p. 250.
79. David Horowitz, *State in the Making*, pp. 300–301.
80. David Bernard Sacher, *David Niles and United States Policy*, pp. 71–72.
81. Harry S. Truman, *Years of Trial and Hope*, pp. 186–87.
82. Celler to Truman, Nov. 26, 1947, HSTL.
83. Truman to Celler, n.d., 1947, HSTL.
84. Sumner Welles, *We Need Not Fail*, p. 63.
85. Note accompanying Nov. 20, 1947 memorandum on Nov. 11 cabinet meeting, President Secretary's File; Cabinet meetings; Box 154, HSTL. See Robert J. Donovan, *Conflict and Crisis*, pp. 329–30.
86. M. Comay to B. Gering, Dec. 3, 1947, *Israel Documents, December 1947–May 1948*, pp. 3–13; quotes are on p. 5–6.
87. Oral history interview with Loy W. Henderson, June 14, 1973, HSTL.
88. Niles and Johnson, quoted in Dan Kurzman, *Genesis 1948*, p. 40.
89. British U.N. Delegation to Foreign Office, Nov. 24, 1947, FO 371 61889, PRO.
90. Marshall to Lovett, Nov. 28, 1947, *FRUS*, vol. 5, pp. 1289–99.
91. Emanuel Neumann, *In the Arena*, p. 254.
92. Ruth Gruber, *Witness*, p. 158.
93. Weizmann to Ginsburg, December 1947, *The Letters and Papers of Chaim Weizmann*, vol. 23, *Aug. 1947–June 1952*, p. 67.
94. "Jubilant Zionists Hold Rally Here," *The New York Times*, Dec. 3, 1947, p. 1.
95. Celler to Truman, Dec. 3, 1947, Box 23, Celler Papers, LOC.

96. Frank J. Adler, *Roots in a Moving Stream*, p. 207.
97. Eddie Jacobson, diary, n.d., 1947, cited in Joel Levitch and Laurel Vlock, "The Diary of Eddie Jacobson," *The Washington Post*, May 6, 1973.
98. Jacobson to Truman, Dec. 12, 1947; quoted in Frank J. Adler, *Roots in a Moving Stream*, p. 207; "Behind the Scenes of the U.N. Decision," *National Jewish Monthly*, January 1948, p. 163.

CHAPTER 11: A "BIG CONSPIRACY" BREWS IN WASHINGTON

1. Abba Eban, *Personal Witness*, p. 114.
2. Eliahu Epstein to Lillie Shultz, Dec. 4, 1947, CZA L35/76.
3. Henderson to Marshall, Nov. 10, 1947, *FRUS*, vol. 5, p. 1249.
4. "U.S. Embargoes Arms to Mid-East," *The New York Times*, Dec. 6, 1947, p. 3.
5. Michael Cohen, *Truman and Israel*, p. 174.
6. Leonard Slater, *The Pledge*, p. 132.
7. Ibid., p. 128.
8. Elath to Kirchwey, Dec. 12, 1947, CZA L35/76.
9. Gene Currivan, "Arabs Make Roads," *The New York Times*, Dec. 5, 1947, p. 1.
10. "Arab States Call Meeting," *The New York Times*, Dec. 2, 1947, p. 1.
11. Macatee to Marshall, Dec. 31, 1947, *FRUS*, vol. 5, pp. 1322–28.
12. Evan M. Wilson, *Decision on Palestine*, pp. 133–34.
13. David Horowitz, *State in the Making*, pp. 323–24.
14. "Palestine Will Not View U.S. Troops, Truman Says," *The New York Times*, Jan. 16, 1948, p. 4.
15. Michael Bar-Zohar, *Ben-Gurion: A Biography*, pp. 147–50; quote is on p. 150.
16. Dan Kurzman, *Ben-Gurion*, p. 276.
17. Leonard Slater, *The Pledge*, p. 151.
18. George F. Kennan to Marshall and Lovett, Jan. 20, 1948, *FRUS*, vol. 5, pp. 545–46. The lengthy memo is attached to Kennan's letter. It may be found in "Report by the Policy Planning Staff on Position of the United States With Respect to Palestine," Jan. 19, 1948, *FRUS*, vol. 5, pp. 546–54. The discussion of the memo in the text refers to this report.
19. Truman to Marshall, Feb. 22, 1948, *FRUS*, vol. 6, p. 645.
20. Rusk to Lovett, Jan. 26, 1948, *FRUS*, vol. 5, pp. 556–62.
21. James Forrestal, *The Forrestal Diaries*, ed. Walter Mills, p. 300.
22. James Reston, "Bipartisan Policy on Holy Land Seen," *The New York Times*, Jan. 27, 1948, p. 8.
23. Eleanor Roosevelt to Truman, Jan. 29, 1948, *Eleanor and Harry*, ed. Steve Neal, pp. 123–24.
24. Truman to Eleanor Roosevelt, Feb. 2, 1948, *Eleanor and Harry*, ed. Steve Neal, p. 124.
25. Wadsworth to Truman, n.d. [February 1948], *FRUS*, vol. 5, pp. 596–99.
26. Memo by Ambassador George Wadsworth to Loy Henderson, Feb. 4, 1948, *FRUS*, vol. 5, pp. 592–95.
27. Oral history interview with Oscar R. Ewing, conducted on May 2, 1969 by J. R. Fuchs, HSTL.
28. Ibid.

29. Evan M. Wilson, *Decision on Palestine*, p. 115. Wilson cites an appendix appearing in Ian Bickerton, "President Truman's Recognition of Israel," *American Jewish Quarterly*, December 1968, pp. 229–40.
30. Andie Knutson to Phileo Nash, Aug. 6, 1951, Phileo Nash Papers, HSTL; cited in Michael T. Benson, *Harry S. Truman and the Founding of Israel*, pp. 94–95.
31. Truman to Pepper, Oct. 20, 1947, Confidential File, Box 59, HSTL.
32. "Big Aid to Wallace Is Seen in Victory of Bronx Protégé," *The New York Times*, Feb. 19, 1948, p. 1.
33. James Reston, "Palestine Issue Put High in Bronx Wallace Victory," *The New York Times*, Feb. 20, 1948, p. 14.
34. Truman to Marshall, Feb. 22, 1948, *FRUS*, vol. 5, p. 645.
35. Statement made by Warren Austin before the Security Council, Feb. 24, 1948, *FRUS*, vol. 5, pp. 651–54.
36. Statement by Warren Austin to the UN Security Council, Feb. 25, 1948, *FRUS*, vol. 5, pp. 657–58.
37. Draft of Ben Cohen's article, March 1, 1948. Ben Cohen Papers, Box 8, LOC. Attached to it is a note from Felix Frankfurter: "This is *very* good–many Thanks!" Cohen's revised article appeared in the *New York Herald Tribune*, March 16 and 17, 1948.
38. Francis J. Meyers to Truman, March 4, 1948, Subject File, Box 160, HSTL.
39. Truman to Meyers, n.d., Subject File, Box 160, HSTL.
40. "Truman Says Issue Is of Deep Concern," *The New York Times*, Feb. 25, 1948, p. 1.
41. Celler to Truman, Feb. 25, 1948, Celler Papers, LOC.
42. Kirchwey to Truman, Feb. 25, 1958, CZA L35/137.
43. Marshall to Truman and Cabinet, March 5, 1948, *FRUS*, vol. 5, pp. 678–79.
44. Marshall to Austin, March 5, 1948, *FRUS*, vol. 5, pp. 679–81.
45. Marshall to Austin, March 5, 1948, *FRUS*, vol. 5, pp. 682–85.
46. Marshall to Austin, March 8, 1948, *FRUS*, vol. 5, p. 697.
47. McClintock to Lovett, March 8, 1948, *FRUS*, vol. 5, pp. 697–99.
48. Michael J. Cohen, *Truman and Israel*, p. 189.
49. Clark Clifford, *Counsel to the President*, p. 7; interview with David Ginsburg by the authors, Aug. 3, 2006.
50. Memorandum by Clifford, "Proposed Program on the Palestine Problem," March 6, 1948, *FRUS*, vol. 5, pp. 687–89.
51. Memorandum by Clifford to Truman, March 8, 1948, *FRUS*, vol. 5, pp. 690–96.
52. Austin to Marshall, March 11, 1948, *FRUS*, vol. 5, pp. 707–11.
53. Trygve Lie, *In the Cause of Peace*, p. 169.
54. Austin to Marshall, March 13, 1948, *FRUS*, vol. 5, pp. 712–19; quote is on p. 719.
55. Marshall to Austin, March 16, 1948, *FRUS*, vol. 5, pp. 728–29.
56. Memorandum of telephone conversation by McClintock, March 17, 1948, *FRUS*, vol. 5, pp. 729–31.
57. McClintock to Humelsine, March 17, 1948, *FRUS*, vol. 5, pp. 731–32.
58. Harry S. Truman, *Years of Trial and Hope*, p. 188.
59. Vera Weizmann, *The Impossible Takes Longer*, pp. 221–22; Weizmann to Truman, Feb. 10, 1948, HSTL; Connelly to Weizmann, Feb. 12, 1948, Box 10, Max Lo-

wenthal Papers; Chaim Weizmann, *The Letters and Papers of Chaim Weizmann*, vol. 2, p. 85.

60. "Zionist Leaders Turned to Kansas Citian," *The Kansas City Times*, May 13, 1965; Eddie Jacobson, "Two Presidents and a Haberdasher–1948, *American Jewish Archives* 20, no. 1 (April 1968): 3–15.
61. Jacobson to Truman, February 18, 1948, Jacobson Papers, HSTL.
62. Truman to Jacobson, Feb. 27, 1948, Jacobson Papers, HSTL.
63. Daniel J. Fellman, *An American Friendship*, p. 138.
64. Abba Eban, *Personal Witness*, p. 134.
65. Jacobson to Josef Cohn, March 27, 1952, A. J. Granoff Papers, Box 2, HSTL. The letter is reprinted in full under the title "Two Presidents and a Haberdasher–1948," *American Jewish Archives* 20, no. 1 (April 1968): 3–14. The sections of the text that follow this citation are all drawn from Jacobson's account in this letter
66. Harry S. Truman, *Years of Trial and Hope*, pp. 189–90.
67. Vera Weizmann, *The Impossible Takes Longer*, pp. 228–29.
68. Statement of Warren Austin before the Security Council of the U.N., March 19, 1948; *FRUS*, vol. 5, pp. 742–44.
69. Julius Haber, *The Odyssey of an American Zionist*, p. 323.
70. Statement by Rabbi Abba Hillel Silver to the Security Council, March 19, 1948, Zionist Organization of American Files, American Jewish Historical Society, New York.
71. Entry on Truman's calendar, March 19, 1948, in Margaret Truman, *Harry S. Truman*, pp. 424–25.
72. "Clifford Sets the Record Straight: Talk delivered before the American Jewish Historical Society and the American Historical Association," *Near East Report*, Dec. 29, 1976; Loy Henderson Papers, LOC. Also see Jonathan Daniels, interview with Clark Clifford, Oct. 24, 1949, Jonathan Daniels Papers, HSTL. In this version, Clifford has Truman saying to Weizmann, "He must think I'm a shitass," rather than a liar, which is what Clifford used when he spoke about it publicly.
73. Truman had done the same thing in 1946, when Secretary of Commerce Henry A. Wallace gave a speech in New York calling for policies friendly to the Soviet Union. Secretary of State James F. Byrnes was in Europe at a meeting of the Council of Foreign Ministers, and Wallace's speech, which he said the president had approved in advance, harmed Byrnes's ability to negotiate with the Soviets and to maintain a tough stance with them. Truman responded by saying he had approved only Wallace's giving a speech, not its contents. He then fired Wallace and sent him back to private life.
74. Clifford, memorandum, March 19, 1948, Box 13, Clark Clifford Papers, HSTL.
75. Clark Clifford, "Recognizing Israel," *American Heritage*, April 1977, p. 7.
76. Oral history interview of Clark Clifford by Richard Holbrooke and Brian VanDeMark, May 4, 1988, Richard Holbrooke Papers, HSTL.
77. Jonathan Daniel, notes on his interview with Clark Clifford, Oct. 24, 1949, Jonathan Daniels Papers, HSTL.
78. Charles Ross, memorandum, March 29, 1948, Box 6, Ross Papers, HSTL.

79. Eben A. Ayers, diary entry, March 25, 1948 (Jan. 1–May 31 Folder) Eben A. Ayers Papers, HSTL; "Clifford Sets the Record Straight," *Near East Report*, Dec. 29, 1976, enclosed with Henderson to Norman F. Dacey, n.d., January 1977, in Loy Henderson Papers, Box 11, Folder: Israel-Palestine correspondence, LOC.
80. Statement by President Truman, White House Press Release, March 25, 1948, David Niles Papers, Box 29, HSTL.
81. Truman to Eleanor Roosevelt, March 25, 1948, in Steve Neal, ed., *Eleanor and Harry*, pp. 133–34.
82. Trygve Lie, *In the Cause of Peace*, pp. 170–71.
83. "A Land of Milk and Honey," *The New York Times*, March 21, 1948, p. E8.
84. *New York Herald Tribune*, March 21, 1948.
85. *P.M.*, March 21, 1948.
86. T. O. Thackrey, "Betrayal Reaffirmed," *New York Post* March 21, 1948.
87. Eddie Jacobson to Josef Cohn, March 30, 1952, A. J. Granoff Papers, Box 2, HSTL.
88. Eliahu Elath, "Samuel Irving Rosenman and his Contribution Before the Establishment of the State of Israel" (Hebrew), *Molad* 7, no. 37–38 (247–248) (Spring 1976): 448–54. We would like to thank Tuvia Friling for translating this article for us
89. Weizmann to Doris May, March 23, 1948, Letter 128, *The Letters and Papers of Chaim Weizmann*, Series A, August 1947–1952, pp. 91–92.
90. Weizmann to Truman, April 9, 1948, Letter 138, *The Letters and Papers of Chaim Weizmann*, Series A, August 1947–1952, pp. 99–101.

CHAPTER 12: A NEW COUNTRY IS BORN

1. Henderson to Lovett, March 27, 1948, *FRUS*, vol. 5, pp. 767–69.
2. Memo Prepared by Dept. of State, April 2, 1948, *FRUS*, vol. 5, pp. 778–96.
3. Julian Louis Meltzer, "Arabs Fire Shells at Zionist Village," *The New York Times*, April 6, 1948, p. 11; Mallory Browne, "Jew and Arab Meet in U.N. Truce Move," *The New York Times*, April 8, 1948, p. 1.
4. "Soviet Charges U.S. Wrecks Partition for Oil and a Base," *The New York Times*, March 31, 1948, p. 1.
5. *FRUS*, vol. 5, pp. 776–78. Shertok's remarks to the Security Council are printed in an Editorial Note.
6. "Silver Exhorts US to Back Partition," *The New York Times*, April 1, 1948, p. 8.
7. Mallory Browne, "Truce Talks Fail; Arab Shuns Agency," *The New York Times*, April 9, 1948, p. 1; and "Arab Terms Reported," *The New York Times*, April 10, 1948, p. 6.
8. Thomas J. Hamilton, "U.N. Morale Sagging Under Heavy Strains," *The New York Times*, April 28, 1948, p. E4.
9. Howard M. Sachar, *The History of Israel from the Rise of Zionism to Our Time*, pp. 304–5.
10. Abba Eban, *Personal Witness*, pp. 136–37; Howard M. Sachar, *The History of Israel*, pp. 305–9.

11. Michael J. Cohen, *Palestine and the Great Powers*, p. 335.
12. Ibid., p. 337.
13. Wasson to Marshall, April 13, 1948, *FRUS*, vol. 5, p. 817.
14. Dana Adams Schmidt, "34 Jews Are Slain in Hospital Convoy," *The New York Times*, April 14, 1948, p. 6.
15. Letter by 81 doctors to Truman, April 21, 1948, American Jewish Historical Society, New York.
16. "Dr. Louis Dublin's Report on his Meeting with President Truman," April 22, 1948, box 42, American Jewish Historical Society.
17. Mrs. Samuel W. Halprin, National President, Hadassah, to Joseph C. Satterthwaite, Deputy Director of Near Eastern Division of the State Department, April 26, 1948; enclosing Hadassah Press Release, April 26, 1948, Box 42, Folder 1, American Jewish Historical Society.
18. Text of speech by Austin to U.N. Special Session on Palestine, April 19, 1948, Subject File, Box 160, HSTL.
19. Evan M. Wilson, *Decision on Palestine*, p. 138.
20. Trygve Lie, *In the Cause of Peace*, pp. 171–73.
21. Douglas to Marshall, April 20, 1948, *FRUS*, vol. 5, p. 837.
22. Douglas to Marshall, April 22, 1948, *FRUS*, vol. 5, p. 847.
23. McClintock to Lovett, April 22, 1948, *FRUS*, vol. 5, pp. 845–46.
24. Marshall to the U.K., May 5, 1948, *FRUS*, vol. 5, pp. 915–16.
25. John C. Ross to Marshall, May 6, 1948, *FRUS*, vol. 5, pp. 917–20.
26. Austin to Marshall, May 9, 1948, *FRUS*, vol. 5, pp. 949–53.
27. Jacobson to Dr. Josef Cohn, March 30, 1952, published as Eddie Jacobson, "Two Presidents and a Haberdasher–1948," *American Jewish Archives*, April 1968, p. 11.
28. Vera Weizmann, *The Impossible Takes Longer*, p. 231. The authors wish to note that many books and memoirs offer different versions of what happened on the twenty-third and the events surrounding it. We have evaluated all of the different accounts and present it to the best of our ability.
29. Ibid., p. 232.
30. Abba Eban, *Personal Witness*, pp. 140–41.
31. Ibid., pp. 142–43.
32. "Arab King Warns Palestine Invasion . . . Is Set," *The New York Times*, April 27, 1948, p. 1.
33. Evan Wilson, *Decision on Palestine*, p. 139.
34. Wasson to Marshall, May 3, 1948, *FRUS*, vol. 5, pp. 889–91.
35. John E. Horner, memorandum, May 4, 1948, *FRUS*, vol. 5, pp. 898–901.
36. Memorandum for the President, May 11, 1948, attached to proposed State Department resolution to be brought before the United Nations that would call for a truce in Palestine. Clifford scribbled on it "Lovett brought this in May–in high excitement–but this whole scheme flopped at Lake Success and no action was ever taken." See Eliahu Epstein to Shertok, May 14, 1948, Weizmann Archives, Rehovoth, Israel. Epstein had cabled Shertok that Clifford had telephoned friends in Washington advising that U.S. recognition was imminent, but they needed a request. He had consulted with Cohen and Ginsburg and drafted the request to the president and secretary of state.

37. Shertok to Rusk, May 4, 1948, David Niles Papers, Israel File, HSTL.
38. Rusk to Lovett, May 4, 1948, *FRUS*, vol. 5, pp. 894–96.
39. Austin to Marshall, enclosing Rusk to Lovett, May 3, 1948, *FRUS*, vol. 5, pp. 886–89.
40. Niles to Clifford, May 6, 1948, cited in Zvi Ganin, *Truman, American Jewry and Israel*, p. 218.
41. Dean Alfange to Truman, May 5, 1948, Subject File, Box 160, HSTL.
42. Clark Clifford, notes, May 4, 1948, Clark Clifford Papers, Box 13, HSTL.
43. Interview with Clark Clifford by Richard Holbrooke and Brian VanDeMark, "On State Department Antipathy Towards Israel Recognition Debate," May 12, 1948 Showdown, Lovett vis-à-vis Marshall and Clifford, May 14, 1948; "HST's Sympathy Toward Jews," in Richard C. Holbrooke Papers, Chronological File, Feb. 10, 1988, to Nov. 14, 1988, Box 2, HSTL. Richard Holbrooke gave the transcripts of his interviews with Clifford, used for preparation of *Counsel to the President*, to the Truman Library. Unless otherwise noted, all dialogue and accounts of the meeting are from the Holbrooke Papers and from Clark Clifford with Richard Holbrooke, *Counsel to the President*, pp. 3–18. According to Max Lowenthal, the meeting was called "at Clifford's urging"; diary entry of May 12, 1948, Max Lowenthal Papers.
44. Austin to Marshall, May 4, 1948, enclosing Rusk to Jessup, *FRUS*, vol. 5, p. 897.
45. Unsigned report and letter to Clifford from State Department, May 7, 1948, Clifford Papers, Box 13, HSTL.
46. Rusk to Marshall, May 8 or 7, 1948, *FRUS*, vol. 5, pp. 930–35.
47. Report of meeting between Shertok and Epstein with Marshall, Lovett, and Rusk, May 8, 1948, Israel Documents, Document 483, pp. 757–69. The paragraphs on this meeting are based on this summary, prepared by Shertok.
48. Vera Weizmann, *The Impossible Takes Longer*, p. 231; Dan Kurzman, *Ben-Gurion*, p. 282; Abba Eban, *Personal Witness*, p. 144. The story appears in all three books in a slightly different versions. Vera Weizmann based her recollection on her personal diary, which she wrote the day her husband phoned Shertok at the airport.
49. Dan Kurzman, *Ben-Gurion*, p. 282–85. Kurzman asserts that upon returning to Palestine, Shertok told Ben-Gurion that he favored postponement and Ben-Gurion told him forcefully that he had to change his position. Michael Bar-Zohar, in his biography *Ben-Gurion*, says only that Shertok had "profound inner disquiet and grave uncertainties" and that during the flight home, he had decided to advocate postponement. When he told Ben-Gurion, the future prime minister told him not to tell the leadership he thought Marshall was right; Michael Bar-Zohar, *Ben Gurion: A Biography*, pp. 151–52. Sharett's biographer, Gabriel Sheffer, disputes the accounts by Kurzman and Bar-Zohar, which emanated from Ben-Gurion, and writes, "this version of what went on at that secret meeting should be taken with more than a grain of salt." Sheffer argues that Shertok always and forcefully urged creation of the new state and rejected Marshall's advice. He cites as evidence Shertok's talk to the Mapai Central Committee, in which Shertok said that even during the last days of the U.N. debate he had argued that its creation could not be postponed; Gabriel Sheffer, *Moshe Sharett: Biography of a Political Moderate*, pp. 325–26, 332.

50. Max Lowenthal, diary entry, May 12, 1948, Max Lowenthal Papers, University of Minnesota, Minneapolis.
51. Oral history interviews with Max Lowenthal, conducted by Jerry N. Hess, Sept. 20, 1967, and Nov. 29, 1967, HSTL.
52. Lowenthal, statement, May 9, 1948, Clark Clifford Papers, Box 13, HSTL. The discussion of Lowenthal's views is taken entirely from this memorandum.
53. Lowenthal, memorandum, "Palestine: What Are the Alternatives Before the President at This Moment?," May 11, 1948, Clark Clifford Papers, Box 13, HSTL.
54. Lowenthal, memorandum, May 12, 1948, Max Lowenthal Papers, University of Minnesota, Minneapolis.
55. Interview with Clark Clifford by Richard Holbrooke, Nov. 14, 1988. The discussion of the May 12 meeting is taken from this interview and Clark Clifford's memoir *Counsel to the President.*
56. George C. Marshall, memorandum of conversation, May 12, 1948, *FRUS*, vol. 5, pp. 972–76.
57. Clark Clifford, "Factors Influencing President Truman's Decision to Support Partition and Recognize the State of Israel," in Clifford, Eugene Rostow, and Barbara Tuchman, *The Palestine Question in American History*, p. 47. For a discussion of Clifford's role at the May 12 meeting, also see, Henry D. Fetter, " 'Showdown in the Oval Office': 12 May 1948 in History," *Israel Affairs*, vol. 14, no. 3 (July 2008): pp. 499–518.
58. The attempt of the State Department to spread British disinformation about purported Soviet infiltration of the Yishuv is discussed in Elihu Bergman, "Unexpected Recognition: Some Observations of a Last-Gasp Campaign in the U.S. State Department to Abort a Jewish State," *Modern Judaism* 19 (No. 1, Feb. 1999): pp. 133–71. Bergman writes, "NEA bought into a somewhat desperate British scheme to portray illegal Jewish immigration as a dangerous conduit for introducing a substantial Communist cadre into Palestine, which could later be activated in support of Soviet expansionist aims in the Near East" (p. 156). He describes it as a "Red Scare scenario" meant by the British to capture American attention and change the U.S. policy against recognition (p. 158).
59. Ibid., p. 975.
60. Clark Clifford, *Counsel to the President*, p. 13.
61. Max Lowenthal, diary entry, May 12, 1948, Max Lowenthal Papers, University of Minnesota, Minneapolis.
62. Max Lowenthal, memorandum, May 15, 1948, on Clifford's conversation with Lowenthal at his office on the issue of recognition of a new Jewish state, Max Lowenthal Papers, University of Minnesota, Minneapolis.
63. Presidential Press Conference, May 13, 1948, HSTL.
64. Diary entry, May 13, 1949 (*sic*), Max Lowenthal Papers, University of Minnesota, Minneapolis. Lowenthal inadvertently entered the wrong year for his entry.
65. Arthur Krock, "What Is Meant by 'De Facto' Recognition?," *The New York Times*, May 20, 1948, p. 28.
66. Author's interview with David Ginsburg, Aug. 3, 2006, Washington, D.C.
67. Lovett, memorandum of conversations, May 17, 1948, *FRUS*, vol. 5, pp. 1005–7.

68. Max Lowenthal, memorandum, May 14, 1948, Max Lowenthal Papers, University of Minnesota, Minneapolis. Lowenthal relates what Clifford told him about the events of the previous day.
69. Epstein to Marshall, May 14, 1948, Loy Henderson Papers, Box 11, HSTL.
70. Vera Weizmann, *The Impossible Takes Longer*, p. 234.
71. David Bernard Sacher, "David K. Niles and United States Policy," Senior Honors Thesis, Harvard University, 1959, p. 1.
72. Dean Rusk to William N. Franklin, June 13, 1974, in Editorial Note, *FRUS*, vol. 5, p. 993. Also see Max Lowenthal, memorandum, May 18, 1948, Max Lowenthal Papers, University of Minnesota, Minneapolis. Lowenthal had been told of Austin's surprise the next day by Ben Cohen.
73. Granoff to Truman, May 15, 1948, Official File, Box 772, HSTL.
74. Harry S. Truman, *Years of Trial and Hope*, pp. 193–94.

CONCLUSION

1. Marshall Newton, "Rally Held Here," *The New York Times*, May 17, 1948, p. 1.
2. Max Lowenthal, memorandum, May 18, 1948; diary entries, Monday, May 17, and May 20, 1948, Max Lowenthal Papers, University of Minnesota, Minneapolis.
3. Max Lowenthal, diary entry, May 21, 1948.
4. Frank Adler, *Roots in a Moving Stream*, p. 212.
5. Max Lowenthal, diary entry, May 22, 1948.
6. Anthony Leviero, "Weizmann Visits Truman," *The New York Times*, May 26, 1948, p. 1.
7. Vera Weizmann, *The Impossible Takes Longer*, p. 240.
8. Anthony Leviero, "Weizmann Visits Truman," *The New York Times*, May 26, 1948, p. 1.
9. Frank Adler, *Roots in a Moving Stream*, p. 213.
10. Weizmann to Truman, May 26, 1948, Subject File, Box 160, HSTL.
11. James G. MacDonald, *My Mission to Israel*, pp. 3–7.
12. Oral history interview with Matthew Connelly, Aug. 21, 1968, HSTL.
13. Eliahu Elath, *Israel and Elath*, pp. 23–24.
14. Interview with Clark Clifford by Richard Holbrooke, May 4, 1988, Holbrooke Papers, HSTL.
15. Alfred Lilienthal, *Washington Report on Middle East Affairs*, June 1999, pp. 49–50.
16. Ibid., p. 50.
17. Eliahu Elath, "Harry S. Truman–The Man and the Statesman," First Annual Harry S. Truman Lecture, May 18, 1977, Hebrew University, p. 48.
18. Quoted in Clark Clifford, *Counsel to the President*, p. 25; Dan Kurzman, *Ben-Gurion*, p. 416.
19. David B. Sacher, *David K. Niles and United States Policy*, p. 96.
20. Truman to Weizmann, Nov. 29, 1948, in Harry S. Truman, *Years of Trial and Hope*, pp. 197–99.
21. Vera Weizmann to Jacobson, December 1952, Eddie Jacobson Papers, HSTL.
22. Interview with Herb Jacobson by Daniel Fellman, July 26, 2004, in Daniel J. Fellman, *An American Friendship*, pp. 169–75.

23. Truman to Jacobson, June 30, 1955, Jacobson Papers, HSTL.
24. Truman, address at Jacobson Memorial Dinner, Nov. 26, 1952, Jacobson Papers, HSTL.
25. Daniel J. Fellman, *An American Friendship*, p. 64.
26. "Eddie Jacobson, Truman Partner," *The New York Times*, Oct. 26, 1955, p. 31.
27. Quoted in Joseph B. Schechtman, *The United States and the Jewish State Movement*, p. 424.
28. Clark Clifford, "Recognizing Israel," *American Heritage*, April 1977, p. 11.
29. Clifford, memorandum to the president, Nov. 19, 1947, Clark Clifford Papers, HSTL. Also see Clark Clifford, *Counsel to the President*, pp. 189–94.
30. Bowles to Clifford, Sept. 23, 1948, Box 12, Clifford Papers, HSTL.
31. Diary of Eban A. Ayers, Sept. 9, 1948, Box 20, Eban A. Ayers Papers, HSTL.
32. The points about Lincoln are made in an essay by Eric Foner, "The President and the Prophet," which appeared in *The Nation*, Feb. 5, 2007. He bases his argument on James Oakes, *The Radical and the Republican: Frederick Douglass, Abraham Lincoln, and the Triumph of Antislavery Politics* (New York: Norton, 2007).
33. Eliahu Elath, "Harry S. Truman–The Man and the Statesman," p. 53.

BIBLIOGRAPHY

ARCHIVES AND LIBRARIES

United States

Harry S. Truman Library, Independence, Missouri (HSTL)
Papers as U.S. Senator and Vice President (SV)

Presidential papers:
Confidential File (CF)
Official File (OF)
President's Personal File (PPF)
President's Secretary's File (PSF)
White House Office of the President's Correspondence Secretary Files (PSF)
White House Scrapbooks
Postpresidential Papers

Oral history interviews:
Dean Acheson, Clark M. Clifford, Matthew J. Connelly, George M. Elsey, Oscar Ewing, Abraham Feinberg, A. J. Granoff, Averell Harriman, Loy Henderson, Robert Lovett, Max Lowenthal, Ted Marks, Edward D. McKim, Philleo Nash, Mary Ethel Noland, Harry Rosenfeld, Samuel I. Rosenman, Harry H. Vaughan, Fraser Wilkins.

The papers of:
Dean Acheson, Eben A. Ayers, Clark M. Clifford, Matthew J. Connelly, Jonathan Daniel, Democratic National Committee, George M. Elsey, A. J. Granoff, Richard C. Holbrooke (interviews with Clark Clifford, February to November 1988), Edward Jacobson, Charles F. Knox, Jr., Howard J. McGrath, Frank McNaughton, Charles S. Murphy, Philleo Nash, David K. Niles, Mary Ethel Noland, Harry Rosenfeld, Samuel I. Rosenman, Charles G. Ross, Joel D. Wolfsohn.

The Library of Congress, Washington, D.C. (LOC)
The papers of:
Emanuel Celler, Ben Cohen, Felix Frankfurter, Loy Henderson, Harold Ickes, Philip Jessup, Robert A. Taft

Columbia University, New York, New York (CU)
The papers of James G. McDonald

Schlesinger Library, Radcliffe Institute, Harvard University, Cambridge, Massachusetts (SL)
The papers of Freda Kirchwey

National Archives, College Park, Maryland (NA)
Department of State, Record Group 59: Palestine Files
Records of the Joint Chiefs of Staff

University of Minnesota Archives, Minneapolis, Minnesota (UM)
The papers of Max Lowenthal

Center for Jewish History, New York City
American Jewish Historical Society and YIVO Collections

Great Britain

Public Records Office, Kew Gardens, London (PRO)
Foreign Office (FO) FO 371
Cabinet Papers (CAB)

Israel

Central Zionist Archives, Jerusalem (CZA)
Jewish Agency, Political Department, American Section

Chaim Weizmann Archives, Rehovot (WA)
The papers of Chaim Weizmann

OFFICIAL DOCUMENTS

Foreign Relations of the United States (FRUS). Vols. from 1944 to 1949. Washington, D.C.: U.S. Government Printing Office.

Israel Documents, December 1947 to May 1948 (ID). Ed. Gedalia Yogev. Jerusalem: Israel Government Printing Office, 1979.

Public Papers of the Presidents of the United States: Harry S. Truman, 1945–1948, 1949–1952. Washington, D.C.: U.S. Government Printing Office, 1961, 1965.

United Nations, General Assembly. *Official Records*, First Special Session, 1947.

——. *Official Records*, Second Session, 1947. *Report of the United Nations Special Committee on Palestine*.

ARTICLES AND SPEECHES

Bergson, Eliahu. "Unexpected Recognition: Some Observations of a Last-Gasp Campaign in the U.S. State Department to Abort a Jewish State." *Modern Judaism* 19 (1999): 133–71.

Clifford, Clark. "Clifford Sets the Record Straight." Talk delivered before the American Jewish Historical Society and the American Historical Association. *Near East Report*, Dec. 29, 1976.

——. "Recognizing Israel." *American Heritage* 28 (April 1977): 4–11.

Cohen, Michael J. "Truman and the State Department: The Palestine Trusteeship Proposal, March 1948." *Jewish Social Studies* (Spring 1981): 165–78.

Elath, Eliahu. "Harry S. Truman–The Man and Statesman," First Annual Harry S. Truman Lecture, May 18, 1977, Hebrew University of Jerusalem; Harry S. Truman Research Institute.

——. "Samuel Irving Rosenman and His Contribution Before the Establishment of the State of Israel" [Hebrew]. *Molad* 8, no. 37–38 (Spring 1976): 448–454. Translated by Tuvia Friling.

Foner, Eric. "The President and the Prophet." *The Nation*, Feb. 5, 2007.

Ganin, Zvi. "The Limits of American Jewish Political Power: America's Retreat from Partition, November 1947–March 1948." *Jewish Social Studies* 39 (Winter-Spring 1977): 1–36.

Halperin, Samuel, and Irvin Oder. "The United States in Search of a Policy: Franklin D. Roosevelt and Palestine." *The Review of Politics* 24, no. 3 (July 1962): 336–37.

Klieman, Aaron S. "In the Public Domain: The Controversy over Partition for Palestine." *Jewish Social Studies* (Spring 1980): 147–64.

Mayerberg, Samuel S. "Two Presidents and a Haberdasher–1948." *American Jewish Archives* 20 (April 1968): 3–15.

Rucker, Laurent. "Moscow's Surprise: The Soviet-Israeli Alliance of 1947–1949." Working Paper 48, Woodrow Wilson International Center for Scholars, Washington, D.C.

Slonim, Shlomo. "The 1948 American Embargo on Arms to Palestine." *Political Science Quarterly* 94, no. 3 (Fall 1979): 495–513.

Wilson, Evan M. "The Palestine Papers, 1943–1947." *Journal of Palestine Studies* 2, no. 4 (Summer 1973): 34–37.

BOOKS, MEMOIRS, AND PUBLISHED DOCUMENTARY COLLECTIONS

Acheson, Dean. *Present at the Creation: My Years in the State Department*. New York: Norton, 1969.

Adler, Frank. *Roots in a Moving Stream*. Kansas City, Mo.: The Temple, Congregation of B'nai Jehudah, 1972.

Alpern, Sara. *Freda Kirchwey: A Woman of the Nation*. Cambridge, Mass.: Harvard University Press, 1987.

Avineri, Shlomo. *The Making of Modern Zionism*. New York: Basic Books, 1981.

Ayers, Eben. *Truman in the White House: The Diary of Eben Ayers*. Ed. Robert Ferrell. Columbia: University of Missouri Press, 1991.

Baram, Philip J. *The Department of State in the Middle East, 1919–1945*. Philadelphia: University of Pennsylvania Press, 1978.

Bar-Zohar, Michael. *Ben-Gurion: A Biography*. New York: Delacorte Press, 1978.

Beir, Robert L., with Joseph, Brian. *Roosevelt and the Holocaust*. Fort Lee, N.J.: Barricade Books, 2006.

Beisner, Robert L. *Dean Acheson: A Life in the Cold War*. New York: Oxford University Press, 2006.

Ben-Gurion, David. *Israel: A Personal History*. New York: Funk and Wagnall's, 1971.

Benson, Michael T. *Harry S. Truman and the Founding of Israel*. Westport, Conn.: Praeger, 1997.

Beschloss, Michael. *Presidential Courage: Brave Leaders and How They Changed America*. New York: Simon and Schuster, 2007.

Bisgyer, Maurice. *Challenge and Encounter*. New York: Crown, 1976.

Bohlen, Charles E. *Witness to History 1929–1969*. New York: Norton, 1973.

Brands, H. W. *Inside the Cold War: Loy Henderson and the Rise of the American Empire, 1918–1961*. New York: Oxford University Press, 1991

Bullock, Alan. *Ernest Bevin: Foreign Secretary, 1945–1951*. New York: Norton, 1983.

Celler, Emanuel. *You Never Leave Brooklyn: The Autobiography of Emanuel Celler*. New York: John Day Company, 1953.

Clifford, Clark, with Richard Holbrooke. *Counsel to the President: A Memoir*. New York: Random House, 1991.

Clifford, Clark M., Eugene V. Rostow, and Barbara W. Tuchman. *The Palestine Question in American History*. New York: Arno Press, 1978. (Contains papers given on December 28, 1976, at a joint meeting of the American Jewish Historical Society and the American Historical Association.)

Cochran, Bert. *Harry Truman and the Crisis Presidency*. New York: Funk and Wagnall's, 1973.

Cohen, Michael J. *Palestine and the Great Powers, 1945–1948*. Princeton, N.J.: Princeton University Press, 1982.

——. *Truman and Israel*. Berkeley: University of California Press, 1990.

Cohen, Naomi. *Not Free to Desist: The American Jewish Committee, 1906–1966*. Philadelphia: Jewish Publication Society of America, 1972.

Crossman, Richard. *Palestine Mission*. London: Hamish Hamilton, 1947.

Crum, Bartley. *Behind the Silken Curtain: A Personal Account of Anglo-American Diplomacy in Palestine and the Middle East*. New York: Simon and Schuster, 1947.

Daniels, Jonathan. *The Man of Independence*. Philadelphia: Lippincott, 1950.

Dinnerstein, Leonard. *America and the Survivors of the Holocaust*. New York: Columbia University Press, 1982.

Donovan, Robert. *Conflict and Crisis: The Presidency of Harry S. Truman, 1945–1948*. New York: Norton, 1977.

Eban, Abba. *An Autobiography*. Glasgow: William Collins, 1977.

——. *Personal Witness: Israel Through My Eyes*. New York: Putnam, 1992.

——. *Voice of Israel*. New York: Horizon Press, 1957.

Elath, Eliahu. *Israel and Elath: The Political Struggle for the Inclusion of Elath in the Jewish State*. London: Weidenfeld, 1966.

——. *Zionism at the UN: A Diary of the First Days*. Philadelphia: Jewish Publication Society of America, 1976.

Elsey, George McKee. *An Unplanned Life: A Memoir.* Columbia: University of Missouri Press, 2005.

Ferrell, Robert H. *Harry S. Truman and the Modern American Presidency.* Boston: Little, Brown, 1983.

Fink, Reuben. *America and Palestine.* New York: American Zionist Emergency Council, 1944.

Feingold, Henry L. *The Politics of Rescue: The Roosevelt Administration and the Holocaust, 1938–1945.* New Brunswick, N.J.: Rutgers University Press, 1970.

Feis, Herbert. *The Birth of Israel: The Tousled Diplomatic Bed.* New York: Norton, 1969.

Forrestal, James. *The Forrestal Diaries.* Ed. Walter Mills. New York: Viking, 1951.

Frankfurter, Felix. *Felix Frankfurter Reminisces.* Recorded in talks with Dr. Harlan B. Phillips. New York: Reynal and Company, 1960.

Friling, Tuvia. *Arrows in the Dark: David Ben-Gurion, the Yishuv Leadership, and Rescue Attempts During the Holocaust.* Vol. 2. Madison: University of Wisconsin Press, 2005.

Gal, Alon. *David Ben-Gurion and the American Alignment for a Jewish State.* Jerusalem: Magnes Press, 1991.

Ganin, Zvi. *Truman, American Jewry, and Israel, 1945–1948.* New York: Holmes and Meir, 1979.

García Granados, Jorge. *The Birth of Israel: The Drama as I Saw It.* New York: Knopf, 1948.

Gilbert, Martin. *Churchill and the Jews: A Lifelong Friendship.* New York: Henry Holt, 2007.

——. *Israel: A History.* New York: William Morrow, 1998.

Gilead, Zerubavel, and Dorothea Krook. *Gideon's Spring: A Man and His Kibbutz.* New York: Ticknow and Fields, 1985.

Goldmann, Nahum. *The Autobiography of Nahum Goldmann: Sixty Years of Jewish Life.* New York: Holt, Rinehart and Winston, 1969.

Grose, Peter. *Israel in the Mind of America.* New York: Knopf, 1983.

Gruber, Ruth. *Destination Palestine: The Story of the Haganah Ship* Exodus 1947. New York: Current Books, 1948.

——. *Inside of Time: My Journey from Alaska to Israel.* New York: Carroll and Graf, 2003.

——. *Witness.* New York: Schocken Books, 2007.

Haber, Julius. *The Odyssey of an American Zionist.* New York: Twayne Publishers, 1956.

Halperin, Samuel. *The Political World of American Zionism.* Detroit: Wayne State University Press, 1961.

Hamby, Alonzo. *Man of the People: A Life of Harry S. Truman.* New York: Oxford University Press, 1995.

Hassett, William D. *Off the Record with F.D.R., 1942–1945.* New Brunswick, N.J.: Rutgers University Press, 1958.

Hazony, Yoram. *The Jewish State: The Struggle for Israel's Soul.* New York: Basic Books, 2000.

Hertzberg, Arthur, ed. *The Zionist Idea.* New York: Atheneum, 1975.

Herzl, Theodor. *The Jewish State.* New York: Dover, 1988. (Originally published in the United States by the American Zionist Emergency Council, 1946.)

Horowitz, David. *State in the Making.* Translated from the Hebrew by Julian Meltzer. New York: Knopf, 1953.

Hull, Cordell. *Memoirs*. Vol. 2. New York: Macmillan, 1948.

Hurewitz, J. C. *The Struggle for Palestine*. New York: Schocken, 1950.

Huthmacher, J. Joseph. *Senator Robert F. Wagner and the Rise of Urban Liberalism*. New York: Atheneum, 1971.

Isaacs, Stephen D. *Jews and American Politics*. New York: Doubleday, 1974

Issacson, Walter, and Evan Thomas. *The Wise Men: Six Friends and the World They Made*. New York: Simon and Schuster, 1986.

Judt, Tony. *PostWar: A History of Europe Since 1945*. New York: Penguin Press, 2005.

Kaplan, Robert D. *The Arabists: The Romance of an American Elite*. New York: Free Press, 1993.

Kaufman, Menahem. *An Ambiguous Partnership: Non Zionists and Zionists in America, 1939–1948*. Jerusalem: Magnes Press, 1991.

Kenen, I. L. *Israel's Defense Line: Her Friends and Foes in Washington*. Buffalo, N.Y.: Prometheus, 1981.

Klausner, Abraham J. *A Letter to My Children from the Edge of the Holocaust*. San Francisco: Holocaust Center of Northern California, 2002.

Kochavi, Arieh J. *Post-Holocaust Politics: Britain, the United States, and Jewish Refugees, 1945–1948*. Chapel Hill: University of North Carolina Press, 2001.

Koestler, Arthur. *Promise and Fulfillment: Palestine 1917–1949*. New York: Macmillan, 1949.

Kuntzel, Matthias. *Jihad and Jew-Hatred: Islamism, Nazism and the Roots of 9/11*. New York: Telos Press, 2007.

Kurzman, Dan. *Ben-Gurion: Prophet of Fire*. New York: Simon and Schuster, 1983.

———. *Genesis 1948: The First Arab-Israeli War*. New York: Signet, 1970.

Laqueur, Walter. *A History of Zionism: From the French Revolution to the Establishment of the State of Israel*. New York: Schocken, 1972.

Lash, Joseph P. *Eleanor: The Years Alone*. New York: Norton, 1972.

Lelyveld, Joseph. *Omaha Blues*. New York: Farrar, Straus and Giroux, 2005.

Lie, Trygve. *In the Cause of Peace: Seven Years with the United Nations*. New York: Macmillan, 1954.

Litvinoff, Barnet. *Weizmann: Last of the Patriarchs*. New York: G. P. Putnam's Sons, 1976.

Louis, William Roger. *The British Empire in the Middle East, 1945–1951*. Oxford, England: Clarendon Press, 1984.

Lowdermilk, Walter Clay. *Palestine: Land of Promise*. New York: Harper and Brothers, 1944.

Makovsky, Michael. *Churchill's Promised Land: Zionism and Statecraft*. New Haven, Conn.: Yale University Press, 2007.

Mandel, Daniel. *H. V. Evatt and the Establishment of Israel: The Undercover Zionist*. London: Frank Cass, 2004.

McCullough, David. *Truman*. New York: Simon and Schuster, 1992.

McDonald, James. *My Mission to Israel*. New York: Simon and Schuster, 1951.

Miller, Aaron David. *Search for Security: Saudi Arabian Oil and American Foreign Policy, 1939–1949*. Chapel Hill: University of North Carolina Press, 1980.

Miller, Merle. *Plain Speaking: An Oral Biography of Harry S. Truman*. New York: Berkley-Putnam, 1973.

Miscamble, Wilson D. *From Roosevelt to Truman: Potsdam, Hiroshima, and the Cold War*. New York: Cambridge University Press, 2007.

Morris, Benny. *The Birth of the Palestinian Refugee Problem Revisited*. Cambridge, England: Cambridge University Press, 2004.

——. *1948: A History of the First Arab-Israeli War*. New Haven, Conn.: Yale University Press, 2008.

Mosley, Leonard. *Marshall: Hero for Our Times*. New York: Hearst Books, 1982.

Murphy, Bruce Allen. *The Brandeis/Frankfurter Connection*. New York: Anchor Press, 1983.

Neal, Steve. *Eleanor and Harry: The Correspondence of Eleanor Roosevelt and Harry S. Truman*. New York: Citadel Press, 2002.

Neumann, Emanuel. *In the Arena: An Autobiographical Memoir*. New York: Herzl Press, 1976.

Oren, Michael B. *Power, Faith, and Fantasy: America in the Middle East, 1776 to the Present*. New York: Norton, 2007.

Pawel, Ernst. *The Labyrinth of Exile: A Life of Theodor Herzl*. New York: Farrar, Straus and Giroux, 1989.

Pogue, Forrest C. *George C. Marshall: Statesman, 1945–1959*. New York: Viking, 1987.

Proskauer, Joseph. *A Segment of My Times*. New York: Farrar Straus, 1950.

Raphael, Marc Lee. *Abba Hillel Silver: A Profile in American Judaism*. New York: Holmes and Meier, 1989.

Reinharz, Jehuda. *Chaim Weizmann: The Making of a Zionist Leader*. New York: Oxford University Press, 1985.

——. *Chaim Weizmann: The Making of a Statesman*. New York: Oxford University Press, 1993.

Ro'I, Yaacov. *From Encroachment to Involvement: A Documentary Study of Soviet Policy in the Middle East, 1945–1973*. New York: John Wiley, 1974.

Rose, Norman. *Chaim Weizmann: A Biography*. New York: Viking Penguin, 1986.

Rosen, Robert N. *Saving the Jews: Franklin D. Roosevelt and the Holocaust*. New York: Thunder's Mouth Press, 2006.

Roosevelt, Eleanor. *My Day*. Vol. 11: *The Post War Years*. New York: Pharos Books, 1990.

Roosevelt, Franklin D. *FDR: His Personal Letters, 1928–1945*. Ed. Elliott Roosevelt. Vol. 2. New York: Duell, Sloan and Pearce, 1950.

Rosenman, Samuel, and Dorothy Rosenman. *Presidential Style: Some Giants and a Pygmy in the White House*. New York: Harper and Row, 1976.

Sachar, Howard M. *The History of Israel from the Rise of Zionism to Our Time*. New York: Alfred A. Knopf, 1996.

Sacher, Abram. *The Redemption of the Unwanted: The Post-Holocaust Years*. New York: St. Martin's Press, 1983.

Saint John, Robert. *Eban*. New York: Doubleday, 1972.

Schechtman, Joseph B. *The United States and the Jewish State Movement: The Crucial Decade, 1939–1949*. New York: Herzl Press, 1966.

Sheffer, Gabriel. *Moshe Sharett: Biography of a Political Moderate*. Oxford, England: Clarendon Press, 1996.

Silver, Abba Hillel. *Vision and Victory: A Collection of Addresses by Dr. Abba Hillel Silver*. New York: Zionist Organization of America, 1949.

Snetsinger, John. *Truman, the Jewish Vote, and the Creation of Israel*. Stanford, Calif.: Hoover Institution Press, 1974.

Spalding, Elizabeth Edwards. *The First Cold War: Harry Truman, Containment, and the Remaking of Liberal Internationalism.* Levington: University of Kentucky Press, 2006.
Stark, Freye. *Dust in the Lion's Paw.* London: Century Publishing, 1961.
Stettinius, Edward R., Jr. *The Diaries of Edward R. Stettinius, Jr., 1943–1946.* Ed. Thomas M. Campbell and George C. Herring. New York: Franklin Watts, 1975.
Stone, I. F. *This Is Israel.* New York: Boni and Gaer, 1948.
——. *Underground to Palestine.* New York: Boni and Gaer, 1946.
Sykes, Christopher. *Cross Roads to Israel: Palestine from Balfour to Bevin.* London: Collins, 1965.
Teveth, Shabtai. *Ben-Gurion: The Burning Ground, 1886–1948.* Boston: Houghton Mifflin, 1987.
Truman, Harry S. *Dear Bess: The Letters from Harry to Bess Truman, 1910–1959.* Ed. Robert H. Ferrell. New York: Norton, 1983.
——. *Memoirs.* Vol. 1: *Years of Decisions.* New York: Doubleday, 1955.
——. *Memoirs.* Vol. 2: *Years of Trial and Hope, 1946–1952.* New York: Doubleday, 1956.
——. *Mr. Citizen.* New York: Popular Library, 1953.
——. *Off the Record: The Private Papers of Harry S. Truman.* Ed. Robert H. Ferrell. New York: Harper and Row, 1980.
Truman, Margaret. *Bess W. Truman.* New York: Macmillan, 1986.
——. *Harry S. Truman.* New York: Morrow, 1972.
Tuchman, Barbara W. *Bible and Sword: England and Palestine from the Bronze Age to Balfour.* New York: Ballantine Books, 1956.
Urofsky, Melvin. *We Are One!* Garden City, N.Y.: Anchor Press, 1978.
Urquhart, Brian. *Ralph Bunche: An American Odyssey.* New York: Norton, 1993.
Vandenberg, Arthur. *The Private Papers of Senator Vandenberg.* Ed. Arthur H. Vandenberg, Jr. Boston: Houghton Mifflin, 1952.
Wallace, Henry A. *The Price of Vision: The Diary of Henry A. Wallace.* Ed. John Morton Blum. Boston: Houghton Mifflin, 1973.
Weisgal, Meyer. *Meyer Weisgal . . . So Far.* New York: Random House, 1971.
Weizmann, Chaim. *Trial and Error: The Autobiography of Chaim Weizmann.* New York: Harper and Brothers, 1949.
——. *The Letters and Papers of Chaim Weizmann.* Vol. 1: *1898–1931.* Ed. Barnet Litvinoff. New Brunswick, N.J.: Transaction Books, 1983.
——. *The Letters and Papers of Chaim Weizmann.* Vol. 2: *1931–1952.* Ed. Barnet Litvinoff. New Brunswick, N.J.: Transaction Books, 1984.
Weizmann, Vera (as told to David Tutaev). *The Impossible Takes Longer: The Memoirs of Vera Weizmann.* New York: Harper, 1967.
Welles, Sumner. *We Need Not Fail.* Boston: Houghton Mifflin, 1948.
Williams, Francis. *Ernest Bevin: Portrait of a Great Englishman.* London: Hutchinson, 1952.
——. *A Prime Minister Remembers: The War and Post-War Memoirs of the Rt. Hon. Earl Attlee.* London: Heinemann, 1961.
Wilson, Evan M. *Decision on Palestine: How the U.S. Came to Recognize Israel.* Stanford, Calif.: Hoover Institution Press, 1979.
Wise, Stephen. *Challenging Years.* London: East and West Library, 1951.
Wyman, David. *The Abandonment of the Jews.* New York: Pantheon, 1984.

Wyman, David, and Rafael Medoff. *A Race Against Death: Peter Bergson, America, and the Holocaust.* New York: New Press, 2002.
Zaar, Isaac. *Rescue and Liberation: America's Part in the Birth of Israel.* New York: Bloch Publishing, 1954.

UNPUBLISHED SOURCES

Fellman, Daniel J. "An American Friendship: A Critical Examination of the Life of Eddie Jacobson and His Relationship with President Harry S Truman." Thesis for Ordination, Hebrew Union College–Jewish Institute of Religion, March 2005.
Sacher, D. B. "David Niles and American Policy." Senior honors thesis, Harvard University, 1959.

INDEX